T0178296

Lecture Notes in Bioinformatics 13234

Subseries of Lecture Notes in Computer Science

More information about this subseries at https://link.springer.com/bookseries/5381

Lingling Jin · Dannie Durand (Eds.)

Comparative Genomics

19th International Conference, RECOMB-CG 2022
La Jolla, CA, USA, May 20–21, 2022
Proceedings

 Springer

Editors
Lingling Jin (ID)
University of Saskatchewan
Saskatoon, SK, Canada

Dannie Durand (ID)
Carnegie Mellon University
Pittsburgh, PA, USA

ISSN 0302-9743 ISSN 1611-3349 (electronic)
Lecture Notes in Bioinformatics
ISBN 978-3-031-06219-3 ISBN 978-3-031-06220-9 (eBook)
https://doi.org/10.1007/978-3-031-06220-9

LNCS Sublibrary: SL8 – Bioinformatics

This Springer imprint is published by the registered company Springer Nature Switzerland AG
The registered company address is: Gewerbestrasse 11, 6330 Cham, Switzerland

Preface

With the advent of high-throughput DNA sequencing and chromosome conformation capture technologies, more and more high quality genome sequences and genome-related data are available. The challenges for computational comparative genomics are also increasing. The comparison of related genomes provides a great inferential power for analyses of genome evolution, gene function, ancestral genome organization, cellular processes, mechanisms of evolution, and the evolution of cancer genomes.

This volume contains the papers presented at the 19th Annual RECOMB Satellite Workshop on Comparative Genomics, held at La Jolla, USA, during May 20–21, 2022. RECOMB-CG is a forum that brings together leading researchers in the mathematical, computational, and life sciences to discuss cutting edge research in comparative genomics, with an emphasis on computational approaches and novel experimental results.

There were 28 submissions from authors in 15 countries. Each manuscript was reviewed by at least three, but usually four, members of the Program Committee (PC) as well as sub-reviewers who were sought based on their expertise for specific papers. The review process was single blind. After careful consideration, 18 papers (64%) were accepted for oral presentation at the meeting and inclusion in this proceedings.

RECOMB-CG 2022 would like to thank the invited speakers: Aoife McLysaght (University of Dublin, Trinity College, Ireland), Rachel Dutton (University of California at San Diego, USA), and Nandita Garud (University of California at Los Angeles, USA).

We express our appreciation to the PC members and sub-reviewers for their diligent work reviewing the manuscripts and providing thorough discussion and review reports that informed the review process.

We also thank the other members of the Steering Committee, Marília Braga, Jens Lagergren, Aoife McLysaght, Luay Nakhleh, and David Sankoff for their guidance and helpful discussions.

Special thanks to the local organizer, Siavash Mirarab, and to the Jacobs School of Engineering, University of California, San Diego, for organizing and providing financial support for the conference.

We used the EasyChair conference system to handle submissions, reviews, and formatting.

In closing, we would like to thank all the scientists who submitted papers and posters and those who enthusiastically attended RECOMB-CG 2022.

May 2022

Lingling Jin
Dannie Durand

Organization

Program Chairs

Lingling Jin	University of Saskatchewan, Canada
Dannie Durand	Carnegie Mellon University, USA

Steering Committee

Marília Braga	Bielefeld University, Germany
Dannie Durand	Carnegie Mellon University, USA
Jens Lagergren	KTH Royal Institute of Technology, Sweden
Aoife McLysaght	University of Dublin, Trinity College, Ireland
Luay Nakhleh	Rice University, USA
David Sankoff	University of Ottawa, Canada

Local Organizer

Siavash Mirarab	University of California at San Diego, USA

Program Committee

Max Alekseyev	George Washington University, USA
Nikita Alexeev	Saint Petersburg National Research University of Information Technologies, Mechanics and Optics (ITMO University), Russia
Lars Arvestad	Stockholm University, Sweden
Mukul Bansal	University of Connecticut, USA
Sèverine Bérard	University of Montpellier, France
Anne Bergeron	Université du Québec à Montréal, Canada
Paola Bonizzoni	University of Milano-Bicocca, Italy
Marília Braga	Bielefeld University, Germany
Broňa Brejová	Comenius University in Bratislava, Slovakia
Cedric Chauve	Simon Fraser University, Canada
Leonid Chindelevitch	Imperial College London, UK
Miklós Csűrös	University of Montreal, Canada
Daniel Doerr	Heinrich Heine University Düsseldorf, Germany
Mohammed El-Kebir	University of Illinois at Urbana-Champaign, USA
Nadia El-Mabrouk	University of Montreal, Canada
Oliver Eulenstein	Iowa State University, USA

Yong Wang	Academy of Mathematics and Systems Science, China
Tandy Warnow	University of Illinois at Urbana-Champaign, USA
Rohan Williams	National University of Singapore, Singapore
Yufeng Wu	University of Connecticut, USA
Louxin Zhang	National University of Singapore, Singapore
Xiuwei Zhang	Georgia Tech, USA
Fa Zhang	Chinese Academy of Sciences, China
Jie Zheng	ShanghaiTech University, China

Many thanks to our sub-reviewers: Laurent Bulteau, Diego P. Rubert, Lianrong Pu, Lina Herbst, Ke Chen, Lucas Oliveira, Alexey Markin, Xinnan Dai, Wandrille Duchemin, Dominique Lavenier, Luca Denti, Anton Zamyatin, Raffaella Rizzi.

Previous Meetings

The 1st RECOMB-CG, Minneapolis, USA, 2003
The 2nd RECOMB-CG, Bertinoro, Italy, 2004
The 3rd RECOMB-CG, Dublin, Ireland, 2005
The 4th RECOMB-CG, Montréal, Canada, 2006
The 5th RECOMB-CG, San Diego, USA, 2007
The 6th RECOMB-CG, Paris, France, 2008
The 7th RECOMB-CG, Budapest, Hungary, 2009
The 8th RECOMB-CG, Ottawa, Canada, 2010
The 9th RECOMB-CG, Galway, Ireland, 2011
The 10th RECOMB-CG, Niterói, Brazil, 2012
The 11th RECOMB-CG, Lyon, France, 2013
The 12th RECOMB-CG, New York, USA, 2014
The 13th RECOMB-CG, Frankfurt, Germany, 2015
The 14th RECOMB-CG, Montréal, Canada, 2016
The 15th RECOMB-CG, Barcelona, Spain, 2017
The 16th RECOMB-CG, Québec, Canada, 2018
The 17th RECOMB-CG, Montpellier, France, 2019
The 18th RECOMB-CG, virtual, 2021

Sponsors

Jacobs School of Engineering, University of California, San Diego

Contents

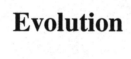

Evolution

On the Comparison of Bacteriophage Populations

Anne Bergeron[1]([✉]), Marie-Jean Meurs[1], Romy Valiquette-Labonté[1],
and Krister M. Swenson[2]

[1] Université du Québec à Montréal, Montreal, Canada
bergeron.anne@uqam.ca
[2] LIRMM, Université de Montpellier, CNRS, Montpellier, France

Abstract. The production of cheese and other dairy products relies on the constant monitoring of viruses, called bacteriophages, that attack the organisms responsible for the fermentation process. Bacteriophage species are characterized by a stable core genome, and a 'genetic reservoir' of gene variants that are exchanged through recombination. Phylogenetic analysis of phage populations are notably difficult due not only to extreme levels of horizontal exchange at the borders of functional modules, but also inside of them.

In this paper we present the first known attempt at directly modeling gene flux between phage populations. This represents an important departure from gene-based alignment and phylogenetic reconstruction, shifting focus to a genetic reservoir-based evolutionary inference. We present a combinatorial framework for the comparison of bacteriophage populations, and use it to compute recombination scenarios that generate one population from another. We apply our heuristic, based on this framework, to four populations sampled from Dutch dairy factories by Murphy [14]. We find that, far from being random, these scenarios are highly constrained. We use our method to test for factory-specific diversity, and find that there was likely a large amount of recombination in the ancestral population.

Find instructions for reproducing the results at:
https://bitbucket.org/thekswenson/phage_population_comparison
The code is publicly available at:
https://bitbucket.org/thekswenson/phagerecombination

1   Introduction

Bacteriophages – or simply *phages* – are viruses that infect bacteria. They are the most abundant and diverse organisms on the planet, and are found in every community where bacteria thrive: soil, water, air, lungs, guts, sewers, plants, and milk [9]. Where their presence intersects human activity, they can be beneficial, when they are used in therapies to combat bacterial infections [4], or detrimental, when they destroy batches of dairy fermentation in artisanal or industrial food factories [13]. Due to their economic impacts, dairy bacteriophage populations

© The Author(s), under exclusive license to Springer Nature Switzerland AG 2022
L. Jin and D. Durand (Eds.): RECOMB-CG 2022, LNBI 13234, pp. 3–20, 2022.
https://doi.org/10.1007/978-3-031-06220-9_1

have been extensively sequenced in the last few years. These populations can be separated by geography [5, 6, 14], by time [10, 11], or by their bacterial host [3].

Bacteriophages are divided into *species*, characterized by a common *core* genome distributed along their single chromosome, where genes appear in the same order for each member of the species. Between these regions of core genome, there is a *variable* genome composed of regions that are shared by some members of the family, but not all of them. In an individual phage, the variable region between two consecutive regions of core genome may be empty, or may have one or more *variants* that are presumed to perform the same biological function, but with different proteins [3, & references therein].

Kupczok [10] sequenced 34 dairy phages from a single German dairy factory, sampled over three decades. Their analyses concluded that, over such a period of time, point mutations were *"[...] unlikely to constitute the major driver of phage genome evolution"*. However, the variable genome of the sequenced phages changed considerably over time: *"The frequent gene loss and regain suggest the existence of a pangenome (i.e. genetic reservoir) that is accessible by genetic recombination."*

1.1 Recombinations and Mosaicism in Phage Genomes

Genetic *recombination* allows two phages to exchange or borrow significant parts of their genomes, creating novel viruses. These exchanges take place inside a single cell, and are presumed to occur either between two co-infecting phages, or by an infecting phage and a *prophage* (*i.e.* a phage genome that inserted itself into a host bacterium). When the exchanges occur between similar sequences, recombinations are called *homologous*, and when they occur between unrelated sequence they are called *illegitimate*.

A striking feature of the comparison of phage genomes is their extreme *mosaicism*, where regions of unrelated sequences alternate with regions of very high similarity, as illustrated in Fig. 1.

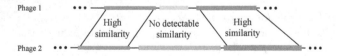

Fig. 1. Alignments of phage genomes exhibit alternating regions of high local similarity and unrelated regions.

A few decades ago, Botstein [2] proposed a *theory of modular evolution* for bacteriophages based on homologous recombinations. In this model, recombinations are mediated by flanking regions of high sequence similarity. DePaepe [7] characterized biological mechanisms that could be responsible for such rearrangements, calling them *relaxed homologous recombinations*, and qualifying them as *"...strangely dependent on the presence of sequence homology, but highly tolerant to divergence"*.

Recombination that does not follow the Botstein model also likely plays an important role in phage evolution. Pedulla [15] found three recent recombinations in Mycobacteriophages that have no flanking regions of similarity. Yahara [18] analyzed recombination within *soft-core* genes (*i.e.* the genes that exist in at least 90% of the phages they studied) of *Helicobacter pylori* prophages, showing extreme sequence divergence that they attribute to recombination within genes.

Thus, phages are particularly ill suited to traditional evolutionary analyses using multiple sequence alignment. The extraordinary mosaicism implies that not any set of genes with the same function can be used for phylogenetic inference, since in this case shared function does not imply homology. Instead, Brussow [3] recommends that evolutionary histories should be established using only the homologous sequences belonging to the same functional module (*i.e.* what we call a *variant* in the present article). Yet, to make matters worse, the high dissimilarity attributed to pervasive recombination within genes makes phylogenetic reconstruction on individual variants very difficult [18].

1.2   Recombination Between Phage Populations

The conditions of high mosaicism in bacteriophages motivate a fresh perspective on evolutionary history inference, one that uses the fact that a population of phages represent a genetic reservoir that is constantly testing combinations. To this end, in a previous work we inferred a recombination scenario within a population of phages while explicitly using the Botstein model of recombination with flanking homologous regions [17]. The present article differs from our previous work in two important ways:

1. here we infer recombination scenarios *between* phage populations instead of *within* them, and
2. our present model can accomodate flanking homology or ignore it.

When building our modules we assume the existence of flanking homology "anchors" between every adjacent module (see Sect. 3.1 for details on how we constructed our modules).

Our work is timely, given that *"little is known about genetic flux by recombination between populations"* [18]. While it represents a first attempt at reconstructing the evolution of phages in such a global sense, we expect this perspective to increase in importance as phage sequencing becomes more prevalent.

Dutch Dairy Factories. In Murphy [14], phages are sampled across geographic regions. They sequenced 38 phage genomes from four Dutch dairy factories, and added to their dataset phages of the same species from various countries and continents (Australia, Canada, Denmark, France, Germany, Ireland, Italy, Poland, United Kingdom, United States, and New Zealand). By using hierarchical clustering on protein families presence/absence data, they were able to – mostly – separate continents and countries. However, this technique was not able to separate the four Dutch factories.

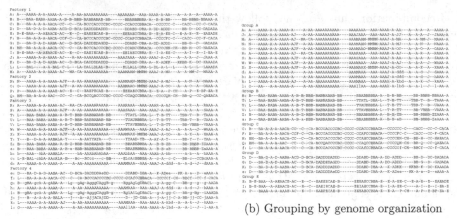

(a) Grouping by factories

(b) Grouping by genome organization

Fig. 2. (a) The variable parts of the 38 phage genomes of Murphy [14], color-coded by factory. Each letter stands for a variant spanning the interval between two regions of the core genome. (b) Highly similar genome organization is observed in 31 of the 38 phage genomes. Using the color-coding of panel (a), we see that each of the 5 groups has a representative in at least two different factories.

Figure 2a shows the variable parts for the 38 phage genomes of [14], color-coded by factory. Each letter stands for a variant spanning the interval between two consecutive *anchors* (*i.e.* regions of core genome), and dashes represent empty variants. The phages were grouped by similar genome organizations and variants. Seven of these phages were in groups occurring in only a single factory, having few shared modules with the other factories, so were removed from consideration. Figure 2b shows the 31 remaining phages. The comparison of Figs. 2a and 2b implies that each factory hosts a crew of different phages, that is more or less conserved across factories, suggesting a common ancestral population.

We apply the algorithm described in this paper to the populations from each pair of the Dutch factories. The lengths of the calculated recombination scenarios are used to infer relative properties of population diversity between the factories. Experiments are performed to determine if factories have specific qualities, and to demonstrate a likely high amount of ancestral recombination within phage populations.

Paper Outline. In this paper, we introduce the concept of recombination scenarios, describing how a phage population can be derived from another.

The next section gives the basic definitions and properties, Sect. 2.2 presents the theoretical basis of the greedy heuristics, and Sect. 2.3 derive lower bounds adapted to specific characteristics of biological data. Finally in Sect. 3, our heuristic is used to compare the four dairy factories of Fig. 2a.

2   Methods

2.1   Basic Definitions and Properties

A *phage species* is a set of phage genomes whose core genome contains the same number m of distinct regions, called *anchors*, thus the same number m of variable regions called *modules*. Each module has two or more *variants* within the species. In this paper, we work with circularized versions of phage genomes[1].

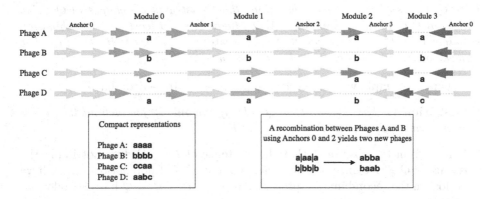

Fig. 3. Each phage in a set can be represented by the sequence of its variants. For each module, an arbitrary symbol is assigned to each variant, including empty variants. This compact representation captures the different assortments of modules within the population. Intervals are sequences of consecutive modules in the circular order. A recombination exchanges intervals between two parents, creating two new phages.

By representing variants of a module by single symbols such as a, b, c, . . ., it is possible to represent individual members of a species by the sequence of their variants, as in Fig. 3. Anchors are numbered from 0 to $m-1$ in the clockwise direction, where $m \geq 2$ is the number of modules.

More formally, given sets of variants $\mathcal{V}_i$ for each module i, a phage p can be represented by $p = p_0 p_1 \ldots p_{m-1}$ where $p_i \in \mathcal{V}_i$. The *recombination* operation at anchors a and b between two phages p and q yields new phages c and d:

$$p = p_0 p_1 \cdots p_{a-1} | p_a \cdots p_{b-1} | p_b \cdots p_{m-1}$$
$$q = q_0 q_1 \cdots q_{a-1} | q_a \cdots q_{b-1} | q_b \cdots q_{m-1}$$

yields

---

[1] After invading a cell, linear phage genomes are often circularized. This is due to a variety of mechanisms: a circular configuration may protect phages from degradation by the defense mechanisms of the bacteria; it may allow the phage genome to be duplicated as a plasmid, or to be integrated in the host genome; or it may be used to initiate a *rolling circle* replication procedure that leads to a concatenamer [16].

$$c = p_0 p_1 \ldots p_{a-1} | q_a \ldots q_{b-1} | p_b \ldots p_{m-1}$$
$$d = q_0 q_1 \ldots q_{b-1} | p_a \ldots p_{b-1} | q_b \ldots q_{m-1}.$$

The recombining phages are called *parents*, and the newly constructed phages, their *children* or *descendants*, the pair $\{c, d\}$ is a pair of *twins*. Comparisons between phage populations are based on the following relation, which is the main focus of this paper:

Definition 1. *Given two populations P and Q, a recombination scenario from P to Q is a sequence of recombinations that constructs all phages of Q using phages of P and their descendants. When there exists at least one recombination scenario from P to Q, we say that P generates Q, and we write $P \to Q$. The number of recombinations in a shortest scenario is ℓ_{PQ}.*

There is a simple way to check whether $P \to Q$. Indeed we have:

Proposition 1. *The relation $P \to Q$ holds if and only if, for each module, every variant that appears in Q also appears in P.*

If follows from Proposition 1 that the existence of $P \to Q$ does not imply the existence of $Q \to P$, since certain variants of modules of population P may have been lost in the recombination process. The operation $P \to Q$ has the following properties:

Proposition 2. *For any population P, $P \to P$ and $\ell_{PP} = 0$. If $P \to Q$ and $Q \to R$, then $P \to R$, and $\ell_{PR} \leq \ell_{PQ} + \ell_{QR}$.*

Since the measure ℓ_{PQ} is not symmetric, it is not a distance, but it is always possible to convert it to a distance by considering the sum $\ell_{PQ} + \ell_{QP}$. However, with actual biological data, it turns out to be much more interesting to compare ℓ_{PQ} and ℓ_{QP}.

The central problem that we address in this paper is the following:

Problem 1. Given phage populations P and Q such that $P \to Q$, compute a recombination scenario of length ℓ_{PQ}.

The computational complexity of Problem 1 is currently unknown, even when $|Q| = 1$. The main theoretical hurdle is that the notion of *breakpoint*, which is central to most genome rearrangement problems, is not well-defined: it is often impossible to determine *a priori* the number, nature, or positions of breakpoints.

In order to develop approximate solutions, we need objective functions that are guaranteed to decrease at each iteration of the process, these are developed in Sect. 2.2, where a first greedy heuristics is outlined. Evaluating the performance of the heuristics requires theoretical lower bounds for the length of a recombination scenario. These bounds are derived in Sect. 2.3.

2.2 Minimum Covers

We define a *circular interval* $(s..t)$ as the subset of integers $\{s, s+1, \ldots, t\}$, where additions are done modulo m. An *interval* in a phage p is denoted by $p(s..t)$; a phage interval $p(s..t)$ is *contained* in a phage interval $q(s'..t')$ if the circular interval $(s..t)$ is contained in $(s'..t')$, and $p_k = q_k$ for all k in $(s..t)$.

In particular, the equality of phage intervals $p(s..t) = q(s..t)$ implies that they share the same modules with the same variants.

Definition 2 (Covers and minimum covers)
Let P be the population of parents, and Q the population of children. A cover of a child c is a set $\mathcal{C}(c) = \{p^1(s_1..t_1), p^2(s_2..t_2), \ldots, p^n(s_n..t_n)\}$ of n intervals, where each phage $p^k \in P$, and such that:

1. *The union $\bigcup_{k \in \{1..n\}} p^k(s_k..t_k)$ is equal to c;*
2. *No interval in $\mathcal{C}(c)$ is contained in another interval of $\mathcal{C}(c)$.*
3. *Each interval $p^k(s_k..t_k)$ is maximal, in the sense that neither $p^k(s_k - 1..t_k)$, nor $p^k(s_k..t_k + 1)$ is contained in phage c;*

A minimum cover is a cover with the smallest number of intervals.

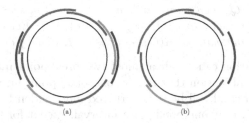

Fig. 4. (a) A cover of a circular phage, in black, by intervals of four potential parents. (b) A minimum cover extracted from cover (a).

Figure 4 gives an example of a cover and a minimum cover. The condition that no interval of a cover is contained in another implies that, in a cover, all left bound s_k are distincts and all right bounds t_k are distincts. Thus the intervals of a cover can be ordered along the circle by their distinct and increasing left bounds s_k, and we can refer without ambiguity to a pair of *consecutive intervals* of a cover $p(s_k..t_k)$ and $q(s_{k+1}..t_{k+1})$. These can be used to propose a first definition of breakpoints induced by covers:

Definition 3 (Breakpoint interval). *Given two consecutive intervals of a cover $p(s_k..j - 1)$ and $q(i..t_{k+1})$, where $i \le j$, of a child c, then the interval $(i..j)$ of anchors is called a breakpoint interval.*

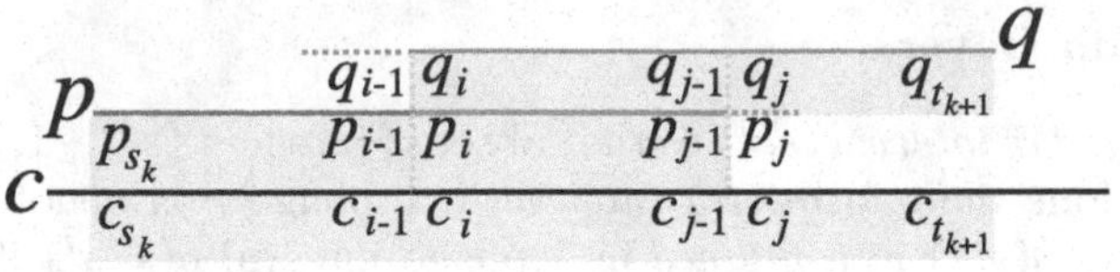

$$p \quad \begin{array}{cccc} & q_{i-1}\,q_i & q_{j-1}\,q_j & q_{t_{k+1}} \end{array} \quad q$$

Fig. 5. Breakpoint interval. All variants of modules in shaded areas are equal. A recombination of p and q with anchors $a \in (i..j)$ and b outside the interval $(s_k..t_{k+1})$ will create the interval $c(s_k..t_{k+1})$ of child c. Note that the interval $(i..j)$ corresponds either to the overlap of parents p and q, or, when $i = j$, is the single anchor shared by p and q.

Note that when $i = j$, there is a single anchor in the interval, and this corresponds to the classical notion of a breakpoint. However, breakpoint intervals can be arbitrary wide in the general case. Figure 5 illustrates the concept. Breakpoint intervals correspond to anchors that can be used to construct the union of two consecutive intervals. Indeed, a recombination of phages p and q, with anchors $a \in (i..j)$ and b outside the union of $p(s_k..j - 1)$ and $q(i..t_{k+1})$, that is $b \in (t_{k+1} + 1..s_k)$, will create the interval $c(s_k..t_{k+1})$ of child c. Such a recombination is said to *repair* the breakpoint interval.

Proposition 3 (Upper bound). *Let $s(c)$ be the size of a minimum cover by P for each child $c \in Q$, then there exists a recombination scenario from P to Q of length less than or equal to $R(Q) = \sum_{c \in Q}(s(c) - 1)$. Each recombination in the scenario lowers the value of $R(Q)$ by at least 1, and by at most $2|Q|$.*

Proof. A recombination can repair at most two breakpoint intervals, and there always exists a recombination that repairs one breakpoint interval of a child. In the worst case, each child will be reconstructed independently, and with recombinations that repair only one breakpoint interval, except for the last one, since a child with a minimum cover of size 2 can always be constructed in one recombination. In the best case, all children share the same two breakpoint intervals in their minimum covers, implying that $R(Q)$ may decrease by as much as $2|Q|$.

From Proposition 3, we can sketch a greedy heuristics that tests all candidate recombinations and selects an optimal one. Performance, in terms of computer resources, is not a priority, unless some steps have the potential to lead to combinatorial explosions.

Greedy Heuristics
Input: Two populations P, Q such that $P \to Q$, and $P \cap Q = \emptyset$
While $Q \neq \emptyset$

1. Compute the size $s(c)$ of a minimum cover for each phage $c \in Q$, and $R(Q)$.
2. If there exists c such that $s(c) = 2$, remove c from Q. Add the two children of the recombination that creates c to P.
3. Otherwise, find a recombination that maximizes $R(Q) - R'(Q)$, where $R'(Q)$ is computed after simulating each possible candidate recombination. Add the children of this recombination to P.

Fortunately, due to applications in surveillance systems, the problem of computing minimum covers of circles has received a lot of attention. Lee and Lee [12] gave a solution in $\mathcal{O}(|S|)$ if the intervals of S are maximal elements, in terms of inclusion, and the intervals are sorted by their increasing starting points. In this case, there is at most one interval in S that begins, or ends, at any point in $(0..m-1)$. Let I be an interval in S that ends at t, its successor $succ(I)$ is the – unique – interval that contains $(t \bmod m)$, and has the largest starting point. The algorithm described in [12] finds a minimum cover by iterating the function $succ(I)$. When the iteration begins with the largest interval of S, a minimum cover is found after at most $2m$ iterations.

However, in general, minimum covers are far from unique: many pairs of parents can repair a breakpoint interval, and a child may have alternate minimum covers that do not share any breakpoint interval. Moreover, there exists the theoretical possibility that a shortest scenario must use breakpoint intervals of covers that are not minimal (see Example 1 of Annex 1).

Our first experiments with phage data appeared to yield "pretty good" results. Many recombinations were shared by two or more children, and many children were constructed using exactly $\left\lfloor \frac{s(c)+1}{2} \right\rfloor$ recombinations, where $s(c)$ is the size of a minimal cover of child c, which is a strict lower bound for a single child.

Quantifying "pretty good" without relying on minimum covers is discussed in the next section.

2.3 Lower Bounds

In this section, we explore under which conditions some breakpoints are *mandatory*, in the sense that they belong to any possible cover. As we will show, it is easy to construct an example with no mandatory breakpoints. Thus, a lower bound based on mandatory breakpoints has the potential to be useless in the general case, but, in datasets that come from the comparison of phages populations, they are sufficiently abundant to provide practical lower bounds.

The formal definition is based of the detection of *useful breakpoints*, that identify intervals that no single parent can cover, but that are covered by two consecutive intervals of a cover.

Definition 4. *A useful breakpoint is an interval $c(i-1,j)$ of a child c in Q, that is not contained in any parent of P, yet both $c(i-1,j-1)$ and $c(i,j)$ are. A useful breakpoint is thus the interval spanning a breakpoint interval together with its two flanking modules.*

A useful breakpoint can be *erased* by simply adding to the set of parents P a new parent that contains the interval $c(i-1..j)$: Example 1 of Annex 1 shows a population whose only child has 9 breakpoint intervals, but only two of them are useful. On the other hand, in our phage comparisons, almost all breakpoint intervals are useful.

```
          Parents
          A:  a a a a a a a a a a a a a
          B:  b b b b b b b b b a b a a
          C:  c c c c c c c c c c b c c
          D:  d d d d d d c d d d d a a
       Children
          X:  a|d|b|c|d|c|c|d|b a b|a|a|
          Y:  a a|b|c c|d|c|c|a a b c|a
      Anchors: 0  1 2 3 4 5 6 7 8 9 0 1 2 0
```

Fig. 6. In this example child Y has 6 mandatory breakpoint. Five have a single anchor: a|b with anchor {2}, b|c with anchor {3}, c|d with anchor {5}, c|a with anchor {8}, c|a with anchor {12}; and one has two anchors d|c|c with anchors {6, 7}. One region, in gray, contains two overlapping useful breakpoint, a|a|b with anchors {9, 10}, and a|b|c with anchors {10, 11}. Breakpoints in blue are exclusive to Y, breakpoints in red are exclusive to X, and the three breakpoints in violet are *shared*.

Two sets *overlap* if their intersection is non-empty, and neither is contained into the other. When the sets of anchors of two or more useful breakpoint overlap, then there exists alternative covers of different lengths for that region. We want to eliminate this possibility:

Definition 5 (Mandatory breakpoint). *A mandatory breakpoint of a child c in Q is a useful breakpoint whose set of anchors does not overlap any other such set of anchors.*

In Fig. 6, all breakpoints of child X are mandatory, and we can deduce a cover from these breakpoints by constructing the sequence of parents that are exchanged at each breakpoint. In this case the only possible cover is ADBCDCDBA.

However, child Y = aabccdccaabca that has two overlapping useful breakpoint: a|a|b with anchors {9, 10}, and a|b|c with anchors {10, 11}. This yield two different covers of Y, one of length 9, ABCDCABCA, and one of length 8, ABCDCACA.

We say that two mandatory breakpoints $c(i-1, j)$ and $d(i-1, j)$ are *shared*, if they have the same set of anchors and a common pair of parents (p, q) that can repair them. Then either $c(i-1, j) = d(i-1, j)$, or $c(i-1, j)$ and $d(i-1, j)$ are a pair of twins constructed by the same recombination.

In Fig. 6, there is one pair of equal mandatory breakpoints between children X and Y at anchor {3} with parents B and C, and two pair of twins, one with anchor {5}, and one with anchors {6, 7}, both with parents C and D.

Being based on equality, the 'shared' relation is an equivalence relation on the set of mandatory breakpoints. In the example of Fig. 6, child Y has 6 mandatory breakpoints and X has 8. Three of them are shared, thus there are 11 breakpoints that must be repaired, yielding a minimum of 6 recombinations, since a recombination repairs at most two shared mandatory breakpoints (see Example 2 of Annex 1 for a recombination scenario of length 6).

In general, we have:

Proposition 4. *Let P and Q be two populations such that P → Q, let M be the number of shared mandatory breakpoints of children in Q with respect to P, then the length of a shortest recombination scenario is at least* $r(Q) = \lfloor \frac{M+1}{2} \rfloor$.

3 Experiments

3.1 Dataset Construction

In order to identify core and variable genomes, we aligned the 38 genomes of the Murphy study [14] using the Alpha aligner [1]. This aligner identifies the core and variable genomes, the core becoming the anchors between the variable regions that comprise the modules. For this experiment, we used the default values of the software. Alpha was the preferred alternative to a painstaking and time consuming breakpoint analysis done by hand, where sequence similarity is queried using a tool such as BLAST. Alpha is adapted specifically to phages, and identifies major breakpoints and variants in a single automated step.

The size of the 38 genomes varied between 29 097 and 31 049 bps, with a core genome of size 7722 bps distributed across 84 anchors.

Each phage p_j from the collection receives an identifier that is a single unique letter, and a new variant for module i in phage p_j receives the identifier of p_j as its variant identifier. This yields the set of strings displayed in Fig. 2a, which is processed by the greedy algorithm. Each phage belongs to one of four factories, identified by F1, F2, F3 and F4 in the original paper. The number of phages per factory are, respectively, 13, 4, 13 and 8.

When computing a recombination scenario from a source factory to a target factory, we may encounter a variant that occurs only in the target and not in the source. We call such a variant a *missing* variant. The algorithm deals with missing variants in two ways: if stretches of missing variants are shared by two or more children in the target factory, supplementary chromosomes that contain these stretches are added to population P; if stretches of missing variants are unique to a child, the importation of each stretch counts as one recombination, and these recombinations can be done at the end of the scenario. These strategies simulate the 'genetic reservoir', and should be sound as long as the stretches are not too long.

3.2 Comparing Factories

Each factory was compared with the three others, resulting in 12 recombination scenarios whose length varied from 29 to 158. Table 1 presents various statistics of these scenarios, and a *gap ratio* that compares the performance of the heuristics to both the lower bound $r(Q)$, given by Proposition 4 and the upper bound $R(Q)$, given by Proposition 3. If $L \leq R(Q)$ is the length of an actual scenario, we score it with the formula:

$$\text{Gap ratio} = \frac{(R(Q) - L)}{(R(Q) - r(Q))}.$$

Gap ratios range from 0, for the worst scenario, to 1 for a scenario whose length is equal to $r(Q)$.

Since ℓ_{PQ} depends on the number of phages to reconstruct, we also give in Table 1 the average length needed to reconstruct one child.

Table 1. Comparisons of four Dutch dairies. Scenario F1 → F2 computed by the greedy algorithm has 29 recombinations, and theoretical lower bound of 24, and upper bound of 47, yielding a gap ratio of 0.78; among the 29 recombinations, 20 were used to import missing data; the size of target factory F2 is 4 phages, thus the average length of the scenario is 7.25 recombinations per phage.

Factories	Scenario Length	Lower Bound $r(Q)$	Upper Bound $R(Q)$	Gap ratio	Missing Data	Target Dairy Size	Average Length
F1 → F2	29	24	47	0.78	20	4	7.25
F1 → F3	112	88	213	0.81	77	13	8.62
F1 → F4	84	60	161	0.76	53	8	10.50
F2 → F1	158	111	358	0.81	190	13	12.15
F2 → F3	149	107	402	0.86	105	13	11.46
F2 → F4	109	80	231	0.81	79	8	13.63
F3 → F1	152	90	361	0.77	101	13	11.69
F3 → F2	53	39	90	0.73	40	4	13.25
F3 → F4	110	81	209	0.77	73	8	13.75
F4 → F1	151	109	282	0.76	100	13	11.62
F4 → F2	54	42	90	0.75	36	4	13.50
F4 → F3	151	103	370	0.82	94	13	11.62

Gap ratios range from 0.73 to 0.86, which is promising, given our very conservative lower bounds that assume that all recombinations occur between pairs of mandatory breakpoints; and given the fact that the upper bounds, based on minimum covers, cannot be lowered in the general case. Indeed, upper bounds are reached on recombination scenarios in which children do not share breakpoints, and on parents who contribute at most a single interval of contiguous modules in a child. In our dataset, specialized subpopulations were apparent in Fig. 2, implying that children of these subpopulations did not share breakpoints, and parents that contribute a single interval of contiguous modules in a child are used to model the "genetic reservoir".

Another interesting aspect of these comparisons is that Factory F1 is obviously the best 'constructor', with an average of 8.79 recombinations to reconstruct all other factories, compared to 12.90 for Factory F3, for example. On the other hand, F4 is the hardest to construct, being the highest result for each of the other factories.

3.3   Shared Evolution

We tested how much shared evolution there was when creating the phages from one factory using the phages from another. To do this, we compared the scenario lengths from Table 1 to the scenario lengths obtained separately from each factory to each phage individually. The results are shown in Table 2. The table shows that, by considering the phages in the target factories simultaneously, we economize 12, 1.75, 15.7, and 7.6 recombinations per phage for target factories F1, F2, F2, and F4 respectively.

Table 2. Creating each phage individually instead of creating all phages from a factory at once. Each individual phage for a factory was created independently of the others. The sum of the lengths of the recombination scenarios creating all phages from F1, using the phages from F2, is in the first row: 61 more recombinations were required to do this as compared to creating all phages of F1 with shared recombinations. Overall, to create F1 from each of the other factories we use 13.63 recombinations more per phage.

Factories	Scenario length	Sum of individual lengths	Difference	Economy per phage
F2 → F1	158	219	61	13.23
F3 → F1	152	230	78	
F4 → F1	151	184	33	
F1 → F2	29	31	2	2.75
F3 → F2	53	55	2	
F4 → F2	54	61	7	
F1 → F3	112	131	19	15.85
F2 → F3	149	247	98	
F4 → F3	151	244	89	
F1 → F4	84	100	16	10.25
F2 → F4	109	150	41	
F3 → F4	110	135	25	

Due to the limited numbers of phages sampled from each factory, the variants from a target factory was not always present in a source factory. These missing variants played a role in our comparison, as they favored single-recombination replacement of longer stretches in the single-target comparisons. As described in Sect. 3.1, each stretch of missing variants that did not exist in any phage of a target factory were counted as a single recombination. Consider one of these stretches defined on a target factory with a single phage. Adding phages to this factory can only fragment these stretches, implying more stretches and more recombinations.

This gives single phage targets a significant advantage, in that they will have longer stretches of missing variants that each can be repaired with a single recombination. Despite this advantage enjoyed by the single phage targets, we observe large savings in all cases.

3.4 Population Structure

How distinctive are the populations within the factories? We approached this question by conducting experiments to test how random the structure is within the factories. The test statistic that we used was the sum of the scenario lengths over all pairs of factories.

The first null hypothesis was that the phages were randomly partitioned into factories of sizes 13, 4, 13, and 8. We tested this hypothesis by repeatedly (n times) partitioning the data uniformly at random into the prescribed sizes and rerunning the experiment of Sect. 3.2. This gave us the null distribution on the test statistic. A single sample T-test against this null distribution yielded a p-value lower then 10^{-6} even with $n = 20$. For larger n the p-value dropped precipitously.

The second null hypothesis we tested took into account the group structure as defined in Fig. 2b. The null hypothesis was that the phages were randomly distributed to the factories while preserving the group structure within each factory. That is, define a group structure vector [A, B, C, D, E] for a factory, containing the frequency of each group in the factory. We construct four random factories with the following frequency vectors:

$$F1 = [3, 1, 3, 3, 2],$$
$$F2 = [2, 0, 1, 0, 1],$$
$$F3 = [5, 5, 0, 0, 0], \text{ and}$$
$$F4 = [3, 0, 1, 1, 0].$$

We run the experiment of Sect. 3.2 on n of these randomly constructed factories to obtain the null distribution on our test statistic. A single sample T-test against this null distribution for $n = 100$ gives a p-value is less than 10^{-11}.

4   Discussion and Conclusion

In this paper, we described the first combinatorial framework for the comparison of bacteriophage populations using recombinations. Our work represents a shift to a more global perspective of evolutionary analysis for bacteriophages, since the alignment-centric view breaks down in the presence of large amounts of recombination [3, 18].

Our experiments show that the populations of phages sampled within the factories are not random, suggesting isolated evolution within individual factories. They also illuminate the potentially large amount of shared evolution in the recombination histories leading to the factories that we see today, supporting the genetic reservoir hypothesis.

While the application of our methods to the Dutch factories gives us insight into the relative diversity of phages in one factory with respect to another, our application is "still seriously data limited" due to the disproportionately sparse sampling of phage populations [3]. We expect the gene reservoir paradigm of variant sharing between populations to gain tremendous significance as better samplings become possible. There are signs that this time could be near, as a recent survey estimates the number of different phage species in the ocean to be more than 195,000 [8].

The problem of recombination scenario inference is a tricky one, whose computational complexity remains unknown. However, with some realistic assumptions on real data, we were able to give good solutions to comparisons arising from biological data. There are still many combinatorial problems that arise from this framework, which include increasing the lower bound—since we used a very conservative approach—and the development of better algorithms to find recombination scenarios.

Dataset

Bacteriophage genomes used in this paper identified by their one letter code, their name, and their accession number.

A	Phi17	KP793114	B	Phi13.16	KP793116
C	Phi19	KP793103	D	PhiJF1	KP793129
E	Phi43	KP793110	F	Phi4	KP793101
G	PhiG	KP793117	H	PhiA.16	KP793102
I	PhiD.18	KP793107	J	PhiL.18	KP793120
K	PhiF.17	KP793113	L	Phi5.12	KP793108
M	PhiM.16	KP793128	N	Phi109	KP793121
O	Phi93	KM091443	P	Phi129	KP793112
Q	PhiLj	KP793133	R	Phi155	KP793130
S	Phi16	KP793135	T	Phi44	KP793124
U	Phi114	KP793115	V	Phi15	KM091442
W	Phi40	KP793127	X	Phi145	KM091444
Y	Phi10.5	KP793119	Z	Phi19.3	KP793105
a	Phi19.2	KP793111	b	PhiL.6	KP793122
c	Phi4.2	KP793123	d	PhiM.5	KP793126
e	PhiF0139	KP793118	f	PhiA1127	KP793106
g	PhiC0139	KP793109	h	Phi91127	KP793125
i	PhiB1127	KP793104	j	PhiS0139	KP793134
k	PhiE1127	KP793131	l	PhiM1127	KP793132

Funding Information. AB is partially supported by Canada NSERC Grant number 05729-2014. MJM is partially supported by Canada NSERC Grant number 06487-2017. RVL acknowledges the support of the Natural Sciences and Engineering Research Council of Canada (NSERC) and the Fonds de recherche du Québec - Nature et technologies (FRQNT). KMS is partially supported by VIROGENESIS (EU H2020-PHC-32-2014 #634650) and the Labex NUMEV (ANR-10-LABX-20, 2017-2-46).

Annex 1

Example 1. A shortest scenario is not necessarily tied to a minimum cover.

```
        Parents:              Recombination 1
        A: Xoooooooo          G: ooXX|oo|XXo
        B: oXXooooooo         H: oooo|XX|ooo
        C: oooXXoooo
        D: oooooXXoo          1: ooXX|XX|XXo
        E: oooooooXX          2: oooo|oo|ooo
        F: oXooooooX
        G: ooXXooXXo          Recombination 2
        H: ooooXXooo          F: oX|oooooo|X
                              1: oo|XXXXXX|o
        Child:
        J: XXXXXXXXX          3: oX|XXXXXX|X
                              2: oo|oooooo|o

                              Recombination 3
                              A: X|oooooooo|
                              3: o|XXXXXXXX|

                              J: X|XXXXXXXX|  *
                              2: o|oooooooo|
```

Child J has a minimum cover of size 5, namely ABCDE. Thus a shortest scenario must have at least 3 recombinations. Using the minimum cover, there is a trivial scenario of length 4, but there is an alternate one of length 3 that uses the cover AFGHGF, which is not a minimal cover.

Example 2. A recombination scenario of length 6 for the example of Fig. 6.

```
Parents
  A: aaaaaaaaaaaaa
  B: bbbbbbbbbabaa
  C: cccccccccccbcc
  D: dddddcdddaa
Children
  X: adbcdccdbabaa
  Y: aabccdccaabca
```

Recombination 1
```
C: ccccc|c|ccccbcc
D: ddddd|d|cddddaa

1: ccccc|d|ccccbcc
2: ddddd|c|cddddaa
```

Recombination 2
```
A: aaaaaaa|aa|aaa
1: ccccdcc|cc|bcc

3: aaaaaaa|cc|aaa
4: ccccdcc|aa|bcc
```

Recombination 3
```
B: bb|bbbbbbbab|aa
3: aa|aaaaaacca|aa

5: aa|bbbbbbbab|aa
6: bb|aaaaaacca|aa
```

Recombination 4
```
4: ccc|ccdccaabc|c
5: aab|bbbbbbaba|a

Y: aab|ccdccaabc|a *
7: ccc|bbbbbbaba|c
```

Recombination 5
```
2: d|ddddccd|dddaa
5: a|abbbbbb|babaa

8: a|ddddccd|babaa
9: d|abbbbbb|dddaa
```

Recombination 6
```
Y: aa|bc|cdccaabca
8: ad|dd|dccdbabaa

X: ad|bc|dccdbabaa *
10: aa|dd|cdccaabca
```

References

1. Bérard, S., Chateau, A., Pompidor, N., Guertin, P., Bergeron, A., Swenson, K.M.: Aligning the unalignable: bacteriophage whole genome alignments. BMC Bioinform. **17**(1), 30 (2016). https://doi.org/10.1186/s12859-015-0869-5
2. Botstein, D.: A theory of modular evolution for bacteriophages. Ann. N. Y. Acad. Sci. **354**(1), 484–491 (1980). https://doi.org/10.1111/j.1749-6632.1980.tb27987.x
3. Brüssow, H.: Population genomics of bacteriophages. In: Polz, M.F., Rajora, O.P. (eds.) Population Genomics: Microorganisms. PG, pp. 297–334. Springer, Cham (2018). https://doi.org/10.1007/13836_2018_16
4. Cafora, M., et al.: Phage therapy against pseudomonas aeruginosa infections in a cystic fibrosis zebrafish model. Sci. Rep. **9**(1), 1527 (2019). https://doi.org/10.1038/s41598-018-37636-x
5. Castro-Nallar, E., et al.: Population genomics and phylogeography of an Australian dairy factory derived lytic bacteriophage. Genome Biol. Evol. **4**(3), 382–393 (2012). https://doi.org/10.1093/gbe/evs017
6. Chmielewska-Jeznach, M., Bardowski, J.K., Szczepankowska, A.K.: Molecular, physiological and phylogenetic traits of lactococcus 936-type phages from distinct dairy environments. Sci. Rep. **8**(1), 12540 (2018). https://doi.org/10.1038/s41598-018-30371-3

7. De Paepe, M., Hutinet, G., Son, O., Amarir-Bouhram, J., Schbath, S., Petit, M.A.: Temperate phages acquire DNA from defective prophages by relaxed homologous recombination: the role of Rad52-like recombinases. PLoS Genet. **10**(3), e1004181 (2014)

8. Gregory, A.C., et al.: Marine DNA viral macro- and microdiversity from pole to pole. Cell **177**(5), 1109–1123.e14 (2019)

9. Hatfull, G.F.: Dark matter of the biosphere: the amazing world of bacteriophage diversity. J. Virol. **89**(16), 8107–8110 (2015). https://doi.org/10.1128/JVI.01340-15

10. Kupczok, A., et al.: Rates of mutation and recombination in siphoviridae phage genome evolution over three decades. Mol. Biol. Evol. **35**(5), 1147–1159 (2018). https://doi.org/10.1093/molbev/msy027

11. Lavelle, K., et al.: A decade of streptococcus thermophilus phage evolution in an Irish dairy plant. Appl. Environ. Microbiol. **84**(10), e02855-17 (2018). https://doi.org/10.1128/AEM.02855-17

12. Lee, C., Lee, D.: On a circle-cover minimization problem. Inf. Process. Lett. **18**(2), 109–115 (1984)

13. Marcó, M.B., Moineau, S., Quiberoni, A.: Bacteriophages and dairy fermentations. Bacteriophage **2**(3), 149–158 (2012). https://doi.org/10.4161/bact.21868. pMID 23275866

14. Murphy, J., et al.: Comparative genomics and functional analysis of the 936 group of lactococcal siphoviridae phages. Sci. Rep. **6**, 21345 (2016)

15. Pedulla, M.L., et al.: Origins of highly mosaic mycobacteriophage genomes. Cell **113**(2), 171–182 (2003)

16. Skalka, A.M.: DNA replication-bacteriophage lambda. In: Arber, W., et al. (eds.) Current Topics in Microbiology and Immunology, pp. 201–237. Springer, Heidelberg (1977). https://doi.org/10.1007/978-3-642-66800-5_7

17. Swenson, K.M., Guertin, P., Deschênes, H., Bergeron, A.: Reconstructing the modular recombination history of staphylococcus aureus phages. BMC Bioinform. **14**, S17 (2013)

18. Yahara, K., Lehours, P., Vale, F.F.: Analysis of genetic recombination and the pan-genome of a highly recombinogenic bacteriophage species. Microb. Genom. **5**(8) (2019)

Syntenic Dimensions of Genomic Evolution

Zhe Yu and David Sankoff[(⊠)]

Department of Mathematics and Statistics, University of Ottawa,
Ottawa, ON K1N 6N5, Canada
sankoff@uottawa.ca

Abstract. We compare several types of evolutionary divergence of synteny blocks: sequence level divergence of the genes in the block, loss of genes affecting the length and structure of the blocks and spatial position of the block in relation to the chromosomal centromere, and suggest other dimensions, such as the predominant functional characteristic of the genes in the block. We focus on the evolutionary history of the allotetraploid *Coffea arabica* genome and its two progenitor genomes through three major genomic events spanning 120 million years.

Keywords: Synteny block · Coffee fractionation · Sequence divergence · Chromosomal structure · Principal components

1 Introduction

Sequence-level divergence is the universal criterion for the degree of genomic evolution. There is a widely-assumed need to align this divergence to chronological time, whether by fossil dating or strictly comparative techniques. The divergence of sequences, however, is measured in many different ways. The K_s score is essentially an assessment of the amount of evolutionary drift due to synonymous, i.e. translationally neutral, mutations in the codons of a gene, a way of avoiding biases due to selection acting on proteins. In contrast, total sequence difference $Dist$ includes mutations which change protein, especially as synonymous coding positions become saturated over time. The fourfold degeneracy transversion rate 4DTv [1] measures not the evolutionary drift as does K_s, and not non-synonymous changes as does $Dist$, but how far evolution has proceeded in the direction of increased transversions over transitions in mutations at synonymous sites (e.g. [2]). The three measures K_s, $Dist$ and 4DTv measure three different aspects of sequence evolution, somewhat independant of each other, and are thus of more biological interest than merely competing approaches to the inference of chronological time.

In Sects. 4 and 5, we introduce a number of other ways of quantifying the degree of evolution and investigate to what extent these new ways, as well as K_s and $Dist$ developed in Sect. 3, proceed independently of each other. These new approaches are based not on individual genes, but on syntenic blocks, sets

© The Author(s), under exclusive license to Springer Nature Switzerland AG 2022
L. Jin and D. Durand (Eds.): RECOMB-CG 2022, LNBI 13234, pp. 21–30, 2022.
https://doi.org/10.1007/978-3-031-06220-9_2

containing several (default at least 5) collinear homologous genes in two genomes or two subgenomes, created by evolutionary events such as speciation and polyploidization. In particular we assess the length of the synteny blocks, the rate of fractionation (loss of one of a pair of homologous genes), gap size between two unfractionated pairs of genes, and position of the synteny block on the chromosome in relation to the centromere. For comparability, we include $Dist$ and K_s in terms of their mean values over all genes in a synteny block. In Sect. 2 we discuss how synteny blocks may be constructed or inferred from comparative genomic data.

We study these properties of syntenic blocks in the context of a soon-to-be published project on the sequence, evolution and population history of the *Coffea arabica* genome[1]. We access the annotated sequence of this allotetraploid genome, with 11 pairs ($n = 22$) of homologous chromosomes, as well as the two 11-chromosome progenitor genomes *C. canephora* and *C. eugenioides*. We focus on three evolutionary events occurring at three points of time, namely the gamma hexaploidization at the root of the core eudicot clade 1.2×10^8 Mya, the speciation event giving rise to the ancestors of *C. canephora* and *C. eugenioides* $\approx 10^7$ Mya, and the allotetraploidization event creating *C. arabica* < 1 Mya, as schematized in Fig. 1.

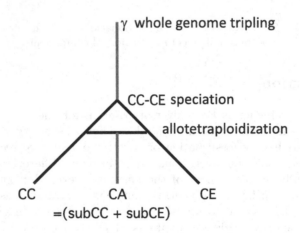

Fig. 1. Three genomic events in the evolutionary history of *C. arabaca*: the gamma hexaploidization at the root of the core eudicots, the speciation giving rise to *C. canephora* (CC) and *C. eugenioides* (CE), and the allotetraploidization event producing *C. arabica* (CA), composed of subgenomes subCC and subCE.

2   The Construction and Biological Significance of Synteny Blocks

Synteny blocks are in the first instance analytical constructs. In the genomes of two species such as CC *versus* CE or CC *versus* subCE, the two sets of collinear

---

[1] J. Salojärvi, personal communication.

genes in a block are of course "unaware" of each other, and what happens to one is independent of what happens to the other. In the subgenomes of a polyploid, on the other hand, such as subCC *versus* subCE, there is likely some degree of interaction, at least in the early days (measured in My), due to dosage considerations, often leading to fractionation, the loss of one of a pair of homologous genes in the block, usually not both.

The methodology we employ to infer synteny blocks is contained in the widely-used and well-documented SYNMAP package on the COGE platform [3,4]. This combines a syntenically-validated version of BLAST with a gap-size constrained search for collinear genes, DAGCHAINER [5], which is followed by the QUOTAALIGN algorithm [6] for piecing together neighbouring synteny blocks.

This pipeline is designed to balance the search for all blocks in the data, and their full extents, against coincidence, random effects and other noise. Thus, for example, short blocks may be missed and long blocks with gaps may appear as two blocks.

For the four *Coffea* genomes and subgenomes CC, CE, subCC and subCE (denoted just subCE), we carried out six SYNMAP comparisons of pairs, and four self-comparisons, as listed in Table 1. Each comparison of two genomes reveals synteny blocks for all the events that have affected both. All ten comparisons contain synteny blocks created by the core eudicot hexaploidization gamma. Four these comparisons, those involving either CC or subCC *versus* CE or subCE, also contain synteny blocks created by the speciation of CC and CE, while CC *versus* subCC and CE *versus* subCE contain synteny blocks caused by the tetraploidization event.

Whereas estimates of K_s or $Dist$ for a pair of homologous genes may have a large variance, the average of these quantities in a synteny block containing many pairs, all of which were created at the same time, is much more precise. This makes it fairly easy to assign each of the hundreds of synteny blocks encountered in a genomic comparison to one of the three genomic events.

3   Review of Sequence Divergence

The distinctions between the measures of sequence divergence are based on their different focus on synonymous and non-synonymous mutations and on the different quantities that are inferred from the mutational counts.

The aim of this section is to emphasize the multiplicity of evolutionary directions followed by genomes. Thus although we model the connection between each measure of sequence divergence and elapsed time, this does not detract from the conceptual independence of the biological quantities that underlie them.

For our purposes, the sequence distance $Dist$ between two homologous genes, normalized by the total number of nucleotides in the coding sequence (CDS), is most simply modelled as

$$E(Dist) = E(Dist_1 + Dist_2)$$
$$= E(Dist_1) + E(Dist_2)$$
$$= \frac{3}{4}\theta_1(1 - e^{-\lambda_1 t}) + \frac{3}{4}\theta_2(1 - e^{-\lambda_2 t}),$$

(1)

where t measures the time since the divergence of the two aligned genes, and θ_1 and θ_2 are the proportions of non-synonymous and synonymous positions, respectively, in its codons. The factor $\frac{3}{4}$ is due to mutations back to the original state. $\lambda_1 < \lambda_2$ are rate parameters. $Dist$ can be predicted by a linear combination of the Jukes-Cantor distances [8] for non-synonymous and synonymous sites. The proportions θ_1 and θ_2 are often taken to be $\frac{2}{3}$ and $\frac{1}{3}$, respectively. For larger t, such as that for the gamma hexaploidization, the synonymous regions tend toward saturation, so $Dist$ effectively becomes a measure of protein evolution.

Again, for the purposes of our model, we set

$$E(Dist_2) = \frac{3}{4}[1 - e^{-\lambda_2 E(K_s)}],$$
$$E(K_s) = \frac{-\log[1 - \frac{4}{3}E(Dist_2)]}{\lambda_2}$$

(2)

so that K_s depends only on the synonymous positions. Even if $Dist_1$ and $Dist_2$ are inferred from the sequence data in much the same way, i.e., as counts of mutations, K_s is formulated as a direct estimate of t, while $Dist$ is understood to approach a limiting value of $\frac{3}{4}$ as t increases, so is only an indirect measure of elapsed time.

In contrast to $Dist$, the quantity K_s does not reflect any evolution in the protein produced by the gene.

Since some synonymous sites can have fourfold degeneracy while others have two fold degeneracy, a more accurate model would have

$$E(K_s) = E(K_{s4} + K_{s2})$$
$$E(K_{s4}) = \frac{-\log[1 - \frac{4}{3}E(Dist_{2,4})]}{\lambda_2}$$
$$E(K_{s2}) = \frac{-\log[1 - 2E(Dist_{2,2})]}{\lambda_2}$$

(3)

where $s4$ and $s2$ refer to mutations in the fourfold and twofold synonymous sites, respectively.

Yet another measure of sequence divergence is the so-called 4DTv transversion ratio [7], observed in the fourfold degenerate positions. Although this involves compiling all the observed mutations in these positions, it is not interpreted directly in terms of time elapsed as in Eq. (3). Instead, mutations are classified into transitions (purine-purine or pyrimidine-pyrimidine mutations) or transversions (purine-pyrimidine or pyrimidine-purine), and it is the proportion of transversions or the ratio of transversions *versus* transitions that is used as a statistic for gauging elapsed time. The rationale is that transversions are rarer, for DNA-structural reasons, but as time progresses the transitions become saturated so that the transversions reach equity with them. Modelling this process, undertaken by Hasegawa *et al.* [7], is complicated by the fact that an observed mutation may be the result of a series of transitions and transversions before the moment of observation.

What is important is that all three measures we have discussed, $Dist$, K_s and 4DTv, pertain to distinct biological processes, protein evolution, neutral DNA evolution, and local structural change within the class of neutral mutations, respectively, although $Dist$ is of course affected by neutral change, especially at early times.

Besides $Dist$, K_s and 4DTv, there are several other ways of measuring sequence divergence, but in the analysis in Sects. 6 and 7 below, we will focus on two representatives, K_s and 4DTv, to compare with the other measures of evolution we develop.

4 Fractionation and Gap Size

The evolutionary process of gene loss from a genome, such as through DNA excision or pseudogenization [9], counterbalances novel gene acquisition through processes such as tandem duplication, gene family expansion and whole genome doubling. Loss serves a number of functions, not least in compensating for the energetic, material and structural costs of gene complement expansion.

A measure of evolution reflecting gene loss is the decreasing length of synteny blocks. Two processes erode gene length. One is genome rearrangement, which simply breaks blocks into two shorter ones. The other is gene loss, creating gaps within the block, leading eventually to the fragmentation of a block.

One type of gene loss widespread in the lineages of plant genomes, and also occurring in a few yeast, fish and amphibian genomes, is "fractionation" after whole genome doubling or tripling, where over time one of a pair or triplet of paralogous genes in parallel syntenic contexts is discarded. A second measure of evolution is then the proportion of fractionated gene pairs (i.e., singletons) within a synteny block.

It is not always clear whether gene pairs are fractionated randomly or whether whole chromosomal fragments are excised in a single event [10]. Our third measure reflects this, the mean of the frequency distribution of gap lengths within syntenic blocks calculated during the comparison of chromosomes from two genomes or subgenomes.

A somewhat surprising result from [9] is that the dynamics of gene loss from syntenic blocks of orthologous genes after speciation parallels the process of fractionation after polyploidization. This includes the fate of deleted genes, the erosion of synteny blocks, the increase in fractionation rate and the size of gaps in the synteny blocks. Indeed, in many ancient polyploids, we cannot assume that fractionation has involved interaction between the two subgenomes, since the same statistics pertain equally well to gene loss from orthologous synteny blocks in the genomes of independent species.

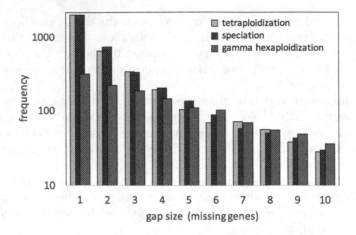

Fig. 2. Gap size frequencies (log scale) in syntenic blocks generated by three events.

Figure 2 shows the frequency distribution of gap lengths within syntenic blocks calculated during the comparison of chromosomes from two genomes or subgenomes.

All three measures related to gene loss are tabulated in Table 1.

5   Spatial Evolution

The distribution of genes on chromosomes is not homogeneous. Heterochromatin, particularly constitutive heterochromatin, including the subtelomeric region and the pericentromeric heterochromatin around the centromere are gene poor.

Part of this relative gene sparsity in the pericentromeric region in *Coffea* is due to the high rate of fractionation there. Figure 3 shows how fractionation rates are highest in this region in almost all chromosomes.

Though we have not explored this in the present research, we know that gene function also tends to differ according to chromosomal position. As with many plant genomes, genes on the chromosomal arms remote from the centromere may be enriched for housekeeping functions whereas more species-specific genes acting in response to external stimuli will be situated preferentially close to the centromere, possibly in relation to elevated tandem duplication in this region [11, 12].

To see whether the position of the synteny block with respect to the centromere is affected by evolution, we tabulated the distance (in bp) between the centromere-containing region and the nearest end of the synteny block (Fig. 3). We can call this measure "distance to centromere", or simply "c-mere distance".

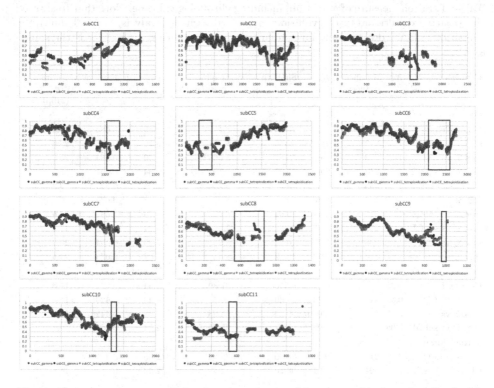

Fig. 3. Gene retention rates (1.0 − fractionation rate) in synteny blocks only, calculated in sliding 100-gene (or gene-pairs) windows along subCC chromosomes; subCE orange points (from tetraploidization) obscuring almost identical trends in subCC grey points (also from teraploidization), subCC from gamma (blue) and subCE from gamma (red). Green box indicates region predicted to contain the centromere on subCC. (Color figure online)

6 Data Summary

Table 1 lists statistics for the hundreds of synteny blocks located in each of the ten SYNMAP comparisons, segregated according to the genomic event that generated them. Note that the same comparison shows up as containing as many as three sets of synteny blocks, one set pertaining to each genomic event.

The age value is $\log_{10}$ tens of millions of years, normalized to span the interval $[0, 1]$.

7 Correlational Analysis

Table 2 contains the matrix of Spearman correlations among the seven measures. It is clear that age is highly correlated with the other six measures, which is not surprising in that they all represent aspects of genome evolution. More important

Table 1. Seven measure of evolution, including chronological time. Note that to assure the positive directionality of evolution in our subsequent analysis, we will multiply length by -1. We do the same for similarity, so that resulting rank order is the same as *Dist*.

Comparison	Age	Number of blocks	K_s	Similarity (%)	Length (pairs/singles)	Gap size	Fract rate	c-mere dist. (Mbp)
Gamma								
CC vs CE	1	253	2.42	70.70	8.48	3.69	0.680	26.4
subCC vs CE	1	232	2.63	70.85	8.26	3.72	0.686	23.6
subCE vs CC	1	262	2.60	70.87	8.56	3.80	0.684	20.0
subCC vs subCE	1	244	2.69	70.97	8.10	3.84	0.691	17.2
CC vs CC	1	263	2.50	70.77	8.79	3.84	0.694	22.2
CE vs CE	1	255	2.88	70.83	8.27	3.68	0.679	29.6
subCC vs subCC	1	215	2.62	70.62	8.47	3.79	0.694	17.5
subCE vs subCE	1	256	2.47	70.78	7.98	3.74	0.683	17.7
subCC vs CC	1	238	2.71	70.82	8.93	3.85	0.694	20.2
subCE vs CE	1	248	2.73	70.82	8.46	3.67	0.675	24.5
Speciation								
CC vs CE	0.05	465	0.0261	97.50	37.89	2.54	0.421	17.0
subCC vs CE	0.05	446	0.0350	97.30	36.43	2.46	0.420	14.4
subCE vs CC	0.05	258	0.0307	97.29	63.98	2.52	0.442	12.3
subCC vs subCE	0.05	240	0.0335	97.07	65.35	2.54	0.442	9.3
Tetraploidization								
CC vs subCC	0	210	0.015	98.40	79.54	2.50	0.435	11.7
CE vs subCE	0	524	0.018	98.20	33.82	2.33	0.393	14.8

is that all the other six are substantially less correlated with each other than with age, confirming that they all represent somewhat independent aspects of evolution.

Two groups, or clusters, have generally higher within-group than between-group correlations, as indicated in the shaded cells in the table.

A principal component analysis of Table 2 produces the configuration in Fig. 4, clearly displaying the distinction between the two clusters discussed in the table. The first principal component contrasts - similarity (or *Dist*) and the measures of fractionation, namely fractionation rate and average gap size, against K_s, distance to centromere and decrease in the length of synteny blocks.

That the two fractionation measures are closely associated stems from the fact that if step-by-step fractionation tends to involve two or more neighbouring genes at a time, this will also increase fractionation rate. On the other hand, for the a block to show many large gaps, and to host a large number of singletons, it must have sufficient length. This explains why -length is widely separated from the two measures of fractionation.

That distance to centromere is associated with -length, turns out to be a consequence of the near-telomeric positioning of the short synteny blocks originally generated by the gamma hexaploidization event. That there are ten comparisons bearing on gamma amplifies this effect.

Table 2. Correlations among the measures. Two distinct groups emerge, indicated by gray cells above and below the diagonal, respectively. in the table

	distance Dist	K_s	-length	gap size	c-mere dist	fract rate	age
Dist		0.668	0.621	0.703	0.709	0.795	0.861
K_s	0.668		0.794	0.671	0.715	0.634	0.861
-length	0.621	0.794		0.591	0.703	0.575	0.820
gap size	0.703	0.671	0.591		0.532	0.936	0.851
c-mere dist	0.709	0.715	0.703	0.532		0.518	0.820
fract rate	0.795	0.634	0.575	0.936	0.518		0.841
age	0.861	0.861	0.820	0.851	0.820	0.841	

The same explanation holds for the proximity of -similarity (and thus $Dist$) to the two measures of fractionation. For K_s on the other hand, the synonymous positions in the synteny blocks generated by gamma will be saturated or nearly so. The longer synteny blocks associated with speciation and tetraploidization will have smaller K_s than the saturated short blocks associated with gamma, explaining the proximity of K_s and -length.

The second component clearly contrasts all six measures with chronological age. This effect is likely due to correspondences between the measures at the gamma event, while there is no correspondence between any of them and age, which is constant.

The third component reveals a secondary association between distance from centromere and -similarity, which explains a non-neglible percentage of the variance.

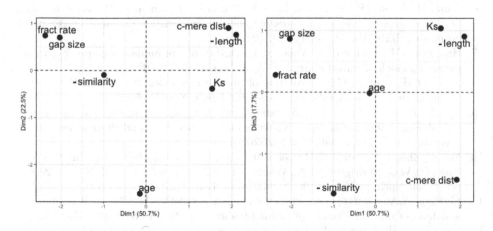

Fig. 4. First three components of principal components analysis

8 Discussion

The picture of syntenic evolution that emerges from this work is that there are at least five types of components of divergence of synteny blocks: neutral and protein-level sequence divergence of the genes in the block, loss of genes affecting the structure of the blocks (two such measures in our data) and their lengths, and spatial position of the block in relation to the chromosomal centromere. We predict that the predominant functional characteristic of the genes in the block will be another such measure. There are certainly other components of genomic evolution, such as genomic rearrangements, retropositional activity, gene movement in and out of blocks, that are not accessible to the analyses of single synteny blocks, but could be analyzed with the same philosophy of genomic evolution we advocate here.

References

1. Kumar, S., Subramanian, S.: Mutation rates in mammalian genomes. Proc. Natl. Acad. Sci. U.S.A. **99**(2), 803–808 (2002)
2. Duchêne, S., Ho, S.Y.W., Holmes, E.C.: Declining transition/transversion ratios through time reveal limitations to the accuracy of nucleotide substitution models. BMC Evol. Biol. **15**, 36 (2015)
3. Lyons, E., Freeling, M.: How to usefully compare homologous plant genes and chromosomes as DNA sequences. Plant J. **53**, 661–673 (2008)
4. Lyons, E., Pedersen, B., Kane, J., Freeling, M.: The value of nonmodel genomes and an example using SynMap within CoGe to dissect the hexaploidy that predates rosids. Trop. Plant Biol. **1**, 181–190 (2008)
5. Haas, B.J., Delcher, A.L., Wortman, J.R., Salzberg, S.L.: DAGchainer: a tool for mining segmental genome duplications and synteny. Bioinformatics **20**, 3643–3646 (2004)
6. Tang, H., Lyons, E., Pedersen, B., Schnable, J.C., Paterson, A.H., Freeling, M.: Screening synteny blocks in pairwise genome comparisons through integer programming. BMC Bioinform. **12**, 102 (2011)
7. Hasegawa, M., Kishino, H., Yano, T.A.: Dating of the human-ape splitting by a molecular clock of mitochondrial DNA. J. Mol. Evol. **22**, 160–74 (1985)
8. Jukes, T.H., Cantor, C.R.: Evolution of protein molecules. In: Munro, H.N. (ed.) Mammalian Protein Metabolism. Academic Press, New York (1969)
9. Yu, Z., Zheng, C., Albert, V.A., Sankoff, D.: Excision dominates pseudogenization during fractionation after whole genome duplication and in gene loss after speciation in plants. Front. Genet. **2020**, 1654 (2020)
10. Yu, Z., Zheng, C., Sankoff, D.: Gaps and runs in syntenic alignments. In: Martín-Vide, C., Vega-Rodríguez, M.A., Wheeler, T. (eds.) AlCoB 2020. LNCS, vol. 12099, pp. 49–60. Springer, Cham (2020). https://doi.org/10.1007/978-3-030-42266-0_5
11. Myburg, A.A., Grattapaglia, D., Tuskan, G.A., Hellsten, U., Hayes, R.D., Grimwood, J., et al.: The genome of Eucalyptus grandis. Nature **510**, 356–362 (2014)
12. Lan, T., et al.: Long-read sequencing uncovers the adaptive topography of a carnivorous plant genome. Proc. Natl. Acad. Sci. **114**, E4435 (2017)

Phylogenetics

Fast and Accurate Branch Support Calculation for Distance-Based Phylogenetic Placements

Navid Bin Hasan[1], Avijit Biswas[1], Metin Balaban[2], Siavash Mirarab[2,3],
and Md. Shamsuzzoha Bayzid[1(✉)]

[1] Computer Science and Engineering,
Bangladesh University of Engineering and Technology, Dhaka 1205, Bangladesh
1605005@ugrad.cse.buet.ac.bd, 1605006@ugrad.cse.buet.ac.bd,
shams_bayzid@cse.buet.ac.bd
[2] Bioinformatics and System Biology Program, UC San Diego,
San Diego, CA 92093, USA
mebalaba@eng.ucsd.edu, smirarab@ucsd.edu
[3] Electrical and Computer Engineering, UC San Diego, San Diego, CA 92093, USA

Abstract. Placing a new sequence onto an existing phylogenetic tree is increasingly used in downstream applications ranging from microbiome analyses to epidemic tracking. Most such applications deal with noisy data, incomplete references, and model misspecifications, all of which make the correct placement uncertain. While recent placement methods have increasingly enabled placement on ultra-large backbone trees with tens to hundreds of thousands of species, they have mostly ignored the issue of uncertainty. Here, we build on the recently developed distance-based phylogenetic placement methodology and show how the distribution of placements can be estimated per input sequence. We compare parametric and non-parametric sampling methods, showing that non-parametric bootstrapping is far more accurate in estimating uncertainty. Finally, we design and implement a linear algebraic implementation of bootstrapping that makes it faster, and we incorporate the computation of support values as a new feature in the APPLES software.

Keywords: Phylogenetic placement · Branch support · Bootstrapping

1 Introduction

Phylogenetic placement of a *query* sequence onto an existing *backbone* tree is increasingly adopted across applications such as microbiome analyses [2,11,22, 31,33,38,49], genome skimming [5,9], and epidemic tracking [28,50]. Phylogenetic placement treats all the queries as independent, resulting in two advantages: the running time scales linearly with the number of queries, and there may be reduced sensitivity to errors in individual sequences as they would not impact the placement of other queries [22]. The price is that the relationship between

L. Jin and D. Durand (Eds.): RECOMB-CG 2022, LNBI 13234, pp. 33–51, 2022.
https://doi.org/10.1007/978-3-031-06220-9_3

queries is not inferred, queries do not help the placement of other queries, and the backbone remains fixed. Notwithstanding these shortcomings, the appeal of having scalable methods has made placement an increasingly popular approach.

A large number of placement methods have been developed based on diverse approaches, including maximum likelihood (ML) [e.g., 6,32,48,52], distance-based estimation [e.g., 3–5], maximum parsimony [e.g., 50], HMMs [e.g., 36,54], supertree approaches [e.g., 30,41], k-mers [e.g., 29], and even machine learning [e.g., 10,24]. The accuracy of these methods depends on the context, but generally, ML methods such as pplacer [32] have been highly accurate [5]. However, these methods do become slow as the backbone grows large (e.g., more than 1000). Meanwhile, denser backbones are known to increase accuracy and improve downstream applications [34,37,39]. This need to scale to large backbone trees has motivated both divide-and-conquer [36,52] methods and the distance-based method APPLES(-2) [3,5]. In particular, APPLES-2 uses divide-and-conquer and dynamic programming to provide sub-linear scaling with the size of the backbone tree when amortized over all queries and provides accuracy levels that match the ML method EPA-ng [6] and comes close to pplacer.

Despite these advantages, the existing distance-based methods lack a key feature: support estimation. The ML placement methods, pplacer and EPA-ng, provide a natural notion of support based on the normalized likelihood of placement across branches. These support values reveal uncertainty, and their incorporation has been known to improve downstream applications of placement such as taxonomic identification [38] and sample comparison [33]. The distance-based method APPLES-2 [3] currently does not output any measure of uncertainty around its placements.

In this paper, we offer parametric and non-parametric methods of measuring support for distance-based placement in the presence of a multiple sequence alignment (MSA). Our non-parametric approach relies on the traditional bootstrapping method, with an efficient implementation that utilizes a linear algebraic formulation to achieve higher speeds. The parametric approach relies on a binomial model paired with the Jukes-Cantor (JC) model [25] or a Poisson approximation of Binomial. We carefully test the accuracy of the support values in extensive simulations, showing that the non-parametric approach provides highly accurate support values. This new feature is fully integrated into APPLES-2 software and is available at https://github.com/navidh86/apples.

2   Approach

We start with some background on APPLES(-2) and the goal of branch support estimation. We then introduce two models for computing support for distance-based placement, one based on a Poisson model approximation and the other based on bootstrapping, along with a particularly optimized implementation of bootstrapping using a linear algebraic formulation.

2.1 Background on APPLES-2

The Least Squares Phylogenetic Placement (LSPP) problem [5] takes as input a reference phylogenetic tree T with n leaves and a vector of distances δ_{qi} between a query taxon q and every taxon i on T. It outputs the placement that minimizes the weighted least squares error:

$$\sum_{i=1}^{n} w_{qi}(\delta_{qi} - d_{qi}(T))^2$$

where $d_{qi}(T)$ is the path distance from q to backbone taxon i on T. The w_{qi} weights reduce the impact of long distances and by default are δ_{qi}^{-2} following Fitch-Margoliash (FM) weighting [17]. APPLES used a dynamic programming algorithm to solve this problem in time that increases linearly with the size of the tree. APPLES-2, a recent improvement [3], is a sub-linear heuristic that solves the LSPP problem on a subset of the leaves. To achieve sub-linear time, APPLES-2 avoids computing all distances to the reference using a heuristic method that clusters backbone leaves. APPLES-2 is faster and more memory-efficient than APPLES, *and* it is also *more* accurate in simulation studies. The increased accuracy likely is due to the reduced impact of long distances, estimation of which is a known major challenge [12,16]. The divide-and-conquer method used in APPLES-2 builds on a rich history of methods with strong theoretical guarantees [14,21,51].

The distances used by APPLES-2 can come from any source. Given an MSA, they can be computed by comparing sequences and calculating a phylogenetically corrected distance. For DNA, the simplest model is JC [25], which requires simply computing the normalized hamming distance h_{qi} (ignoring gap positions) and then correcting it using the simple formula: $-\frac{3}{4}\ln\left(1 - \frac{4}{3}h_{qi}\right)$. For amino acid (AA) alignments, APPLES-2 uses the Scoredist [47] algorithm. Scoredist computes normalized pairwise distances according to the BLOSUM62 [20] matrix and then performs a logarithmic correction.

2.2 Distance-Based Support Estimation: Goals and Background

Branch support estimation aims to show the amount of uncertainty. In the case of placement, we seek a set of placements that collectively have a high probability of including the correct location and a probability assigned to each placement that ideally predicts whether it is correct. We note that the correct interpretation of support is a subtle issue [e.g., 8,35,43,46]. What methods provide (an estimate of the variance of the estimator) is different from what biologists would ideally like to have (an indication of whether a branch is correct). Nevertheless, we can measure the usefulness of a measure of support based on the biologically desired outcome (i.e., probability of correctness).

Standard methods or computing support provide a set of trees. Bayesian analyses approximate the posterior tree distribution. Maximum likelihood and distance-based methods also build a tree distribution but use sampling methods, in particular the bootstrapping procedure [15] that repeatedly resamples sites with replacement. Bootstrapping is a general and valid method for computing distributions around a statistically consistent estimator that asymptotically resembles the distribution of the estimator around the true value if the data generation process was to be repeated [13,45]. Despite being highly parallelizable, repetitive resampling of sites of the input alignment increases the running time, motivating alternative methods [1,18,26,44]. Finally, in phylogenetic placement, support values are computed using the likelihood ratio (i.e., normalizing the likelihood of placement on each branch by the likelihood of other highly likely branches) [6,7] or a Bayesian variation of the same idea [31].

2.3   Non-parametric Bootstrapping

Bootstrapping is adopted to phylogenetic placement as follows. The L sites of the given MSA are subsampled with replacement L times, repeating this process B times to produce B replicate datasets. Each replicate alignment is used for the placement of the query. The placement support on each branch is then set to the fraction of the B replicate placements that fall on that branch. While the backbone topology should ideally be allowed to change for each replicate, this would make it difficult to draw meaningful support values for placements. Thus, we keep the tree topology fixed. We will test if re-estimating branch lengths of the backbone tree for each replicate can improve the exactness of support values.

Noting that the bottleneck in APPLES-2 is computing distances (not the placement step), we improve running time using a linear algebraic formulation. Let M be the MSA of the given reference and query sequences, and let $M_1, M_2, \ldots, M_B$ be the set of B bootstrapped multiple sequence alignments sampled from the original alignment M with replacement. We compute distances between a query and a reference sequence in all of these $B+1$ alignments efficiently using matrix multiplication. Let $\mathbf{H}$ be a $(B+1) \times L$ matrix, where each $\mathbf{H}_{ij}$ is the number of times the j-th site was sampled in the i-th bootstrapped alignment M_i, and $\mathbf{H}_{0j} = 1$ for $0 \leq j \leq L - 1$. We precompute $\mathbf{H}$ once. For each query sequence q, we compute an $L \times n$ matrix $\mathbf{V}$ by setting $\mathbf{V}_{i,j} = 1$ if and only if the i-th sites in the query q and the reference j disagree and neither one is a gap in the original alignment M. We then compute the $(B + 1) \times n$ matrix $\mathbf{P} = \mathbf{H} \cdot \mathbf{V}$. Note that $\mathbf{P}_{i,j}$ gives the number of mismatches between q and reference j in M_i. The normalized Hamming distance is simply computed as $\mathbf{P}_{i,j}/L'_{i,j}$, where $L'_{i,j}$ is the number of sites in M_i where neither the query q nor the reference j is a gap, and is transformed using the JC69 correction. This strategy is faster than the naive approach because 1) string comparisons are made only once (in constructing $\mathbf{V}$) and not once per replicate, 2) site subsampling is implicit in building $\mathbf{H}$ and does not require string operations, and 3) matrix multiplication is a highly optimized operation in most platforms such as NumPy used here.

Considering the two points mentioned earlier, we study two versions of bootstrapping, a "slow" version that does not use the linear algebraic trick and re-estimates branch lengths per replicate, and a "fast" version that uses the linear algebraic formulation and avoids re-estimating branch lengths.

2.4 Parametric Bootstrapping (Binomial and Poisson Models)

We propose a parametric alternative to non-parametric bootstrapping where we sample a collection $\mathcal{D} = \{D_1, D_2, \ldots, D_B\}$ of B new distance matrices from the original distance matrix D_0 computed from original alignment M. These matrices in $\mathcal{D}$ are of the same size as the original distance matrix D_0, each of which can be analyzed to find the placements of a query. The approach does not require repeated distance calculation and thus should be faster.

We model the distribution of distances between each pair of query and reference sequences using the Binomial distribution or its Poisson approximation. Under the JC69 model, the number of mutations *observed* between a query q and a reference r that have l non-gap aligned sites and are separated by a branch of true length t follows a binomial distribution with l tries and success probability $h = \frac{3}{4}(1 - e^{-\frac{4}{3}t})$. Given the observed normalized hamming distance $\hat{h}$, we can use the parametric bootstrapping approach: *i)* draw new values $x_1 \ldots x_B$ from the binomial distribution with l tries and $\hat{h}$ probability of success, *ii)* normalize each x_i by l to get $\hat{h}_1 \ldots \hat{h}_B$, and *iii)* transform each drawn hamming distance to get $\hat{t}_i = -\frac{3}{4} \ln (1 - \frac{4}{3} \hat{h}_i)$. The binomial distribution can be approximated by a Poisson distribution with rate $\lambda = \hat{h} \times l$ if we allow ourselves to forget that h does not diminish as l increases. Importantly, we apply this procedure to each element of the distance matrix independently. This part of the procedure is essentially incorrect since distances are not independent. Thus, the independent draws are based on a consciously incorrect assumption and will be tested empirically.

3 Experimental Study

3.1 Dataset

Simulated Single-Gene RNASim. We use an existing RNASim-VS simulated dataset [5], which contains subsets of a simulated RNASim data set [19]. We used the same data as the original paper [5] who randomly select 1000 queries over 5 replicates (200 queries in each replicate) with various novelty levels. The backbone is a set of 5000 taxa randomly chosen from the full tree (five replicates). Each replicate contains an MSA of 1596 sites of a single gene and the true tree as backbone. The branch lengths have been re-estimated in minimum evolution units using FastTree-2 [40]. In order to assess the performance of our method on fragmentary data and compare it with alternative techniques (e.g., EPA-ng, pplacer) that are specially designed for placing reads, we also created a fragmentary version of the dataset by randomly selecting 200 bp from a random position of the sequence and replacing other letters with a dash.

Web of Life (WoL). We use the Web of Life (Wol) dataset [56] which contains 381 marker genes and a species tree of 10,575 prokaryotic genomes built from ML gene trees using ASTRAL [53]. We reuse the WOL-random subset from a previous publication [3], limited to 1000 species for the backbone (randomly selected) and the *best* $k \in \{5, 10, 25, 30, 40, 50\}$ genes (one replicate), defined as those with the lowest quartet distance to the species tree. We concatenated nucleotide MSAs of the top k genes (one replicate), removing the third codon position (C12), and used this C12 MSA in all our analyses. We use C12 because the third codon significantly misleads the placement accuracy of APPLES-2 on this data [3]. A set of 1000 species not among the 1000 used in the backbone are used as queries. As the backbone, we utilize the ASTRAL tree limited to 1000 backbone species but recalculate its branch lengths using minimum evolution with FastTree-2 [40]. The position of the queries on the full ASTRAL tree is taken as the correct placement because it is obtained using a comprehensive pipeline that uses ML to infer 381 gene trees and ASTRAL to summarize gene trees.

3.2 Measurements

To assess the quality of the branch supports, we first ask how calibrated the support values are by investigating the relationship between bins of branch support and the percentage of correctly placed queries in each bin (accuracy). For example, for branches in the 60–70% support bin, calibrated supports result in roughly 65% accuracy. We report the mean squared error (MSE) of the accuracy versus the median of bins, separately computed for branches with support below or above 70%. However, even when support values are not calibrated, they may be useful in distinguishing correct and incorrect branches (imagine dividing all support values by two). To assess the ability to distinguish correct branches, we use receiver operating characteristic (ROC) curves. ROC curves show the relationship between recall (the percentage of all true branches with support above a threshold t), and false-positive rate (FPR) (the percentage of all false branches with support below t). For thresholds $t \in \{0, 1, \ldots, 100\}$ we classify each branch as: TP : support $\geq t$ and branch is correct, FP : support $\geq t$ and branch is incorrect, TN : support $\leq t$ and branch is incorrect, and FN : support $\leq t$ and branch is correct. We plot $FPR = \frac{FP}{FP+TN}$ versus $Recall = \frac{TP}{TP+FN}$. Finally, we draw the Empirical Cumulative Distribution Function (ECDF) of the support values drawn separately for correct and incorrect branches.

4 Results and Discussion

4.1 Simulated Single-Gene RNASim Dataset: Full-Length Sequences

We start with a comparison of alternative APPLES-2 support measures. The support values obtained by both slow and fast non-parametric bootstrapping are better calibrated with the accuracy (i.e., support values are closer to the expected values) than those obtained by the parametric Poisson and binomial sampling methods (Fig. 1A). The difference between bootstrapping- and Poisson-based methods is more pronounced at lower support levels, where parametric methods tend to underestimate supports. The MSE of the branches with support $\leq 70\%$ is more than 0.17 for parametric methods and only 0.018 for fast bootstrapping. Conversely, parametric sampling over-estimates support at higher ends, leading to at least three times higher MSE than bootstrapping methods.

The slow bootstrapping method produced only slightly better support values than the fast bootstrapping method, especially for high support values where the MSEs are 0.001 and 0.0008 for fast and slow bootstrapping, respectively.

Investigating the distribution of the support values, there is a large gap between the support distribution of correct and incorrect placements, and the gap is more prominent for bootstrapping than the parametric methods (Fig. 1B). Parametric sampling results in many more correct placements with low support and more incorrect placement with high support. With bootstrapping, around 75% of the correct placements have more than 75% support, and around 75% of the incorrect placements have less than 60% support. In contrast, around 60–65% of the correct placements have support above 75% for parametric methods. Parametric methods also have a long tail of incorrect placement with high support. For example, no placement with $\geq 99\%$ bootstrapping support is wrong, but around 4% of such placements are incorrect with parametric sampling.

Examining the predictive power of support (irrespective of their calibration) using ROC curves confirms the superiority of bootstrapping and shows a slight advantage for slow compared to fast bootstrapping (Fig. 1C). The area under ROC curves (AUROC) for the slow and fast bootstrapping is 0.883 and 0.864, respectively, compared to 0.765 and 0.771 for Poisson and Binomial. The AUROCs for parametric methods, while inferior to bootstrapping, nevertheless show fair levels of predictive power. Focusing on bootstrapping, we observe that changing the support threshold from 65% to 100% can lead to FPRs ranging from around 25% to 0%, and recalls ranging from above 80% to 40%, with the 80% support threshold providing a practically useful trade-off (recall: 72% and FPR: 15%).

Finally, examining the top ten highest support placements, we observe that the correct placement is among the top 6 placements for almost all queries with bootstrapping (Fig. 1D). In contrast, going to even the top 10 placements does not capture the correct placement in more than 10% of queries when using parametric models. Furthermore, note that 77%, 75%, 60%, and 55% of the correct placements are within the placements with the highest support obtained by slow bootstrapping, fast bootstrapping, Poisson, and binomial methods, respectively.

Comparison with pplacer and EPA-ng. Comparing fast bootstrapping for APPLES-2 to alternative ML methods EPA-ng and pplacer shows that ML methods result in far higher levels of support than distance-based methods (Fig. 2). For example, close to 76% and 82% of correct placements with pplacer

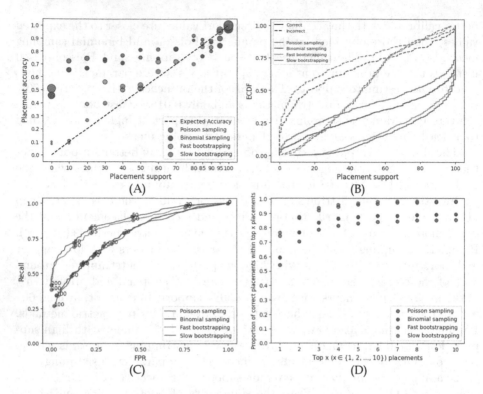

Fig. 1. Results on RNAsim single-gene dataset. (A) Support versus the percentage of correctly placed queries over five replicates. Support values are binned at 0%, 10% ... 80%, 85%, 90%, 95%, and 100% left inclusive (e.g., [0,10)); the last bin only includes 100%. Unity line ($y = x$): fully-calibrated support. Dot sizes are proportional to the number of queries within each bin. The mean squared error (MSE) of the points computed with respect to the unity line and divided between low and high support values are as follows. For low support (support $\leq$ 70%) values, MSE is 0.21 for Poisson, 0.17 for Binomial, 0.018 for Fast BS, and 0.007 for slow BS; for high support (support > 70%) values, MSE is 0.003 for Poisson, 0.004 for Binomial, 0.001 for Fast BS, and 0.0008 for slow BS. (B) Empirical cumulative distribution function (ECDF) of the support for correct/incorrect placements. (C) Receiver operating characteristic (ROC) curves using a range of 0–100 of support thresholds. Selected thresholds are marked. (D) The frequency of the correct placement being among the top $1 \leq x \leq 10$ highest support placements.

and EPA-ng have 100% support, compared to only 38% for APPLES-2. However, such increased support is not universally good. Unlike APPLES-2, pplacer and especially EPA-ng result in high support for incorrect branches often (10% and 40% of their incorrect branches have 100% support, respectively). In other words, both ML methods over-estimate support, but this problem is far worse for EPA-ng. The MSE error of EPA-ng for > 70% branches (0.02) is much higher than pplacer (0.004), which is higher than APPLES-2 (0.001). The high number

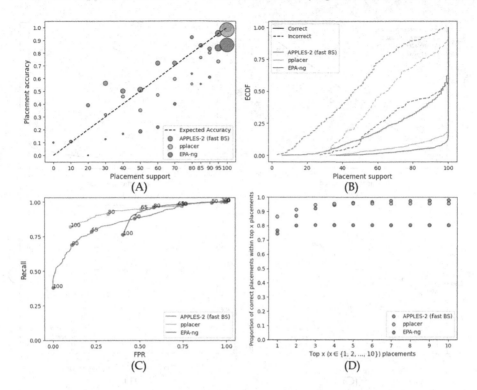

Fig. 2. Comparison to EPA-ng and pplacer on RNASim dataset. Settings are similar to Fig. 1. MSE among support ≤ 70%: 0.02 for APPLES-2, 0.013 for pplacer, and 0.1 for EPA-ng; MSE among support > 70%: 0.001 for APPLES-2, 0.004 for pplacer, and 0.02 for EPA-ng.

of incorrect but confident placements by EPA-ng results in an AUROC of 0.574, which is worse than APPLES-2 (0.861) or pplacer (0.849). For (high) FPR values that both pplacer and APPLES-2 are able to achieve, pplacer has better recalls, showing its better predictive accuracy in that range. Note that even with 100% support, pplacer has substantially higher FPR than APPLES-2.

4.2   Simulated Single-Gene RNASim Dataset: Fragmentary Sequences

Support values dramatically reduce when we examine short fragmentary sequences, especially for APPLES-2 and EPA-ng (Fig. 3). With APPLES-2 bootstrapping, very few placements have support at or close to 100%, and most of those are correct. APPLES-2 support also has high predictive power, with AUROC equal to 0.765, which is lower than full-length sequences but is reasonably high. As we go from 1 to 10 top placements, the number of queries with the correct placement captured among the picks increases from 32% to 72%.

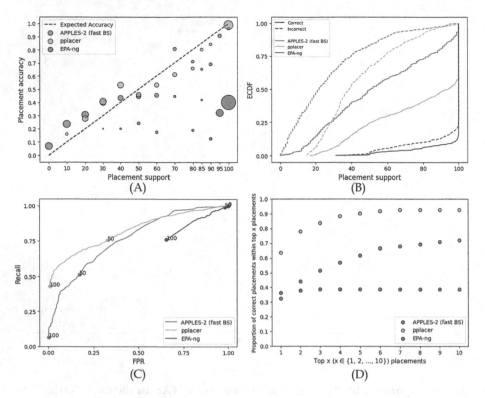

Fig. 3. Results on simulated fragmentary 200 bp dataset. Comparison of APPLES-2 (fast bootstrapping) with EPA-ng and pplacer. MSE among support $\leq 70\%$: 0.01 for APPLES-2, 0.009 for pplacer, and 0.09 for EPA-ng; among support $> 70\%$: 0.02 for APPLES-2, 0.002 for pplacer, and 0.37 for EPA-ng.

We also compare APPLES-2 with EPA-ng and pplacer, which, unlike APPLES-2, are designed explicitly for fragmentary sequences. The behavior of the two ML methods diverges on fragmentary data. The pplacer support values remain highly accurate in terms of calibration with accuracy, the distribution gap between correct and incorrect placements, and ROC. Here, pplacer has the best MSE (0.002 for $> 70\%$ supported branches), followed by APPLES-2 (0.02), and EPA-ng as a distant third (0.37). The AUROC of pplacer is 0.820, which is better than APPLES-2. In particular, at very low FPR values, pplacer is clearly more sensitive than APPLES-2. Note also that pplacer placements (not just their support) are substantially more accurate than APPLES-2 on these fragmentary data. In particular, taking the top 1 to 6 placements from pplacer achieves 63%–91% accuracy, compared with 32%–66% for APPLES-2. Unlike pplacer, EPA-ng is about as accurate as APPLES-2 in terms of placement and has far worse support values. EPA-ng is overtly confident in its placements, producing close to 100% support for almost all queries. The gap between distributions of correct and incorrect branches of EPA-ng is narrow, and the ROC curve confirms a lack of predictive power (AUROC equals 0.308). As a result, unlike APPLES-2, EPA-ng does not benefit from increasing the number of top placements in terms of finding the right branch.

4.3 Multi-gene Web of Life (WoL) Dataset

On the biological multi-gene WoL dataset, the relative accuracy of APPLES-2 support methods is similar to the single-gene RNAsim dataset (Fig. 4). Bootstrapping support values are better calibrated with the frequency of correctness and are more predictive of correctness compared to those obtained by parametric methods. The ROC plots indicate bootstrapping can better distinguish correct and incorrect edges, especially when concatenating fewer genes. However, ROC plots also reveal that on the WoL dataset, the difference between methods is not as substantial as it is in the RNAsim dataset. Across different numbers of genes, both Poisson and binomial methods inferred low supports ($\leq$30) for fairly large numbers of correct placements, especially with fewer genes, in contrast to bootstrapping. However, unlike the single gene dataset where bootstrapping was accurate for high support values, on WoL, all the methods tend to overestimate medium to high support levels (e.g., $\geq$50%). Nevertheless, bootstrapping continues to be more accurate for high support.

Impact of the Number of Genes. Increasing the number of genes results in higher support values for all methods (Fig. 4). Having higher support values, of course, is not always desirable: the support values of the correct and incorrect placements both increase with more genes. Thus, using more genes narrows the gap between the distribution of correct and incorrect results. For example, the proportion of correct placements with $\leq$ 80% bootstrap support goes from 18% with 10 genes to 7% with 50 genes; on the other hand, with only 10 genes, almost half of the incorrect placements have low (<50%) support, whereas with 50 genes, only about a quarter of the incorrect edges have low support. Moreover, as we increase the number of genes, all the methods tend to have longer tails of incorrect placements with high support ($\geq$80%). Around 25% of the incorrect placements have more than 90% support with 10 genes, compared to around 40% of the incorrect placements with 50 genes. The ROC curves show a reduced ability to distinguish correct and incorrect branches with more genes, with AUROC value decreasing (e.g., the AUROC for slow bootstrapping goes from 0.826 with 10 genes to 0.687 with 50 genes). Notably, unlike the RNAsim dataset, no method achieves 0 FPR at any threshold regardless of the number of genes used. The lowest possible FPR is close to 25% with 50 genes and 9% with 10 genes.

Thus, overall, results indicate that increasing the number of genes causes all the methods to over-estimate the branch supports, even when the placements are incorrect. Consequently, the difference between methods shrinks as more genes are concatenated, a pattern most visible in the ROC plots.

Less accurate *support* with increased numbers of genes does not mean less accurate *placements*. The accuracy of the highest support placement is higher

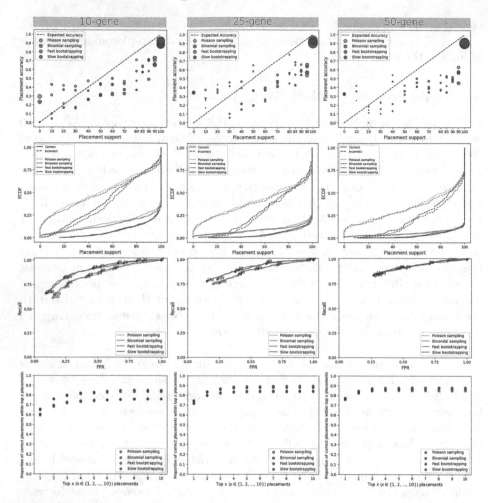

Fig. 4. Results on the WoL multi-gene dataset. We vary the number of genes from 10 to 50. We show the results of four different analyses as in Fig. 1. (i) Relationship between support calculations and frequency of correct placements. Each figure shows the aggregated results on 1000 queries. (ii) Empirical cumulative distribution function (ECDF) plot of the support values, showing the percentage of the correct/incorrect placements (shown in the y-axis) at or below specified placement support levels (in the x-axis). (iii) ROC curves, built using a range of 0–100 of support thresholds. Selected thresholds are shown on the graph. (iv) We show how often the correct placement is among the top x placements (in terms of support values) for different values of x.

for 50 genes than 10 genes. While the correct branch is among the top 4 placements for around 80% of queries 10 with genes, when we go to 50 genes, the correct branch is among the top 2 placements for around 83% of queries for all methods. Thus, placements are more accurate with more genes even if the ability to distinguish correct and incorrect placement diminishes.

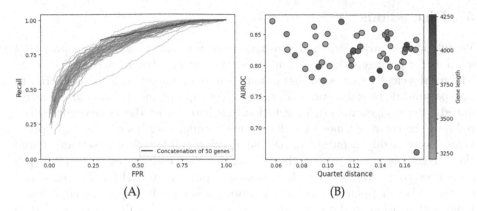

(A)   (B)

Fig. 5. Evaluation of the support values obtained using one gene at a time from the best 50 genes in the WoL dataset. (A) Individual ROC curves for all 50 genes, along with the ROC curve of the concatenation of 50 genes. (B) Quartet distance of the gene trees to the species tree versus AUROC. We color the data points with a color gradient that varies continuously from light blue to dark blue with increasing gene lengths. (Color figure online)

Individual Genes. We also assess the impact of each of the 50 genes separately (Fig. 5) and observe that various genes are substantially different in their ability to distinguish correct and incorrect placements. The AUROC ranges from values as high as 0.87 for the gene that is most similar to the species tree to 0.77 for the second worst gene. An outlier gene has an AUROC of only 0.66. There is a clear decline in AUROC as the quartet distance between the species tree and the gene trees increases (Fig. 5B). Moreover, there is a slight tendency for *decreased* AUROC as the length of alignments increases (most of the long sequences resulted in AUROC values < 0.82). Unlike conditions with multiple genes concatenated, FPR values very close to 0 can be obtained with most genes (Fig. 5A). For any given FPR threshold, we also see recall values that can vary by up to 25%. For example, for 5% FPR, the best gene obtains a recall of 0.59, and the second worst gene has a recall of only 0.24.

4.4   Runtimes

The running time of the slow bootstrapping method is about 200 times slower than the placement without support for 100 replicates. The fast implementation reduces the run time dramatically: a 33X speedup for the RNASim dataset and a 11–15X speedup for WoL (depending on the number of genes). Thus, fast bootstrapping with 100 replicates is 10–20 times slower than no bootstrapping. Since parametric methods were not accurate, we have not tried to optimize them, leaving them slower than non-parametric methods.

5 Discussions

We introduced parametric and non-parametric methods of estimating statistical support for phylogenetic placement using the distance-based method APPLES. The results clearly suggested that our suggested parametric method is inferior to non-parametric bootstrapping. To make bootstrapping more practical, we introduced a linear algebraic implementation that had almost the same accuracy but reduced the running times by orders of magnitude. Fast bootstrapping support values were highly accurate on the single-gene full-length dataset and found the correct placement among the top six or seven candidates with full-length sequences. With fragmentary data, bootstrapping within APPLES-2 remained effective though pplacer was clearly more accurate. On the more challenging biological multi-gene dataset, bootstrap support could still effectively distinguish correct and incorrect branches. Thus, we can reliably estimate support for distance-based placements, addressing one of its main shortcomings compared to slower ML methods.

We only evaluated the method under challenging conditions where the model assumptions of the APPLES-2 were violated. The RNASim dataset is generated under a complex model of sequence evolution [19] and includes many processes not captured by our JC model (e.g., rate variation across sites, selection, and varying substitution rates). Model violations in the multi-gene WoL are even greater. The reference tree used on the WoL dataset, taken as ground truth, is inferred using ASTRAL from a collection of 381 incongruent gene trees inferred under an AA model paired with rate heterogeneity [55]. Instead, we simply concatenate genes and use a JC model with no rate heterogeneity, thereby ignoring not just the complexities of the sequence evolution model but also gene discordance. Recapturing the ASTRAL from 381 gene trees inferred under complex models using concatenation is a difficult task. Nevertheless, while far from perfect, our support values still were informative in detecting incorrect branches (AUROC varied between 69% to 83% with 50 to 10 genes). We chose datasets that violate our model because we believe biological data always do. Nevertheless, our accuracy levels should be interpreted with this point kept in mind.

An interesting pattern was that as the number of concatenated genes increases, so does the support, and not always for better. While more genes increased the *accuracy of the placements*, they also left us with more *positively misleading* results (i.e., incorrect placements with high support). Such a result is not unexpected or new. Concatenation is known to result in high support for incorrect branches in the presence of high levels of discordance [27]. Methods such as bootstrapping are meant to find the variance of statistically consistent estimators on limited data; when the method is used in conditions that violate their assumption, they can be inconsistent or even misleading. In such conditions, increasing the amount of data (e.g., the number of genes) reduces the variance but does not eliminate the bias. Therefore, one gets higher support for correct and incorrect answers alike. Since concatenation is not consistent with highly discordant gene trees [42], when the concatenation is misled, its support values can be misleading, a challenge constantly faced by phylogenomic analyses [e.g.,

23, 43]. We note that other than gene discordance, several factors may lead to the systematic biases in our analyses, including uneven and non-random patterns of missing data, rate heterogeneity across genes, heterotachy, and the use of the simple JC model. These persistent biases can explain the reduced accuracy of support values as the number of genes increases (and variance reduces). Note also that our 10 genes are, by construction, less discordant than 50 genes, and thus, increasing the number of genes may even increase bias.

The fact that Poisson and Binomial parametric sampling methods had low accuracy was also interesting. Two reasons can be considered. First, these methods assumed the JC model generated the data, but our data are generated under more complex models. However, the relatively high accuracy of APPLES-2 placements suggests that the method is not terribly sensitive to model misspecifications. A far more likely problem is that parametric models are used here to estimate a different distance for each pair independently from others. A parametric draw of a distance below (or above) the observed value implies an evolutionary scenario with fewer (or more) mutations *along the path between the two leaves* compared to what is observed. However, if there are more/fewer mutations on any branch on the path, it should affect *all* pairs that connect through that branch. This concept is missing from the parametric method and cannot be easily incorporated. Essentially, we need ways that add noise to all elements of the distance vector for each query jointly, using a model that considers dependence. Such a method would require a covariance matrix, which is not known since the position of the query is not known. However, future work can seek to approximate such a matrix given the optimal placement of the query. We leave the exploration of such an approach for future work.

In this analysis, we focused on the accuracy of support values, not their downstream use. Once calculated, support can be used as weights in downstream applications such as calculating distance between samples using UniFrac, and taxon identification of metagenomic samples (e.g., TIPP [38] uses support to improve taxon id). Such analyses would be most interesting if performed for short sequences where uncertainty is high.

References

1. Anisimova, M., Gascuel, O., Sullivan, J.: Approximate likelihood-ratio test for branches: a fast, accurate, and powerful alternative. Syst. Biol. **55**(4), 539–552 (2006). https://doi.org/10.1080/10635150600755453
2. Asnicar, F., et al.: Precise phylogenetic analysis of microbial isolates and genomes from metagenomes using PhyloPhlAn 3.0. Nat. Commun. **11**(1), 2500 (2020). https://doi.org/10.1038/s41467-020-16366-7. http://www.nature.com/articles/s41467-020-16366-7
3. Balaban, M., Jiang, Y., Roush, D., Zhu, Q., Mirarab, S.: Fast and accurate distance-based phylogenetic placement using divide and conquer. Mol. Ecol. Resour. (2021). https://doi.org/10.1111/1755-0998.13527. https://onlinelibrary.wiley.com/doi/10.1111/1755-0998.13527

4. Balaban, M., Mirarab, S.: Phylogenetic double placement of mixed samples. Bioinformatics **36**(Supplement_1), i335–i343 (2020). https://doi.org/10. 1093/bioinformatics/btaa489. https://academic.oup.com/bioinformatics/article/ 36/Supplement_1/i335/5870522

5. Balaban, M., Sarmashghi, S., Mirarab, S.: APPLES: scalable distance-based phylogenetic placement with or without alignments. Syst. Biol. **69**(3), 566–578 (2020). https://doi.org/10.1093/sysbio/syz063. https://academic.oup.com/ sysbio/advance-article/doi/10.1093/sysbio/syz063/5572672. https://academic. oup.com/sysbio/article/69/3/566/5572672

6. Barbera, P., et al.: EPA-ng: massively parallel evolutionary placement of genetic sequences. Syst. Biol. **68**(2), 365–369 (2019). https://doi.org/10.1093/sysbio/ syy054. https://academic.oup.com/sysbio/article/68/2/365/5079844

7. Berger, S.A., Krompass, D., Stamatakis, A.: Performance, accuracy, and web server for evolutionary placement of short sequence reads under maximum likelihood. Syst. Biol. **60**(3), 291–302 (2011). https://doi.org/10.1093/sysbio/syr010. http://sysbio.oxfordjournals.org/cgi/content/abstract/60/3/291. http://sysbio. oxfordjournals.org/content/60/3/291.abstract. http://sysbio.oxfordjournals.org/ content/60/3/291.full.pdf. http://www.pubmedcentral.nih.gov/articlerender.fcgi? artid=3078422&tool=pmc

8. Berry, V., Gascuel, O.: On the interpretation of bootstrap trees: appropriate threshold of clade selection and induced gain. Mol. Biol. Evol. **13**(7), 999–1011 (1996). https://doi.org/10.1093/molbev/13.7.999. https://academic. oup.com/mbe/article-lookup/doi/10.1093/molbev/13.7.999

9. Bohmann, K., Mirarab, S., Bafna, V., Gilbert, M.T.P.: Beyond DNA barcoding: the unrealized potential of genome skim data in sample identification. Mol. Ecol. **29**(14), 2521–2534 (2020). https://doi.org/10.1111/mec.15507. https:// onlinelibrary.wiley.com/doi/abs/10.1111/mec.15507

10. Brown, D., Truszkowski, J.: LSHPlace: fast phylogenetic placement using locality-sensitive hashing. In: Pacific Symposium on Biocomputing, pp. 310–319 (2013). https://doi.org/10.1142/9789814447973_0031. http://www.ncbi.nlm. nih.gov/pubmed/23424136. http://www.worldscientific.com/doi/abs/10.1142/ 9789814447973_0031

11. Darling, A.E., Jospin, G., Lowe, E., Matsen, F.A., Bik, H.M., Eisen, J.A.: PhyloSift: phylogenetic analysis of genomes and metagenomes. PeerJ **2**, e243 (2014). https:// doi.org/10.7717/peerj.243. https://peerj.com/articles/243

12. Desper, R., Gascuel, O.: Fast and accurate phylogeny reconstruction algorithms based on the minimum-evolution principle. J. Comput. Biol. **9**(5), 687–705 (2002). https://doi.org/10.1089/106652702761034136. http:// www.liebertonline.com/doi/abs/10.1089/106652702761034136. http://www.ncbi. nlm.nih.gov/pubmed/12487758

13. Efron, B.: Bootstrap methods: another look at the jackknife. Ann. Stat. **7**(1), 1–26 (1979). http://www.jstor.org/stable/2958830

14. Erdos, P., Steel, M., Szekely, L., Warnow, T.: A few logs suffice to build (almost) all trees: part II. Theoret. Comput. Sci. **221**(1–2), 77–118 (1999). https://doi.org/ 10.1016/S0304-3975(99)00028-6

15. Felsenstein, J.: Confidence limits on phylogenies: an approach using the bootstrap. Evolution **39**(4), 783–791 (1985). https://doi.org/10.2307/2408678. http://www. jstor.org/stable/2408678

16. Felsenstein, J.: Inferring phylogenies (2003)

17. Fitch, W.M., Margoliash, E.: Construction of phylogenetic trees. Science **155**(3760), 279–284 (1967). https://doi.org/10.1126/science.155.3760.279. https://www.science.org/doi/10.1126/science.155.3760.279

18. Guénoche, A., Garreta, H.: Can we have confidence in a tree representation? In: Gascuel, O., Sagot, M.-F. (eds.) JOBIM 2000. LNCS, vol. 2066, pp. 45–56. Springer, Heidelberg (2001). https://doi.org/10.1007/3-540-45727-5_5

19. Guo, S., Wang, L.S., Kim, J.: Large-scale simulation of RNA macroevolution by an energy-dependent fitness model. arXiv 0912.2326 (2009). http://arxiv.org/abs/0912.2326

20. Henikoff, S., Henikoff, J.G.: Amino acid substitution matrices from protein blocks. Proc. Natl. Acad. Sci. **89**(22), 10915–10919 (1992). https://doi.org/10.1073/pnas.89.22.10915. http://www.pnas.org/cgi/doi/10.1073/pnas.89.22.10915

21. Huson, D.H., Nettles, S.M., Warnow, T.J.: Disk-covering, a fast-converging method for phylogenetic tree reconstruction. J. Comput. Biol. **6**(3–4), 369–386 (1999). https://doi.org/10.1089/106652799318337. http://www.ncbi.nlm.nih.gov/pubmed/10582573

22. Janssen, S., et al.: Phylogenetic placement of exact amplicon sequences improves associations with clinical information. mSystems **3**(3), 00021-18 (2018). https://doi.org/10.1128/mSystems.00021-18. http://msystems.asm.org/lookup/doi/10.1128/mSystems.00021-18

23. Jarvis, E.D., et al.: Whole-genome analyses resolve early branches in the tree of life of modern birds. Science **346**(6215), 1320–1331 (2014). https://doi.org/10.1126/science.1253451. http://www.sciencemag.org/content/346/6215/1320.abstract. http://www.sciencemag.org/cgi/doi/10.1126/science.1253451

24. Jiang, Y., Balaban, M., Zhu, Q., Mirarab, S.: DEPP: deep learning enables extending species trees using single genes. bioRxiv (abstract in RECOMB 2021) (2021). https://doi.org/10.1101/2021.01.22.427808. http://biorxiv.org/content/early/2021/01/24/2021.01.22.427808.abstract

25. Jukes, T.H., Cantor, C.R.: Evolution of protein molecules. In: Mammalian Protein Metabolism, vol. III, pp. 21–132 (1969)

26. Kishino, H., Hasegawa, M.: Evaluation of the maximum likelihood estimate of the evolutionary tree topologies from DNA sequence data, and the branching order in hominoidea. J. Mol. Evol. **29**(2), 170–179 (1989). https://doi.org/10.1007/BF02100115. http://www.springerlink.com/content/ll0lr02023152485

27. Kubatko, L.S., Degnan, J.H.: Inconsistency of phylogenetic estimates from concatenated data under coalescence. Syst. Biol. **56**, 17–24 (2007). http://sysbio.oxfordjournals.org/content/56/1/17.short

28. Libin, P., et al.: PhyloGeoTool: interactively exploring large phylogenies in an epidemiological context. Bioinformatics **33**(24), 3993–3995 (2017). https://doi.org/10.1093/bioinformatics/btx535

29. Linard, B., Swenson, K.M., Pardi, F.: Rapid alignment-free phylogenetic identification of metagenomic sequences. Bioinformatics **35**(18), 3303–3312 (2019). https://doi.org/10.1093/bioinformatics/btz068. https://doi.org/10.1093/bioinformatics/btz068

30. Mai, U., Mirarab, S.: Completing gene trees without species trees in sub-quadratic time. Bioinformatics btab875 (2022). https://doi.org/10.1093/bioinformatics/btab875. https://academic.oup.com/bioinformatics/advance-article/doi/10.1093/bioinformatics/btab875/6493250

31. Matsen, F.A.: Phylogenetics and the human microbiome. Syst. Biol. **64**(1), e26–e41 (2015). https://doi.org/10.1093/sysbio/syu053. http://arxiv.org/abs/1407.1794. https://academic.oup.com/sysbio/article/64/1/e26/2847641

32. Matsen, F.A., Kodner, R.B., Armbrust, E.V.: pplacer: linear time maximum-likelihood and Bayesian phylogenetic placement of sequences onto a fixed reference tree. BMC Bioinform. **11**(1), 538 (2010). https://doi.org/10.1186/1471-2105-11-538.   http://www.pubmedcentral.nih.gov/articlerender.fcgi?artid=3098090&tool=pmcentrez&rendertype=abstract.   http://www.ncbi.nlm.nih.gov/pubmed/21034504.   http://www.pubmedcentral.nih.gov/articlerender.fcgi?artid=PMC3098090

33. Matsen, F.A., IV., Evans, S.N., Matsen, F.A., Evans, S.N.: Edge principal components and squash clustering: using the special structure of phylogenetic placement data for sample comparison. PLoS ONE **8**(3), 1–17 (2013). https://doi.org/10.1371/journal.pone.0056859

34. McDonald, D., Birmingham, A., Knight, R.: Context and the human microbiome. Microbiome **3**(1), 52 (2015). https://doi.org/10.1186/s40168-015-0117-2. http://www.microbiomejournal.com/content/3/1/52

35. Mirarab, S., Bayzid, M.S., Warnow, T.: Evaluating summary methods for multilocus species tree estimation in the presence of incomplete lineage sorting. Syst. Biol. **65**(3), 366–380 (2016). https://doi.org/10.1093/sysbio/syu063. http://sysbio.oxfordjournals.org/content/early/2014/10/13/sysbio.syu063%5Cn. http://sysbio.oxfordjournals.org/content/early/2014/10/13/sysbio.syu063.abstract%5Cn. http://sysbio.oxfordjournals.org/content/early/2014/10/13/sysbio.syu063.full.pdf%5Cn

36. Mirarab, S., Nguyen, N., Warnow, T.: SEPP: SATé-enabled phylogenetic placement. In: Pacific Symposium on Biocomputing, pp. 247–258. World Scientific (2012). https://doi.org/10.1142/9789814366496_0024. http://www.ncbi.nlm.nih.gov/pubmed/22174280. http://www.worldscientific.com/doi/abs/10.1142/9789814366496_0024

37. Nayfach, S., Shi, Z.J., Seshadri, R., Pollard, K.S., Kyrpides, N.C.: New insights from uncultivated genomes of the global human gut microbiome. Nature **568**(7753), 505–510 (2019). https://doi.org/10.1038/s41586-019-1058-x. http://www.nature.com/articles/s41586-019-1058-x

38. Nguyen, N.P., Mirarab, S., Liu, B., Pop, M., Warnow, T.: TIPP: taxonomic identification and phylogenetic profiling. Bioinformatics **30**(24), 3548–3555 (2014). https://doi.org/10.1093/bioinformatics/btu721. http://bioinformatics.oxfordjournals.org/cgi/doi/10.1093/bioinformatics/btu721. https://academic.oup.com/bioinformatics/article-lookup/doi/10.1093/bioinformatics/btu721

39. Pasolli, E., et al.: Extensive unexplored human microbiome diversity revealed by over 150,000 genomes from metagenomes spanning age, geography, and lifestyle. Cell **176**(3), 649–662 (2019). https://doi.org/10.1016/j.cell.2019.01.001. https://linkinghub.elsevier.com/retrieve/pii/S0092867419300017

40. Price, M.N., Dehal, P.S., Arkin, A.P.: FastTree-2 - approximately maximum-likelihood trees for large alignments. PLoS One **5**(3), e9490 (2010). https://doi.org/10.1371/journal.pone.0009490. http://www.pubmedcentral.nih.gov/articlerender.fcgi?artid=2835736&tool=pmcentrez&rendertype=abstract

41. Rabiee, M., Mirarab, S.: INSTRAL: discordance-aware phylogenetic placement using quartet scores. Syst. Biol. **69**(2), 384–391 (2020). https://doi.org/10.1093/sysbio/syz045. https://academic.oup.com/sysbio/advance-article/doi/10.1093/sysbio/syz045/5530610

42. Roch, S., Steel, M.: Likelihood-based tree reconstruction on a concatenation of aligned sequence data sets can be statistically inconsistent. Theor. Popul. Biol. **100**, 56–62 (2015). https://doi.org/10.1016/j.tpb.2014.12.005. http://www.

sciencedirect.com/science/article/pii/S0040580914001075.   https://linkinghub.
elsevier.com/retrieve/pii/S0040580914001075
43. Salichos, L., Rokas, A.: Inferring ancient divergences requires genes with strong
    phylogenetic signals. Nature **497**(7449), 327–331 (2013). https://doi.org/10.
    1038/nature12130.   http://www.nature.com/nature/journal/vaop/ncurrent/full/
    nature12130.html
44. Sayyari, E., Mirarab, S.: Fast coalescent-based computation of local branch support
    from quartet frequencies. Mol. Biol. Evol. **33**(7), 1654–1668 (2016). https://doi.
    org/10.1093/molbev/msw079.   https://academic.oup.com/mbe/article-lookup/
    doi/10.1093/molbev/msw079
45. Singh, K.: On the asymptotic accuracy of Efron's bootstrap. Ann. Stat. **9**(6), 1187–
    1195 (1981)
46. Soltis, P.S., Soltis, D.E.: Applying the bootstrap in phylogeny reconstruction. Stat.
    Sci. **18**(2), 256–267 (2003). http://www.jstor.org/stable/3182855
47. Sonnhammer, E.L., Hollich, V.: Scoredist: a simple and robust protein sequence
    distance estimator. BMC Bioinform. **6**, 1–8 (2005). https://doi.org/10.1186/1471-
    2105-6-108
48. Stark, M., Berger, S.A., Stamatakis, A., von Mering, C.: MLTreeMap-accurate
    maximum likelihood placement of environmental DNA sequences into taxo-
    nomic and functional reference phylogenies. BMC Genomics **11**(1), 461 (2010).
    https://doi.org/10.1186/1471-2164-11-461. http://www.biomedcentral.com/1471-
    2164/11/461
49. Thompson, L.R., et al.: A communal catalogue reveals Earth's multiscale
    microbial diversity. Nature **551**(7681), 457–463 (2017). https://doi.org/10.1038/
    nature24621. http://www.nature.com/doifinder/10.1038/nature24621
50. Turakhia, Y., et al.: Ultrafast Sample placement on Existing tRees (UShER)
    enables real-time phylogenetics for the SARS-CoV-2 pandemic. Nature Genet.
    53(6), 809–816 (2021). https://doi.org/10.1038/s41588-021-00862-7. http://www.
    nature.com/articles/s41588-021-00862-7
51. Warnow, T., Moret, B.M.E., John, K.S.: Absolute convergence: true trees from
    short sequences. In: Proceedings of the Annual ACM-SIAM Symposium on Discrete
    Algorithms (2001)
52. Wedell, E., Cai, Y., Warnow, T.: Scalable and accurate phylogenetic placement
    using pplacer-XR. In: Martín-Vide, C., Vega-Rodríguez, M.A., Wheeler, T. (eds.)
    AlCoB 2021. LNCS, vol. 12715, pp. 94–105. Springer, Cham (2021). https://doi.
    org/10.1007/978-3-030-74432-8_7
53. Zhang, C., Rabiee, M., Sayyari, E., Mirarab, S.: ASTRAL-III: polynomial time
    species tree reconstruction from partially resolved gene trees. BMC Bioin-
    form. **19**(S6), 153 (2018). https://doi.org/10.1186/s12859-018-2129-y. https://
    bmcbioinformatics.biomedcentral.com/articles/10.1186/s12859-018-2129-y
54. Zheng, Q., Bartow-McKenney, C., Meisel, J.S., Grice, E.A.: HmmUFOtu: an HMM
    and phylogenetic placement based ultra-fast taxonomic assignment and OTU
    picking tool for microbiome amplicon sequencing studies. Genome Biol. **19**(1),
    82 (2018). https://doi.org/10.1186/s13059-018-1450-0. https://genomebiology.
    biomedcentral.com/articles/10.1186/s13059-018-1450-0
55. Zhu, Q., et al.: Phylogenomics of 10,575 genomes reveals evolutionary
    proximity between domains Bacteria and Archaea. Nat. Commun. **10**(1),
    5477 (2019). https://doi.org/10.1038/s41467-019-13443-4. http://www.nature.
    com/articles/s41467-019-13443-4
56. Zhu, Q., et al.: WoL: reference phylogeny for microbes (data pre-release) (2019).
    https://biocore.github.io/wol/

The Sackin Index of Simplex Networks

Louxin Zhang$^{(\boxtimes)}$ (iD)

National University of Singapore, Singapore 119076, Singapore
matzlx@nus.edu.sg

Abstract. A phylogenetic network is a simplex network if the child of every reticulation node is a network leaf and each tree node has at most one reticulation node as its child. Simplex networks are a superclass of phylogenetic trees and a subclass of tree-child networks. Generalizing the Sackin index to phylogenetic networks, we prove that the expected Sackin index of a random simplex network is asymptotically $\Theta(n^{7/4})$ in the uniform model.

Keywords: Phylogenetic trees · Tree-child networks · Simplex networks · Sackin index

1 Introduction

Phylogenetic networks have been frequently used for modeling evolutionary history of genomes and genetic flow in population genetics and comparative genomics. Since network models are much more complex than phylgoenetic trees, different classes of phylgoenetic networks have been introduced to investigate different issues of reconstruction of phylogenetic networks [5,8,15,16]. For each special class of phylogenetic networks, algorithmic problems for determining the relationship between phylogenetic trees and networks and for reconstruction of phylogenetic networks from DNA sequences, gene trees and other data have been extensively studied [12,14,17].

The combinatorial and stochastic properties of different classes of phyloge-netic networks have received increasing attention in the study of phylogenetic networks recently. Counting tree-child networks was first studied in [18]. A tight asymptotic value of the number of tree-child networks is given in [10]. Although algorithms are presented for enumerating tree-child networks [4,26], closed for-mulas and even simple recurrence formulas for counting tree-child networks are unknown [6,20]. Counting ranked tree-child networks is studied in [2]. In addi-tion, asymptotic and exact counts of galled trees and galled networks are given in [3] and [9,13], respectively.

The expected height of random binary trees has been known for decades [7]. Recently, the problem of computing the height of random phylogenetic networks is raised in [9,18,24]. The Sackin index of a phylogenetic tree is defined to be

This work was supported by MOE Tier 1 grant R-146-000-318-114.

L. Jin and D. Durand (Eds.): RECOMB-CG 2022, LNBI 13234, pp. 52–67, 2022.
https://doi.org/10.1007/978-3-031-06220-9_4

the sum of the depths of its leaves [21,22]. It is one of the widely-used indices for measuring the balance of phylogenetic trees and testing evolutionary models [1,22,25]. In this paper, we first generalize the Sackin index to phylogenetic networks, where there may be multiple directed paths from the root to each leaf, for measuring the network balance. The Sackin index of a phylogenetic network is defined to be the sum of the lengths of the longest paths from the root to all the leaves. We then prove that the expected Sackin index of a random simplex network on n taxa is $\Theta(n^{7/4})$ in the uniform model, which is significantly larger than the Sackin index of phylogenetic trees on n taxa [23]. As we will see later, simplex networks are the most natural generation of phylogenetic trees, where the child of each reticulation node is a leaf and no two reticulation nodes have a common parent.

The rest of this paper is divided into three parts. In Sect. 2, basic concepts and notation of phylogenetic networks are introduced. In particular, we define the depth of nodes and the Sackin index of phylogenetic networks. In Sect. 3, we present the bound of the expected Sackin index of a random simplex network in the uniform model that is mentioned above. In Sect. 4, we conclude the study with several remarks and open questions.

2 Basic Concepts and Notation

2.1 Tree-Child Networks

For convenience, we consider "planted" phylogenetic networks over taxa (Fig. 1). Such phylogenetic networks over a set X of n taxa are acyclic rooted graphs in which (i) the *root* is of out-degree 1, (ii) there are n nodes of indegree 1 and outdegree 0, called the *leaves*, that are labeled one-to-one by the taxa of X, and (ii) all the other nodes are of degree 3.

Each degree-3 node is called a *tree node* if it is of out-degree 2 and indegree-1; it is a *reticulate node* if it is of indegree 2 and out-degree 1. Note that binary phylogenetic trees are simply phylogenetic networks with no reticulate nodes. An edge (p, q) is a *tree edge* if q is either a tree node or a leaf; it is a *reticulation edge* if q is a reticulation node.

Let N be a phylogenetic network. We use ρ to denote the root of N, $\mathcal{E}(N)$ to denote the set of edges. We also use $\mathcal{L}(N)$, $\mathcal{R}(N)$ and $\mathcal{T}(N)$ to denote the set of the leaves, reticulation and tree nodes, respectively.

Let $u, v \in \{\rho\} \cup \mathcal{L}(N) \cup \mathcal{R}(N) \cup \mathcal{T}(N)$. If $(u, v) \in \mathcal{E}(N)$, u is said to be a *parent* of v and, alternatively, v is a *child* of u. If there is a path from the network root to v that passes u, u is said to be an *ancestor* of v and, alternatively, v is a *descendant* of u.

Let $e' = (u, v) \in \mathcal{E}(N)$ and $e'' = (s, t) \in \mathcal{E}(N)$. The edge e' is said to be *above* e'' if v is an ancestor of s, denoted by $e' \prec e''$. The edges e' and e'' are said to be *parallel*, denoted by $e' \| e''$, if neither of e' and e'' is above the other.

A phylogenetic network is *simplex* if and only if the child of every reticulation is a leaf and every tree node has at most one reticulation node as its child. The

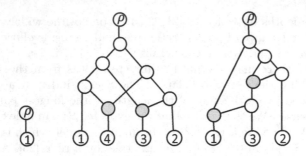

Fig. 1. A phylogenetic network with a single leaf (left), a simplex network on 4 taxa and a tree-child network on 3 taxa that is not simplex (right), where reticulation nodes are represented by filled circles.

middle network in Fig. 1 is simplex, where the child of the reticulations are Leaf 3 and 4.

A phylogenetic network is *tree-child* if every non-leaf node has a child that is either a tree node or a leaf. In Fig. 1, the right phylogenetic network is a tree-child network with 2 reticulation nodes. It is easy to see that a phylogenetic network is tree-child if and only if for every node v, there exists a leaf ℓ such that v and ℓ are connected by a path consisting of tree edges.

It is easy to see that phylogenetic trees are simplex networks, whereas simplex networks are tree-child networks. Simplex networks are also called 1-component tree-child networks [6]. In this paper, we will use the following facts frequently without mention of them, whose proof can be found in [27].

Theorem 1. *Let* $r(N)$, $\ell(N)$ *denote the number of the reticulation nodes and leaves of a tree-child network* N, *respectively. Then,*

(1) $|\mathcal{T}(N)| = \ell(N) + r(N) - 1$.
(2) There are exactly $2\ell(N) + r(N) - 1$ *tree edges and* $2r(N)$ *reticulation edges in* N.

2.2 Node Depth, Network Height and Sackin Index

Let P be a phylogenetic tree and $u \in \mathcal{T}(P) \cup \mathcal{L}(P)$. The *depth* of u is defined to be the number of edges in the unique path from the tree root ρ to u, which is also equal to the number of ancestors of u. Obviously, the depth of the tree root is 0. The *height* of P is defined to be the largest depth from ρ to a leaf of P.

In a phylogenetic network N, there are more than one path from the root ρ to a descendant of a reticulation node. We generalize the concept of node depth to phylogenetic networks as follows.

Let u be a node of N. The *depth* of a node u is defined to be the number of edges in the longest path from the root ρ to u, written as $d_N(u)$. The *ancestor number* of a node u is defined to be the number of the ancestors of the node, written as $\alpha_N(u)$. For example, the depth and ancestor number of Leaf 4 are five and six, respectively, in the right phylogenetic network in Fig. 2.

For a tree-child network N, we define the following parameters:

- The *height* of N is defined to be the largest depth of a leaf, denoted by $h(N)$.
- The *Sackin index* of N is defined to be the sum of the depths of its leaves, denoted by $K(N)$.

Consider a family $\mathcal{F}$ of tree-child networks. The *expected height* of a network of the family $\mathcal{F}$ in the uniform model is defined by:

$$\overline{H}(\mathcal{F}) = \frac{1}{|\mathcal{F}|} \sum_{N \in \mathcal{F}} h(N),$$

where $|\mathcal{F}|$ is the cardinality of the set $\mathcal{F}$. The *expected Sackin index* of a random network in $\mathcal{F}$ in the uniform model is defined by:

$$\overline{K}(\mathcal{F}) = \frac{1}{|\mathcal{F}|} \sum_{N \in \mathcal{F}} K(N).$$

The following results are well known for the class of phylogenetic trees.

Theorem 2. *Under the uniform model,*

(1) [7] The expected height of a random phylogenetic tree over n taxa is asymptotically $2\sqrt{\pi n}$.

(2) [19] The expected Sackin index[1] of a random phylogenetic tree on n taxa is $\frac{2^{2n-2}n!(n-1)!}{(2n-2)!}$, which is asymptotically $\sqrt{\pi}n^{3/2}$.

3 The Expected Sackin Index of Random Simplex Networks

In this section, we will use an enumeration procedure and a simple counting formula for simplex networks that appear in [6] to obtain the asymptotic Sackin index of a simplex network in the uniform model.

3.1 Enumerating Simplex Networks

We first briefly introduce a procedure for enumerating simplex networks appearing in [6]. Let $\mathcal{OC}_n$ denote the class of simplex networks on n taxa and $o_n = |\mathcal{OC}_n|$.

Let $N \in \mathcal{OC}_n$. N may contain 0 to $n-1$ reticulations. Recall that the child of each reticulation is a leaf. All the tree nodes and leaves that are not below

---

[1] The expected Sackin index of a phylogenetic tree on n is different from that reported in literature by n. Here, we work on "planted" phylogenetic trees.

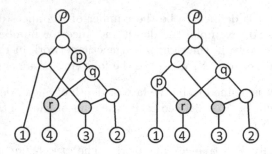

Fig. 2. The Illustration of the procedure for enumerating simplex networks by attaching the grandparents (p, q) of a new leaf (4) below a reticulation node (r) onto either a tree edge (left) or a pair of tree edges (right) in the top tree component of a simplex network.

any reticulations are connected to the root by tree edges, forming a connected subtree, which we call the *top tree component* and denote by $C(N)$ (see [12]). For instance, the top tree component of the simplex network in the middle of Fig. 1 consists of Leaf 1, Leaf 2 and their ancestors, including ρ.

Let $[i, j]$ denote the integer set $\{i, i + 1, \cdots, j\}$, where $0 < i \le j$. For any nonempty $I \subseteq [1, n]$, $\mathcal{OC}_{I,n}$ denotes the subset of simplex networks in which there are $n - |I|$ reticulations whose child are labeled uniquely with the elements of $[1, n] \setminus I$. Clearly, when $I = [1, n]$, $\mathcal{OC}_{I,n}$ is just the set of phylogenetic trees on $[1, n]$.

For simplicity, we write $\mathcal{OC}_{k,j}$ for $\mathcal{OC}_{[1,k],k+j}$ for $j > 1$. The networks of $\mathcal{OC}_{k,j+1}$ can be generated by attaching the two grandparents p and q of Leaf $k + j + 1$ to each network N of $\mathcal{OC}_{k,j}$ in the following two ways [6], where the parent of $k + j + 1$ is denoted by r:

- **(Type-I insertion)** For each tree edge $e = (u, v)$ of $C(N)$, where $u \notin \mathcal{R}(N)$, subdivide e into $(u, p), (p, q)$ and (q, v), and add three new edges $(p, r), (q, r)$ and $(r, k + j + 1)$, This expansion operation is shown on the left network in Fig. 2, where $k + j + 1 = 4$.
- **(Type-II insertion)** For each pair of tree edges $e' = (u, v)$ and $e'' = (s, t)$ of $C(N)$, i.e. $u \notin \mathcal{R}(N)$ and $s \notin \mathcal{R}(N)$, subdivide e' into (u, p) and (p, v) and e'' into (s, q) and (q, t), and add three new edges $(p, r), (q, r)$ and $(r, k + j + 1)$. This insertion operation is shown in the right network in Fig. 2.

Each N of $\mathcal{OC}_{k,j}$ contains exactly $2(k + j) - 1$ tree edges in its $C(N)$. Additionally, all the networks generated using the above method are distinct [6], as the subset of leaves below the lower end of each tree edge is unique in a tree-child network. This implies that $|\mathcal{OC}_{k,j+1}| = (2(k + j) - 1)(k + j)|\mathcal{OC}_{k,j}|$ and thus

$$|\mathcal{OC}_{k,j}| = \frac{(2(k + j) - 2)!}{2^{k+j-1}(k - 1)!}. \tag{1}$$

Note that the simple formula for counting the phylogenetic trees with k leaves is obtained by simply setting j to 0.

By symmetry, $\mathcal{OC}_{I,n} = \mathcal{OC}_{J,n}$ if $|I| = |J|$. Therefore, we obtain [6]:

$$o_n = \sum_{k=1}^{n} \binom{n}{k} |\mathcal{OC}_{k,n-k}| = \frac{(2n-2)!}{2^{n-1}} \sum_{k=0}^{n} \binom{n}{k} \frac{1}{(k-1)!} \tag{2}$$

3.2   The Total Depths of the Nodes in the Top Tree Component

Recall that $\alpha(u)$ denotes the number of ancestors of a node u. Let $\delta_{C(N)}(u)$ denote the number of descendants of u that are in $C(N)$. We also use $\text{ToT}(N)$ to denote the tree edges of $C(N)$. We define:

$$A_C(N) = \sum_{v \in C(N)} \alpha_N(v) = \sum_{u \in C(N)} \delta_{C(N)}(u) \tag{3}$$

and

$$A_C(\mathcal{OC}_{k,j}) = \sum_{N \in \mathcal{OC}_{k,j}} A_C(N). \tag{4}$$

Lemma 1. *Assume $N \in \mathcal{OC}_{k,j}$ and let $n = k + j$. For each e of $\text{ToT}(N)$, we use $N(e)$ to denote the network obtained from N by applying the type-I insertion onto e. Then,*

$$\sum_{e \in ToT(N)} A_C(N(e)) = (2n+3)A_C(N) + (2n-1) \tag{5}$$

Proof. Let $e = (u, v) \in \text{ToT}(N)$. By the description of the Type-I insertion operation, the set of nodes and edges of $N(e)$ are respectively:

$$\mathcal{V}(N(e)) = \mathcal{V}(N) \cup \{p, q, r, n+1\}$$

and

$$\mathcal{E}(N(e)) = \mathcal{E}(N) \cup \{(u,p), (p,q), (q,v), (p,r), (q,r), (r,n+1)\} \setminus \{(u,v)\}$$

(see Fig. 2).

Let $\mathcal{D}(v)$ be the set of descendants of v in $C(N)$. We have the following facts:

$$|\alpha_{N(e)}(x)| = 2 + |\alpha_N(x)|, \ \forall x \in \mathcal{D}(v)$$
$$|\alpha_{N(e)}(y)| = |\alpha_N(y)|, \ \forall y \notin \mathcal{D}(v) \cup \{v, p, q\}$$
$$|\alpha_{N(e)}(p)| = |\alpha_N(v)|,$$
$$|\alpha_{N(e)}(q)| = |\alpha_N(v)| + 1,$$
$$|\alpha_{N(e)}(v)| = |\alpha_N(v)| + 2.$$

By summing above equations, we obtain:

$$A_C(N(e)) = A_C(N) + 2\delta_{C(N)}(v) + 2\alpha_N(v) + 3. \tag{6}$$

Eq. (3) and (6) imply that

$$\sum_{e \in \text{ToT}(N)} A_C(N(e))$$

$$= (2n-1)A_C(N) + 2 \sum_{(u,v) \in \text{ToT}(N)} \delta_{C(N)}(v) + 2 \sum_{(u,v) \in \text{ToT}(N)} \alpha_N(v) + 3(2n-1)$$

$$= (2n-1)A_C(N) + 2 \left(\sum_{v \in C(N)} \delta_{C(N)}(v) - \delta_{C(N)}(\rho) \right) + 2A_C(N) + 3(2n-1)$$

$$= (2n+3)A_C(N) + (2n-1),$$

where $n = k + j$ and we use the fact that $\delta_{C(N)}(\rho) = 2n - 1$ for the network root ρ.

Lemma 2. *Let $N \in OC_{k,j}$ and $N(e', e'')$ denote the network obtained from N by applying type-II insertion to a pair of distinct edges e', e'' of $C(N)$. Then,*

$$\sum_{\{e',e''\} \subset ToT(N)} A_C(N(e', e'')) = (2n^2 + n - 2)A_C(N) - (2n-1), \tag{7}$$

where $n = k + j$.

Proof. After attaching the reticulation node r onto the edges $e' = (u, v)$ and $e'' = (s, t)$, the tree edge e' is subdivided into $e'_1 = (u, p)$ and $e''_1 = (p, v)$; the edge e'' is subdivided into $e'_2 = (s, q)$ and $e''_2 = (q, t)$, where p, q are the parents of r in $N(e', e'')$. We consider two possible cases.

First, we consider the case that $e' \prec e''$. In $N(e', e'')$, for each descendant x of u in $C(N(e', e''))$ such that $x \neq p$ and x is not below s, $\alpha(x)$ increases by 1 because of the subdivision of e'. For each descendant y of s in $C(N(e', e''))$ such that $y \neq q$, $\alpha(y)$ increases by 2 because of the subdivision of e' and e''. Additionally,

$$\alpha_{N(e',e'')}(p) = \alpha_N(v)$$

and because q is below v,

$$\alpha_{N(e',e'')}(q) = \alpha_N(t) + 1.$$

For any tree node or leaves that are not the descendants of u, $\alpha(x)$ remains the same. Summing all together, we have that

$$A_C(N(e', e'')) = A_C(N) + \delta_{C(N)}(v) + \delta_{C(N)}(t) + \alpha_N(v) + \alpha_N(t) + 1 + 2, \tag{8}$$

where 2 is the sum of the total increase of $\alpha(v)$ and $\alpha(t)$.

Summing over all the comparable edge pairs, we have

$$\sum_{e',e''\in\mathrm{ToT}(N):e'\prec e''} A_C(N(e',e''))$$

$$= (A_C(N)+3)|\{(e',e'') \mid e' \prec e''\}| + \sum_{(u,v)\in\mathrm{ToT}(N)} (\delta_{C(N)}(v)+\alpha_N(v))\delta_{C(N)}(v)$$

$$+ \sum_{(u,v)\in\mathrm{ToT}(N)} (\delta_{C(N)}(v)+\alpha_N(v))(\alpha_N(v)-1)$$

$$= (A_C(N)+3)|\{(e',e'') \mid e' \prec e''\}| + \sum_{v\in C(N)\backslash\{\rho\}} (\delta_{C(N)}(v)+\alpha_N(v))^2$$

$$-2A_C(N)+(2n-1), \tag{9}$$

where we use the fact that $\sum_{v\in C(N)\backslash\{\rho\}} \delta_N(v) = A_C(N)-(2n-1)$.

Second, we assume $e'\|e''$. In $N(e',e'')$, for each descendant x of u in $C(N(e',e''))$, $\alpha(x)$ increases by 1 because of the subdivision of e'. For each descendant y of p in $C(N(e',e''))$, $\alpha(y)$ increases by 1 because of the subdivision of e''. Additionally, $\alpha_{N(e',e'')}(p) = \alpha_N(v)$ and $\alpha_{N(e',e'')}(q) = \alpha_N(t)$. For any tree node or leaves that are the descendants of neither u nor p, $\alpha(x)$ remains the same. Hence,

$$A_C(N(e',e'')) = A_C(N) + \delta_{C(N)}(v) + \delta_{C(N)}(t) + \alpha_N(v) + \alpha_N(t) + 2, \tag{10}$$

where 2 counts for the increase by 1 of both $\alpha(v)$ and $\alpha(t)$.

Summing Eq. (10) over parallel edge pairs, we obtain:

$$\sum_{e',e''\in\mathrm{ToT}(N):e'\|e''} A_C(N(e',e''))$$

$$= (A_C(N)+2)\,|\{(e',e''): \ e'\|e''\}|$$

$$+ \sum_{(u,v)\in\mathrm{ToT}(N)} o_N((u,v))(\delta_{C(N)}(v)+\alpha_N(v)), \tag{11}$$

where $o_N((u,v))$ is the number of the edges of $C(N)$ that are parallel to (u,v).

Summing Eq. (9) and (11) and using the facts that $n = k + j$ and $|\{(e',e'') \mid e' \prec e''\}| = A_C(N) - (2n-1)$ and

$$\sum_{(u,v)\in\mathrm{ToT}(N)} (\delta_{C(N)}(v)+\alpha_N(v)) = 2A_C(N) - \delta_{C(N)}(\rho) = 2A_C(N) - (2n-1),$$

we obtain:

$$\sum_{\{e',e''\}\subset\mathrm{ToT}(N)} A_C(N(e',e''))$$

$$= (A_C(N)+2)\binom{2n-1}{2} + |\{(e',e'') \mid e' \prec e''\}| - 2A_C(N) + (2n-1)$$

$$+(2n-1) \sum_{(u,v)\in\mathrm{ToT}(N)} (\delta_{C(N)}(v)+\alpha_N(v))$$

$$= A_C(N)(2n^2 + n - 2) - (2n-1).$$

Theorem 3. *Let $k \geq 1$ and $j \geq 0$. For the subclass of simplex networks with j reticulations whose children are $k+1, \cdots, k+j$ on $n = k+j$ taxa,*

$$A_C(\mathcal{OC}_{k,j}) = \frac{(2n)!}{2^j(2k)!} \left(2^k k! - \frac{(2k-1)!}{2^{k-1}(k-1)!} \right). \tag{12}$$

Proof. Summing Eq. (5) and (7), we obtain

$$\sum_{e \in ToT(N)} A_C(N(e)) + \sum_{\{e',e''\} \subset ToT(N)} A_C(N(e',e'')) = (2n+1)(n+1)A_C(N)$$

for each $N \in \mathcal{OC}_{k,j}$. Summing the above equation over all networks of $\mathcal{OC}_{k,j}$, we have

$$A_C(\mathcal{OC}_{k,j+1}) = (2n+1)(n+1) \sum_{N \in \mathcal{OC}_{k,j}} A_C(N) = (2n+1)(n+1)A_C(\mathcal{OC}_{k,j})$$

or, equivalently, $A_C(\mathcal{OC}_{k,j}) = n(2n-1)A_C(\mathcal{OC}_{k,j-1})$.
  Since it is proved in [26] that

$$A_C(\mathcal{OC}_{k,0}) = 2^k k! - \frac{(2k-1)!}{2^{k-1}(k-1)!},$$

we obtain Eq. (12) by induction.

3.3 The Expected Total C-Depth of Random Simplex Networks

Recall that $\mathcal{OC}_{k,j}$ denotes the set of simplex networks on $k+j$ taxa in which the children of the j reticulations are the leaves labeled with $k+1, k+2, \cdots, k+j$, respectively. Since the children of the k reticulations of a simplex network with j reticulations on $k+j$ taxa can be labelled with any j out of $k+j$ taxa, the *expected total c-depth* of simplex networks with j reticulations on $k+j$ taxa is defined as:

$$D_C(k,j) = \frac{\binom{k+j}{k} \sum_{N \in \mathcal{OC}_{k,j}} A_C(N)}{\binom{k+j}{k} |\mathcal{OC}_{k,j}|} = \frac{A_C(\mathcal{OC}_{k,j})}{|\mathcal{OC}_{k,j}|} \tag{13}$$

in the uniform model. For each $n \geq 1$, the expected total c-depth of simplex networks on n taxa becomes:

$$\overline{D}_C(n) = \left(\sum_{k=1}^{n} \binom{n}{k} A_C(\mathcal{OC}_{k,n-k}) \right) \Big/ \left(\sum_{k=1}^{n} \binom{n}{k} |\mathcal{OC}_{k,n-k}| \right) \tag{14}$$

in the uniform model.

Proposition 1. *For any $k \geq 1$ and $j \geq 0$, the average total c-depth $D_C(k,j)$ has the following asymptotic value*

$$D_C(k,j) = \frac{\sqrt{\pi}n(2n-1)}{\sqrt{k}} \left(1 - \pi^{-1/2}k^{-1/2} + O(k^{-1}) \right), \tag{15}$$

where $n = k+j$.

Proof. Since $\binom{2k}{k} = \frac{2^{2k}}{\sqrt{\pi k}}(1 - (8k)^{-1} + O(k^{-2}))$, by Eq. (1),

$$
D_C(k,j) = \frac{(2n)!}{2^{n-k}(2k)!}\left(2^k k! - \frac{(2k-1)!}{2^{k-1}(k-1)!}\right) / \left(\frac{(2n-2)!}{2^{n-1}(k-1)!}\right)
$$

$$
= \frac{n(2n-1)}{k}\left(\frac{2^{2k}k!k!}{(2k)!} - 1\right)
$$

$$
= \frac{n(2n-1)}{k}\left(\frac{\sqrt{\pi k}}{1 - (8k)^{-1} + O(k^{-2})} - 1\right)
$$

$$
= \frac{\sqrt{\pi}n(2n-1)}{\sqrt{k}}\left(1 - \frac{1}{\sqrt{\pi k}} + \frac{1}{8k} + O(k^{-2})\right).
$$

Thus, Eq. (15) is proved.

Proposition 2. *For any $n > 1$, $\overline{D}_C(n)$ has the following bounds.*

$$
(\sqrt{2} - 1)(1 + O(n^{-1/4})) \leq \frac{\overline{D}_C(n)}{\sqrt{\pi}n^{3/4}(2n-1)} \leq 2(1 + O(n^{-1/4})). \tag{16}
$$

Proof. Here, we use the idea of Laplace method [11] to prove the bounds here. Let $S_k = \binom{n}{k}|\mathcal{OC}_{k,n-k}| = \binom{n}{k}\frac{(2n-2)!}{2^{n-1}(k-1)!}$ and $k_0 = \sqrt{n+1} - 1$.

By considering the ratio S_{k+1}/S_k, we can show (in Appendix A):

- $S_k \leq S_{k+1}$ for $k \in [1, k_0]$;
- $S_k > S_{k+1}$ if $k_0 < k < 2k_0$;
- $S_k > 2S_{k+1}$ if $k \geq 2k_0$.

First, we define

$$
f_1 = \left(\sum_{k \leq k_0} D_C(k, n - k)S_k\right) / \left(\sum_{k \leq k_0} S_k\right).
$$

Since $D_C(k, n - k)$ is a decreasing function (which is proved in Appendix A),

$$
f_1 \geq D_C(\lfloor k_0 \rfloor, n - \lfloor k_0 \rfloor) = \sqrt{\pi}n^{3/4}(2n - 1)(1 + O(n^{-1/4}))
$$

Furthermore, since S_k is increasing on $[1, k_0)$,

$$
f_1 \leq \frac{1}{\lfloor k_0 \rfloor}\sum_{k \leq \lfloor k_0 \rfloor} D_C(k, n - k)
$$

$$
= \frac{\sqrt{\pi}n(2n-1)}{\lfloor k_0 \rfloor}\sum_{k \leq \lfloor k_0 \rfloor}\left(k^{-1/2} - \sqrt{\pi}k^{-1} + (1/8)k^{-3/2} + O(k^{-5/2})\right)
$$

$$
= \frac{\sqrt{\pi}n(2n-1)}{\lfloor k_0 \rfloor}\left[2(\lfloor k_0 \rfloor^{1/2} - 1) - \sqrt{\pi}\ln\lfloor k_0 \rfloor + O(\lfloor k_0 \rfloor^{-1/2})\right]
$$

$$
= 2\sqrt{\pi}n^{3/4}(2n - 1)(1 + O(n^{-1/4}))
$$

Second, define

$$f' = \frac{\sum_{k \in [k_0, 2k_0]} D_C(k, n-k) S_k}{\sum_{k \in [k_0, 2k_0]} S_k}. \tag{17}$$

Since the $D_C(k, n-k)$ is a decreasing function and S_k is decreasing on $[k_0, 2k_0]$,

$$f' \leq D_C(\lceil k_0 \rceil, n - \lfloor k_0 \rfloor) = \sqrt{\pi} n^{3/4} (2n-1)(1 + O(n^{-1/4}))$$

and, by Eq. (15),

$$f' \geq \frac{1}{k_0} \sum_{k \in [k_0, 2k_0]} D_C(k, n-k)$$

$$= \frac{\sqrt{\pi} n (2n-1)}{k_0} \sum_{k \in [k_0, 2k_0]} \left(k^{-1/2} - \sqrt{\pi} k^{-1} + (1/8) k^{-3/2} + O(k^{-5/2}) \right)$$

$$= \frac{\sqrt{\pi} n (2n-1)}{k_0} \left[2(\sqrt{2} - 1) k_0^{1/2} - \sqrt{\pi} \ln 2 + (1/4)(1 - \sqrt{1/2}) k_0^{-1/2} + O(k_0^{-1}) \right]$$

$$= 2(\sqrt{2} - 1) \sqrt{\pi} n^{3/4} (2n-1)(1 + O(n^{-1/4}))$$

Now, we consider

$$f_2 = \frac{\sum_{k > k_0} D_C(k, n-k) S_k}{\sum_{k > k_0} S_k}.$$

For each $k > 2k_0$, $D_C(k, n-k) \leq D_C(2k_0, n - 2k_0) \leq f'$. This implies that

$$\sum_{k > 2k_0} D_C(k, n-k) S_k \leq f' \sum_{k > 2k_0} S_k$$

and thus

$$f_2 = \frac{\sum_{k_0 \leq k \leq 2k_0} D_C(k, n-k) S_k + \sum_{k > 2k_0} D_C(k, n-k) S_k}{\sum_{k > k_0} S_k}$$

$$= \frac{f' \sum_{k_0 \leq k \leq 2k_0} S_k + \sum_{k > 2k_0} D_C(k, n-k) S_k}{\sum_{k > k_0} S_k}$$

$$\leq \frac{f' \sum_{k_0 \leq k \leq 2k_0} S_k + f' \sum_{k > 2k_0} S_k}{\sum_{k > k_0} S_k}$$

$$= f'$$

On the other hand, $S_k \geq 2S_{k+1}$ for $k \geq 2k_0$ and thus $\sum_{k > 2k_0} S_k \leq 2S_{\lceil 2k_0 \rceil}$

$$f_2 \geq \frac{\sum_{k \in [k_0, 2k_0]} D_C(k, n-k) S_k}{\sum_{k > k_0} S_k} = \frac{f' \sum_{k \in [k_0, 2k_0]} S_k}{\sum_{k \in [k_0, 2k_0]} S_k + \sum_{k > 2k_0} S_k}$$

$$\geq \frac{f' \sum_{k \in [k_0, 2k_0]} S_k}{2 \sum_{k \in [k_0, 2k_0]} S_k} = f'/2$$

The bounds on f_1 and f_2 implies that the mean value f over the entire region is between f_1 and f_2, Thus we have proved that

$$(\sqrt{2} - 1)\sqrt{\pi}n^{3/4}(2n - 1)(1 + O(n^{-1/4})) \leq \min(f'/2, f_1) \leq \overline{D}_C(n),$$

and

$$\overline{D}_C(n) \leq \max(f_1, f_2) \leq 2\sqrt{\pi}n^{3/4}(2n - 1)(1 + O(n^{-1/4})).$$

3.4   Bounds on the Sackin Index for a Random Simplex Network

Recall that $K(N) = \sum_{\ell \in \mathcal{L}(N)} d_N(\ell)$ for a network N.

Proposition 3. *For any $N \in \mathcal{OC}_{k,j}$, $K(N) \leq A_C(N) + 1 \leq 2K(N)$, where $k \geq 1, j \geq 0$.*

Proof. Let $N \in \mathcal{OC}_{k,j}$. If $j = 0$, N does not contain any reticulation node and thus every node of N is in the top tree component of N. By definition, $K(N) \leq A_C(N)$. Since N is a phylogenetic tree, N contains the same number of internal nodes (including the root ρ) as the number of leaves. By induction, we can prove that there exists a 1-to-1 mapping $\phi : \mathcal{T}(N) \cup \{\rho\} \to \mathcal{L}(N)$ such that u is an ancestor of $\phi(u)$ for every u (see Appendix B). Noting that $d_N(\rho) = 0$ and $d_N(\phi(\rho)) \geq 1$, we have:

$$A_C(N) + 1 \leq \sum_{\ell \in \mathcal{L}(N)} d_N(\ell) + \sum_{u \in \mathcal{T}(N)} d_N(\phi(u)) = 2K(N).$$

We now generalize the above proof for phylogenetic trees to the general case where $j > 0$ as follows.

We assume that $r_1, r_2, \cdots, r_j$ are the j reticulation nodes and their parents are p'_j and p''_j. Since p'_i and p''_i are both found in the top tree component of N and they cannot be the network root ρ, $d_N(p'_i) \geq 1$ and $d_N(p''_i) \geq 1$ for each $i \in [1, j]$. Since Leaf $(k + i)$ is the child of r_i, by definition, $d_N(k + i) = \max(d_N(p'_i), d_N(p''_i)) + 2$. Now we prove that $K(N) \leq A_C(N) + 1$ by considering two cases.

First, if $d_N(p'_i) \geq 2$ and $d_N(p''_i) \geq 2$ for every $i \in [1, j]$, then $d_N(k + i) = \max(d_N(p'_i), d_N(p''_i)) + 2 \leq d_N(p'_i) + d_N(p''_i)$ for every i. Since N is tree-child, the parents p'_i and p''_i are distinct nodes in the top tree component for different i, $K(N) \leq \sum_{1 \leq i \leq k} d_N(i) + \sum_{1 \leq i \leq j}(d_N(p'_i) + d_N(p''_i)) \leq A_C(N)$.

Second, if $d_N(p'_i) = \min(d_N(p'_i), d_N(p''_i)) = 1$ for some i, then p'_i is the unique child of the root ρ. This implies that there is exactly one index i_0 for which $\min(d_N(p'_{i_0}), d_N(p''_{i_0})) = 1$ and $\max(d_N(p'_{i_0}), d_N(p''_{i_0})) \geq 2$. In this case, $d_N(k + i_0) \leq d_N(p'_{i_0}) + d_N(p''_{i_0}) + 1$ and $d_N(k + i) = \max(d_N(p'_i), d_N(p''_i)) + 2 \leq d_N(p'_i) + d_N(p''_i)$ for every $i \neq i_0$. Therefore, $K(N) \leq A_C(N) + 1$.

The tree-component $C(N)$ contains $k + 2j$ internal tree nodes including ρ. We set

$$\mathcal{T}(C(N)) \setminus \{p'_i, p''_i : 1 \leq i \leq j\} = \{\rho, u_1, \cdots, u_{k-1}\}.$$

Again, there is an 1-to-1 mapping ϕ from $\{\rho, u_1, \cdots, u_{k-1}\}$ to $\mathcal{L}(C(N)) = \{1, 2, \cdots, k\}$ such that the leaf $\phi(u_i)$ is a descendant of u_i. Therefore, since $d_N(\rho) = 0$ and $d_N(\phi(\rho)) \geq 1$,

$$d_N(\rho) + 1 \leq d_N(\phi(\rho));$$
$$d_N(p_i') + d_N(p_i'') \leq 2\max(d_N(p_i'), d_N(p_i'')) < 2d_N(k+i), 1 \leq i \leq j;$$
$$d_N(u_i) + d_N(\phi(u_i)) \leq 2d_N(\phi(u_i)), 1 \leq i \leq k-1.$$

Therefore, $A_C(N) + 1 \leq 2K(N)$.

Theorem 4. *The expected Sackin index $\overline{K}(\mathcal{OC}_n)$ of a simplex network on n taxa is $\Theta(n^{7/4})$.*

Proof. We define

$$K(\mathcal{OC}_{k,j}) = \sum_{N \in \mathcal{OC}_{k,j}} K(N),$$

where $k \geq 1$ and $j \geq 0$. By Proposition 3,

$$K(\mathcal{OC}_{k,j}) \leq A_N(\mathcal{OC}_{k,j}) + |\mathcal{OC}_{k,j}| \leq 2K(\mathcal{OC}_{k,j}).$$

By Eq. (16), we have:

$$\overline{K}(\mathcal{OC}_n) = \frac{1}{\sum_{k=1}^n \binom{n}{k}|\mathcal{OC}_{k,n-k}|} \left(\sum_{k=1}^n \binom{n}{k} K(\mathcal{OC}_{k,n-k}) \right)$$
$$\leq \frac{1}{\sum_{k=1}^n \binom{n}{k}|\mathcal{OC}_{k,n-k}|} \left(\sum_{k=1}^n \binom{n}{k} A_C(\mathcal{OC}_{k,n-k}) \right) + 1$$
$$= \overline{D}_C(n) + 1$$
$$= 4\sqrt{\pi}n^{7/4} + 1.$$

Similarly, using Eq. (16), we have that

$$2\overline{K}(\mathcal{OC}_n) \geq \overline{D}_C(n) = 2(\sqrt{2} - 1)\sqrt{\pi}n^{7/4} + O(1),$$

equivalently,

$$\overline{K}(\mathcal{OC}_n) \geq (\sqrt{2} - 1)\sqrt{\pi}n^{7/4} + O(1).$$

This concludes the proof.

4   Conclusion

What facts about phylogenetic trees remain valid for phylogenetic networks is important in the study of phylogenetic networks. In this paper, a version of the Sackin index is proposed to measure the balance of phylogenetic networks. Here, the asymptotic estimate (up to constant ratio) for the expected Sackin index of a simplex network is given in the uniform model. This is just the first step to develop statistic for imbalance analysis for simplex networks.

This study raises a few research problems. First, the asymptotic value of the expected Sackin index of tree-child networks over n taxa is still unknown. It is also interesting to investigate the Sackin index for galled trees, galled networks and other classes of networks (see [27] for example).

Second, it is even more challenging to estimate the expected height of simplex networks and tree-child networks. Recently, Stufler introduces a branching process method to study the asymptotic global and local shape properties of level-k phylogenetic networks [24]. It is interesting to see whether or not Stufler's method can be used to answer this question.

Third, the phylogenetic trees with the minimum and maximum Sackin index have been characterized [22]. Characterizing the simplex (resp. tree-child) networks with the minimum (resp. maximum) Sackin index is also interesting.

Declaration of Competing Interest

The author declares that he has no known competing financial interests or personal relationships that could have appeared to influence the work reported in this paper.

Acknowledgment. The author thanks Michael Fuchs for reading the first draft of this paper. He also thanks anonymous reviewers for their valuable comments on this work which was submitted to the RECOMB-CG 2020.

Appendix A

Proposition A.1. Let $S_k = \binom{n}{k}\frac{(2n-2)!}{2^{n-1}(k-1)!}$ and $k_0 = \sqrt{n+1}-1$. Then,

- $S_k \leq S_{k+1}$ for $k \in [1, k_0]$;
- $S_k > S_{k+1}$ if $k_0 < k < 2k_0$;
- $S_k > 2S_{k+1}$ if $k \in [2k_0, \infty)$.

Proof. Note that

$$S_{k+1} = \frac{n-k}{k(k+1)}S_k.$$

If $k \leq k_0$, $(k+1)^2 \leq (\sqrt{n+1}-1+1)^2 = n+1$ and, equivalently, $k(1+k) \leq n-k$ and thus $S_{k+1} = \frac{n-k}{k(k+1)}S_k \geq S_k$. Similarly, $S_{k+1} < S_k$ if $k > k_0$.

If $k \geq 2k_0$, $(k+2)^2 \geq 4(n+1)$ and $k^2+k \geq 4(n-k)+k \geq 4(n-k)$ and therefore, $S_{k+1} \leq \frac{1}{4}S_k < \frac{1}{2}S_k$.

Proposition A.2. $D_C(k, n-k)$ is a decreasing function of k on $[1, n-1]$.

Proof. By Eq. (1) and Eq. (12),

$$D_C(k, n-k) = \frac{n(2n-1)}{k}\left(\frac{4^k k! k!}{(2k)!} - 1\right).$$

Hence,

$$n(2n-1)[D_C(k, n-k) - D_C(k+1, n-k-1)]$$
$$= \frac{2^{2k}(k-1)!k!}{(2k)!} - \frac{2^{2k}4k!(k+1)!}{(2k+2)!} - \frac{1}{k(k+1)}$$
$$= \frac{2^{2k}k!(k-1)!}{(2k+1)!} - \frac{1}{k(k+1)}$$
$$= \frac{1}{k(k+1)} \left[\frac{2^{2k}k!(k+1)!}{(2k+1)!} - 1 \right]$$
$$= \frac{1}{k(k+1)} \left[\frac{4 \times 6 \times \cdots 2k \times (2k+2)}{3 \times 5 \times \cdots (2k+1)} - 1 \right]$$
$$> 0.$$

Appendix B

Proposition B.1. Let P be a phylogenetic tree on n taxa. there exists a 1-to-1 mapping $\phi : \mathcal{T}(P) \cup \{\rho\} \to \mathcal{L}(P)$ such that u is an ancestor of $\phi(u)$ for each $u \in \mathcal{T}(P) \cup \{\rho\}$.

Proof. We prove the fact by mathematical induction on n. When $n = 1$, we simply map the root ρ to the only leaf.

Assume the fact is true for $n \le k$, where $k \ge 1$. For a phylogenetic tree P with $k+1$ leaves, we let the child of the root ρ be u and the two grandchildren be v and w. We consider the subtree P' induced by u, v and all the descendants of v and the subtree P'' induced by u, w and all the descendants of w.

Obviously, both T' and T'' have less than k leaves. By induction, there is a 1-to-1 mapping $\phi' : \mathcal{T}(P') \cup \{u\} \to \mathcal{L}(P')$ satisfying the constraints on leaves, and there is a 1-to-1 mapping $\phi'' : \mathcal{T}(P'') \cup \{u\} \to \mathcal{L}(P'')$ satisfying the constraints on leaves. Let $\phi'(u) = \ell$. Then, the function that maps ρ to ℓ, u to $\phi''(u)$ and all the other tree nodes x to $\phi'(x)$ or $\phi''(x)$ depending whether it is in P' or P''. It is easy to verify that ϕ is a desired mapping.

References

1. Avino, M., Ng, G.T., He, Y., Renaud, M.S., Jones, B.R., Poon, A.F.: Tree shape-based approaches for the comparative study of cophylogeny. Ecol. Evol. **9**(12), 6756–6771 (2019)
2. Bienvenu, F., Lambert, A., Steel, M.: Combinatorial and stochastic properties of ranked tree-child networks. arXiv preprint arXiv:2007.09701 (2020)
3. Bouvel, M., Gambette, P., Mansouri, M.: Counting phylogenetic networks of level 1 and 2. J. Math. Biol. **81**(6), 1357–1395 (2020)
4. Cardona, G., Pons, J.C., Scornavacca, C.: Generation of binary tree-child phylogenetic networks. PLoS Comput. Biol. **15**(9), e1007347 (2019)

5. Cardona, G., Rosselló, F., Valiente, G.: Comparison of tree-child phylogenetic networks. IEEE/ACM-TCBB **6**(4), 552–569 (2009). https://doi.org/10.1109/TCBB.2007.70270
6. Cardona, G., Zhang, L.: Counting and enumerating tree-child networks and their subclasses. J. Comput. Syst. Sci. **114**, 84–104 (2020)
7. Flajolet, P., Odlyzko, A.: The average height of binary trees and other simple trees. J. Comput. Syst. Sci. **25**(2), 171–213 (1982)
8. Francis, A.R., Steel, M.: Which phylogenetic networks are merely trees with additional arcs? Syst. Biol. **64**(5), 768–777 (2015)
9. Fuchs, M., Yu, G.R., Zhang, L.: Asymptotic enumeration and distributional properties of galled networks. arXiv preprint arXiv:2010.13324 (2020)
10. Fuchs, M., Yu, G.R., Zhang, L.: On the asymptotic growth of the number of tree-child networks. Eur. J. Comb. **93**, 103278 (2021)
11. Graham, R.L., Knuth, D.E., Patashnik, O.: Concrete Mathematics: A Foundation for Computer Science. Addison-Wesley, Boston (1989)
12. Gunawan, A.D., DasGupta, B., Zhang, L.: A decomposition theorem and two algorithms for reticulation-visible networks. Inf. Comput. **252**, 161–175 (2017)
13. Gunawan, A.D., Rathin, J., Zhang, L.: Counting and enumerating galled networks. Discret. Appl. Math. **283**, 644–654 (2020)
14. Gusfield, D.: ReCombinatorics: The Algorithmics of Ancestral Recombination Graphs and Explicit Phylogenetic Networks. MIT Press, Cambridge (2014)
15. Gusfield, D., Eddhu, S., Langley, C.: Efficient reconstruction of phylogenetic networks with constrained recombination. In: Proceedings of CSB 2003 (2003)
16. Huson, D.H., Klöpper, T.H.: Beyond galled trees - decomposition and computation of galled networks. In: Speed, T., Huang, H. (eds.) RECOMB 2007. LNCS, vol. 4453, pp. 211–225. Springer, Heidelberg (2007). https://doi.org/10.1007/978-3-540-71681-5_15
17. Huson, D.H., Rupp, R., Scornavacca, C.: Phylogenetic Networks: Concepts, Algorithms and Applications. Cambridge University Press, Cambridge (2010)
18. McDiarmid, C., Semple, C., Welsh, D.: Counting phylogenetic networks. Ann. Comb. **19**(1), 205–224 (2015). https://doi.org/10.1007/s00026-015-0260-2
19. Mir, A., Rosselló, F., Rotger, L.A.: A new balance index for phylogenetic trees. Math. Biosci. **241**(1), 125–136 (2013)
20. Pons, M., Batle, J.: Combinatorial characterization of a certain class of words and a conjectured connection with general subclasses of phylogenetic tree-child networks. Sci. Rep. **11**(1), 1–14 (2021)
21. Sackin, M.J.: "Good" and "bad" phenograms. Syst. Biol. **21**(2), 225–226 (1972)
22. Shao, K.T., Sokal, R.R.: Tree balance. Syst. Zool. **39**(3), 266–276 (1990)
23. Steel, M.: Phylogeny: discrete and random processes in evolution. SIAM (2016)
24. Stufler, B.: A branching process approach to level-k phylogenetic networks. Random Struct. Algorithms (2021). https://doi.org/10.1002/rsa.21065
25. Xue, C., Liu, Z., Goldenfeld, N.: Scale-invariant topology and bursty branching of evolutionary trees emerge from niche construction. Proc. Natl. Acad. Sci. **117**(14), 7879–7887 (2020)
26. Zhang, L.: Generating normal networks via leaf insertion and nearest neighbor interchange. BMC Bioinform. **20**(20), 1–9 (2019)
27. Zhang, L.: Clusters, trees, and phylogenetic network classes. In: Warnow, T. (ed.) Bioinformatics and Phylogenetics. CB, vol. 29, pp. 277–315. Springer, Cham (2019). https://doi.org/10.1007/978-3-030-10837-3_12

Phylogenetic Placement Problem: A Hyperbolic Embedding Approach

Yueyu Jiang, Puoya Tabaghi, and Siavash Mirarab[⊠]

Electrical and Computer Engineering, University of California, San Diego 95032, USA
{y5jiang,ptabaghi,smirarb}@ucsd.edu

Abstract. Phylogenetic trees define a *metric space* over their vertices, an observation that underlines *distance-based* phylogenetic inference. Several authors, including Layer and Rhodes (2017), have noted that we can embed leaves of a phylogenetic tree into high-dimensional Euclidean spaces in such a way that it minimizes the distortion of the tree distances. Jiang *et al.* (2021) use a deep learning approach to build a mapping from the space of sequences to the Euclidean space such that the mapped sequences accurately preserve the leaf distances on a given tree. Their tool, DEPP, uses this map to place a new query sequence onto the tree by first embedding it, an idea that was particularly promising for updating a species tree given data from a single gene despite the potential discordance of the gene tree and the species tree. In focusing on Euclidean spaces, these recent papers have ignored the strong theory that suggests hyperbolic spaces are more appropriate for embedding vertices of a tree. In this paper, we show that by moving to hyperbolic spaces and addressing challenges related to non-linearity and precision, we can reduce the distortion of distances for any given number of dimensions. The distortion of distances obtained using hyperbolic embeddings is lower than Euclidean embeddings with the same number of dimensions, both in training (backbone) and testing (query). The low-distortion distances of embeddings result in better topological accuracy in updating species trees using a single gene compared to its Euclidean counterpart. It also improves accuracy in placing queries for some datasets but not all.

Keywords: Phylogenetic placement · Deep learning · Hyperbolic spaces · Tree embedding · Distance-based phylogenetics

1 Introduction

Branch lengths of a phylogenetic tree define distances between species or genes represented at their leaves. Thus, a discrete metric space is defined by the tree topology and branch lengths. Given this observation, we can now ask whether such a discrete metric space can be embedded in a continuous low-dimensional space. The goal is not to embed each tree as a vector, as has been the focus of a rich literature [5]. Rather, the goal is to provide a vector representation for each *tree node* so that certain properties of the original discrete space are

© The Author(s), under exclusive license to Springer Nature Switzerland AG 2022
L. Jin and D. Durand (Eds.): RECOMB-CG 2022, LNBI 13234, pp. 68–85, 2022.
https://doi.org/10.1007/978-3-031-06220-9_5

preserved as much as possible. If such embedding is obtained, continuous optimization methods, widely used in machine learning and elsewhere, can be utilized to analyze phylogenetic trees. For example, a method called Phylo-MCOA uses metric embedding in the Euclidean space paired with an approach called multiple co-inertia analysis to find outlier species among a set of gene trees [30]. Such Euclidean embeddings have the following theoretical underpinning: the N vertices of any tree T can be embedded in $\mathbb{R}^{N-1}$ such that Euclidean distances exactly match with the square root of the tree distances [14].

More recently, metric embedding has been used by Jiang *et al.* [12] to adopt deep learning to the problem of adding novel *query* sequences onto an existing *backbone* tree (i.e., phylogenetic placement). They posed the placement task as a metric learning problem: a deep neural network model can be trained to map observed sequences to the backbone trees. The neural network learns a map from the sequence space into the Euclidean space while aiming to preserve the metric information in the backbone tree. In their proposed method, called DEPP, they train a neural network that maps gene sequences to a high-dimensional Euclidean space while minimizing the difference between Euclidean distances and square roots of the tree distances. The learned map can then be used to embed queries and compute their distances to the backbones species. They then use distance-based phylogenetic placement [3,4] to find the optimal placement of each query on the backbone tree using the computed distances from the query to the backbone species. Embedding tree leaves in a continuous space simplifies the training process because it enables us to define differentiable loss functions that can be minimized using standard back-propagation and stochastic gradient descent operating on the neural network model. Essentially, DEPP relaxes the combinatorial nature of the placement problem by operating in a continuous space. However, a question naturally arises: *are fixed-dimensional Euclidean spaces the best choice for embedding all trees?*

The result that a tree can be perfectly embedded in the Euclidean spaces comes with a major limitation: the number of dimensions has to increase linearly with the number of leaves. Perhaps one can hope that with fewer dimensions, imperfect but low-distortion embeddings can still be obtained. Linial *et al.* [16] have considered the problem of low-distortion Euclidean embedding within a controlled distortion rate and proved that for an isometric embedding of a tree with N vertices, the Euclidean embedding dimension must be at least $d = O(\log(N))$. Thus, the dimensions need to *grow* with the size of the tree to obtain low distortion. Euclidean geometry, however, is not the only option.

Hyperbolic spaces provide better representations than Euclidean spaces for tree-like data structures. Sarkar [25] showed that weighted trees can be embedded with arbitrarily low distortion into a two-dimensional hyperbolic space. Intuitively, the reason is that the volume of a d-dimensional hyperbolic ball of fixed radius grows exponentially with its radius as opposed to Euclidean balls that grow polynomially in volume. The exponential growth enables hyperbolic spaces to embed any weighted tree while almost preserving their metric [10]. Thus, theory suggests hyperbolic spaces are a better fit for working with trees [24].

These advantages have motivated wide-spread adoption of hyperbolic spaces by the machine learning community for problems as varied as classification, manifold learning, and high-dimensional distribution approximation [10,28,29]. In particular, hyperbolic neural networks (HNNs) have been developed to combine the representational power of hyperbolic geometry with the feature extraction capabilities of neural networks [10,26]. More recent advances include hyperbolic graph neural networks [2,6,17], and mixed-curvature variational autoencoders [27].

Despite the theoretical advantages, optimization and embedding in hyperbolic spaces pose their own *practical* challenges, including the impact of bit precision on embeddings [24]. For embedding trees in hyperbolic spaces, Sala *et al.* [24] showed that the representation tradeoff depends on the maximum degree and the longest path length in the tree. For a fixed precision, the required number of bits scales linearly with the length of the longest path, while it also scales logarithmically with the maximum degree of the tree. In an ideal case of infinite bit precision, two-dimensional hyperbolic embeddings should suffice; but in practice, we may need larger embedding dimensions or high precision for representing vertices of trees. In fact, the main challenge in designing HNNs is performing arithmetics in non-Euclidean geometry, a problem approached in various ways in the literature. For instance, Chen *et al.* [7] use the Lorentz model in building computationally stable HNNs.

Recently, hyperbolic embedding has also been adopted for phylogenetics. Matsumoto *et al.* [19] propose an altered version of hyperbolic distance function with the purpose of gaining distance additivity for perpendicular incident (embedded) branches. They evaluated the performance of their proposed distance function for phylogenetic tree distance embedding, embedding internal nodes from partial distance measurements, and integrating the embeddings of multiple phylogenetic trees. Their experiments use small trees ($\approx$100 leaves). Hence it is not clear whether the proposed distance function outperforms the classic hyperbolic distance function in representing additive distances. Corso *et al.* [8] present Neural Distance Embeddings, a framework to embed sequences in geometric spaces—including hyperbolic spaces—to tackle different tasks in bioinformatics, including edit distance approximation, hierarchical clustering, and multiple sequence alignment. They provide extensive numerical experiments on large 16S rRNA gene data and show that embedding methods (data-dependent approaches) outperform classical data-independent approaches with respect to accuracy and inference speed. They also emphasize the importance of using an appropriate geometry for the embedded sequences and how it alone can drastically improve the accuracy of the relevant algorithms.

We consider the problem of discordant phylogenetic placement where gene sequences have evolved on a gene tree but are placed on the backbone species tree. Building on the work of Jiang *et al.* [12], we introduce the Hyperbolic Deep-learning Enabled Phylogenetic Placement (H-DEPP) framework to find low-dimensional hyperbolic embeddings for gene sequences while preserving their phylogenetic distances determined by the backbone tree. The choice of hyperbolic embedding space is motivated by the fact that trees can be accurately represented in low-dimensional hyperbolic spaces, as opposed to high-dimensional

Euclidean spaces utilized in DEPP. This reduction in the dimension of the embedding space leads to i) a simplified neural network with fewer parameters to train and ii) improved performance in estimating the phylogenetic distances for both observed and novel organisms—i.e., small training and testing errors.

2   Background on Hyperbolic Spaces

Hyperbolic space is a connected, negatively curved metric space. There are several isometric models of hyperbolic space, including *Poincaré ball* and *'Loid model*. In the d-dimensional Poincaré space of curvature $C < 0$ (denoted by $\mathbb{I}_C^d$) the points reside in Euclidean ball of radius $\frac{1}{\sqrt{-C}}$, i.e.,

$$\mathbb{I}_C^d = \{x \in \mathbb{R}^d : \sqrt{-C}\|x\|_2 \leq 1\}, \tag{1}$$

where $\|x\|_2$ denotes ℓ_2 norm. The distance between two points $x, y \in \mathbb{I}_C^d$ is

$$d(x,y) = \frac{1}{\sqrt{-C}}\mathrm{acosh}\Big(1 - 2\frac{C\|x - y\|_2^2}{(1 + C\|x\|^2)(1 + C\|y\|^2)}\Big). \tag{2}$$

The points in the 'Loid model of d-dimenaional hyperbolic spaces (with curvature $C < 0$) reside in $\mathbb{L}_C^d = \{x \in \mathbb{R}^{d+1} : x^\top H x = C^{-1}, x_1 > 0\}$, where $H = \mathrm{diag}(-1, 1, 1, \ldots, 1) \in \mathbb{R}^{(d+1)\times(d+1)}$ and x_1 is the first element of vector x. The distance between $x, y \in \mathbb{L}_C^d$ is

$$d(x,y) = \frac{1}{\sqrt{-C}}\mathrm{acosh}(Cx^\top Hy). \tag{3}$$

We can use the exponential map to have a bijection between a tangent space (which is a vector space) and the hyperbolic space, namely the 'Loid model. Let e_1 be the first standard basis for $\mathbb{R}^{d+1}$ and $p_o \overset{\text{def.}}{=} \frac{1}{\sqrt{-C}}e_1$. Then in the 'Loid model, the exponential map is defined for $v \in \mathbb{R}^d$ as follows

$$\exp_{p_o}\left(\begin{bmatrix} 0 \\ v \end{bmatrix}\right) = \cosh(\sqrt{-C}\|v\|_2)p_o + \frac{\sinh(\sqrt{-C}\|v\|_2)}{\|v\|_2}\begin{bmatrix} 0 \\ v \end{bmatrix}. \tag{4}$$

The exponential map (4) provides a way of defining a *hyperbolic embedding function* from a Euclidean embedding function. If $\phi_{\text{Euc.}}$ maps sequences to a d-dimensional Euclidean space, then $\exp_{p_o} \circ \phi_{\text{Euc.}}$ maps them into the hyperbolic space $\mathbb{L}_C^d$. Note that the 'Loid and Poincaré models are equivalent to each other in their embedding capabilities. However, the 'Loid model provides a more stable distance function suitable to perform optimization tasks.

3   Problem Definition

Suppose we are given a backbone species tree and a set of single-gene sequence data, where each sequence corresponds to a leaf on the backbone tree. Sequences have evolved on the gene tree, which generally is not congruent with the backbone tree. Then, we ask: *can we find the correct placement for a new query species using its single gene sequence on the backbone tree?* The reference sequences and the query sequence must all be from the same marker gene and need to be properly aligned (though missing data are allowed).

More formally: let T be the weighted backbone tree with leaf set $V = \{v_1, \ldots, v_N\}$ and d_T be the path length distance between vertex pairs of T. Let $s_1, \ldots, s_N$ be a set of aligned gene sequences (of length L) corresponding to leaves of the backbone tree. We aim to learn an embedding function ϕ from the space of gene sequences to a metric space (M, d) such that for all $i, j \in [N]$, the difference between $d(\phi(s_i), \phi(s_j))$ and $d_T(v_i, v_j)$ is minimized. We call the problem discordant phylogenetic placement when T is the species tree as opposed to the gene tree corresponding to sequences $s_1, \ldots, s_N$. The goal is that for a new query sequence s, the $\phi(s)$ would have distances to all $\phi(s_i)$ embeddings that match the correct placement of s on the (species) tree T.

In practice, given N training sequences, we estimate the function $\widehat{\phi}_N$ as follows

$$\widehat{\phi}_N = \arg\min_{\phi \in \Phi} \sum_{i,j \in [N]} \left(\frac{d(\phi(s_i), \phi(s_j))}{d_T(v_i, v_j)} - 1 \right)^2, \tag{5}$$

where Φ is the set of functions from sequences to the d-dimensional hyperbolic space, $\mathbb{H}^d$, represented by neural networks. The cost function is similar to the weighted least square method in [9], with the only difference being that weights are squared tree distances as opposed to sequence distances. We use the tree distance as the weights because it leads to bounded derivatives in back-propagation. Note that the cost function (5) aims to minimize the deviation of distance ratios from the unit scale and hence it incurs multiplicative errors.

After the training, we use the learned map $\widehat{\phi}_N$ to estimate the distance between the query species and the species in the backbone tree; i.e., $d(\widehat{\phi}_N(s), \widehat{\phi}_N(s_n))$ for $n \in [N]$, where s is the query gene sequence. We then place the query on the backbone tree using the distance-based placement method APPLES-II [3].

4   H-DEPP

The structure of the CNN model used in H-DEPP is identical to DEPP [12]. It encodes each sequence letter as a four-dimensional binary vector and feeds the encodings into three linear convolutional layers of kernel size 1, 5, and 5, respectively. Then, it uses feedforward from the second layer to the third to make the third layer a residual block. The last convolutional layer is followed by a fully-connected layer with d-dimensional outputs. To make the network operate in hyperbolic space, we use two approaches:

Exponential Maps. In this approach, we avoid adding any layers to the network. Instead, we can simply use a Euclidean neural network followed by the exponential map of the 'Loid (or Poincaré) model. Then, the cost function becomes:

$$\text{cost}(\phi_{\text{Euc.}}, s) = \sum_{i,j \in [N]} |s \frac{d(\exp \circ \phi_{\text{Euc.}}(s_i), \exp \circ \phi_{\text{Euc.}}(s_j))}{d_T(v_i, v_j)} - 1|^2,$$

where s is a scale factor (we elaborate later) and $\exp(\cdot)$ is the exponential map for the appropriate model with curvature -1.

HNN++. We add to DEPP a final hyperbolic layer designed by Shimizu *et al.* [26]. This essentially is an all-by-all layer that performs matrix multiplication in the hyperbolic space. Its input is the exponential map of outputs from a Euclidean layer, and its output consists of points in hyperbolic space. These points are used in the cost function (5). We update the parameters according to the hyperbolic back-propagation as implemented by Shimizu *et al.* [26].

Once the method is made hyperbolic, two challenges need to be addressed.

Curvature. We learn the curvature of the hyperbolic space. For any $x_1, x_2 \in \mathbb{I}_C^d$, let $x_1' = \sqrt{-C}x_1$ and $x_2' = \sqrt{-C}x_2$. We can easily verify that x_1' and $x_2' \in \mathbb{I}_{-1}^d$ and $d(x_1, x_2) = \frac{1}{\sqrt{-C}}d(x_1', x_2')$. With this transformation, we keep the domain and distance functions *independent* of curvature (they are defined for the default curvature of -1) and add a scale parameter $s := \frac{1}{\sqrt{-C}}$ to the distance function:

$$d(x, y) = s \cdot d(x_1', x_2') = s \cdot \text{acosh}\left(1 + 2\frac{\|x_1' - x_2'\|_2^2}{(1 - \|x_1'\|_2^2)(1 - \|x_2'\|_2^2)}\right), \quad (6)$$

where $x_1', x_2' \in \{x : \mathbb{R}^d : \|x\|_2 \leq 1\}$. To *jointly* optimize s and the neural network, we alternate between optimizing them, using the alternative updates:

$$s_{k+1} := (1 - \alpha_k)s_k + \alpha_k s_{\phi_k}^*,$$

where ϕ_k, s_k, and α_k are the network, scale, and learning rate at iteration k. Note that since models of hyperbolic space are isometric to each other, the same argument can be made for the 'Loid model as well.

Heterogeneous Distances. The maximum and minimum measured distances vary six orders of magnitude in scale. Hence, any gradient-based training method has to approximate large and small sequences equally accurately; see (5). We exponentially decrease the learning rate to capture and correct for a large range of errors in a fixed number of epochs; i.e., the learning rate at iteration k is $\alpha_k = \alpha_0 p^{-\lfloor \frac{k}{K} \rfloor}$ where p is the decay factor (e.g., $p = 0.95$), $K = 10$ is the number of epoch with a fixed learning rate, and α_0 is the initial learning rate. Moreover, to deal with potential issues with limited precision operations, we normalize the tree distances to be at most 1, and the normalizing factor is absorbed in the scale factor in (6).

5   Experimental Setup

5.1   Datasets

Simulated Data. We first evaluate our method on a published simulated dataset of 201-taxon trees [20]. In this dataset, the species tree and gene trees (simulated by Simphy [18]) are discordant due to incomplete lineage sorting (ILS). We analyze example replicates from two of the six original model conditions corresponding to high and low ILS levels with rate 10^{-6}, each with 50 replicates. From each replicate of both datasets, we chose the first gene that doesn't have identical sequences. The selected true gene trees have on average a normalized [23] (RF) distance of 0.69 and 0.21 to their corresponding species tree, respectively for high and low ILS conditions. We ignore the replicates with all the genes containing identical sequences, which leaves us with 43 replicates for the high ILS condition and 47 replicates for the low ILS condition in total. Note that these simulations include no gaps and thus pose no need for alignment. During the data training, we use the true species tree with branch lengths recalculated using the sites randomly selected from the first 32 genes (with each gene providing 500 sites) as the backbone.

Biological Data. We use the Web-of-life (WoL) dataset by Zhu *et al.* [31] with 10,575 genomes. An ASTRAL [20] tree built from 381 gene trees with substitution-length branch lengths computed using ML from concatenated genes is available. To evaluate the proposed method, we use sequences from the 16S marker gene predicted from the 10,575 genomes using RNAmmer [13]. We align all 16S copies across all the 7,797 species that have at least one copy using UPP [21], which selects a backbone of 1000 sequences and aligns the other sequences independently to those in the backbone.

5.2   Evaluation

We evaluate the method in a leave-out fashion. We remove 5% of the species from the reference species tree to obtain a backbone tree used for training and use the left-out species as queries (i.e., testing). For WoL data, the alignment was not redone after removing queries since UPP aligns most of the sequences independently. A genome can have multiple 16S copies, and we pick one copy randomly to include in the backbone tree for training. For testing, we keep all the copies and test each one separately. Overall, we have 815 query sequences from 319 query species in the WoL dataset.

We also test the ability of our learned model to *update* a species tree using new sequences. Updating, unlike placement, infers a fully resolved tree starting from a given tree and finds the relationship between queries. To do so, we calculate a distance matrix from the learned embeddings including all the backbone and query sequences used in the placement experiment. We then infer the species tree using FastME2 with default options [15]. Note that the relationship between backbone sequences can in fact also change as part of this tree update.

Thus, update here simply means that the information in the starting tree is used (through training) in obtaining the updated tree. We evaluate the performance of the model using both the accuracy of distances and the accuracy of the tree placements or updates.

Distortion: The distortion of the model is measured by the mean squared error between the model prediction and the tree distances. The quantity is computed under two settings. One is the standard mean square error; the other is the average over the square error normalized by the tree distance, as in (5).

Placement or Tree Accuracy. We place each query independently onto the backbone tree and report the placement error—defined as the number of branches between the placement position and the position on the reference tree before removing the query. For tree updating, we compare the full tree inferred using DEPP distances or the tree restricted only to queries (i.e., novel species) against the true species tree (simulated data) or ASTRAL species tree (WoL data) using RF or quartet distance metrics. For WoL data, since query species have multiple copies, we subsample one random copy per species and remove all the other copies from the estimated tree, repeating this process 100 times and reporting the distribution.

6 Results and Discussions

6.1 Comparison of H-DEPP Alternatives

We first compare the various H-DEPP models on the WoL dataset (Fig. 1). Using the HNN++ layer has a higher training and testing distortion error compared to using the exponential maps (Fig. 1(a)–(b)); the HNN layer increases the training error three-folds and the testing distortion close to two-folds. Examining the training distortion versus epochs of learning shows that the HNN layer has made training more challenging (Fig. 1(c)). Even after 9000 epochs, the training error does not seem to have converged for the HNN layer, compared to exponential maps that seem to have converged by 3000 epochs. This pattern may be due to the fact that the exponential maps do not perform any arithmetic operations in the hyperbolic space through the network layers and instead map points from the Euclidean to the hyperbolic space in the loss function. In doing so, they avoid the difficulty of propagating (back) the gradient of the loss function through the more complex arithmetic operations needed in the hyperbolic space.

Comparing the two models realized with exponential maps, 'Loid has lower testing distortion, and its training error converges slightly faster than the Poincaré model; see Fig. 1(a)–(c). While the two models are equivalent mathematically (with infinite precision), these improvements point to different levels of difficulty in optimization given limited precision and the idiosyncrasies of how stochastic gradient descent and learning rates interact in each model.

The differences in training and testing distortion errors, however, do not translate to dramatically varied placement errors (Fig. 1-(d)). The placement of the exponential map approach is slightly lower than the HNN+ approach. Mean placement errors are 1.88, 1.77, 1.78 for HNN++, Poincaré (Exp. map), and 'Loid (Exp. map). The lack of improvement in placement accuracy when distances become far more accurate may seem surprising. However, we note that these levels of error are very low to begin with, leaving relatively little room for improvement. Moreover, note that APPLES-2 does not use *all* of the distances; to achieve high accuracy and scalability, APPLES-2 only considers small distances (in our runs, only five smallest distances). Thus, the reduced distortion among high distances does not directly benefit placement accuracy.

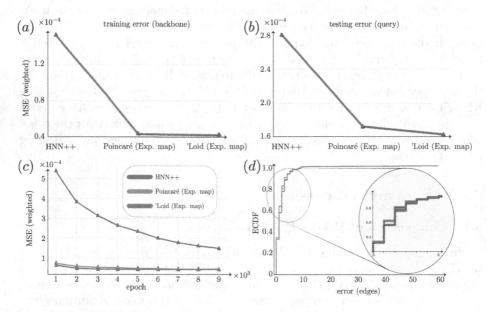

Fig. 1. Comparison of Poincare model and Loid model. Fixing $d = 128$, we show (a, b) the median of mean squared error of predicted distances on backbone and query species (i.e., training and testing distortion), respectively. (c) Convergence of the training loss versus the number of epochs of training. (d) Empirical cumulative distribution function (ECDF) of placement errors.

6.2   Comparison to Euclidean Embedding

The theoretical advantages of hyperbolic spaces compared to Euclidean space translate to much better training and testing distortion error, regardless of the number of dimensions d used; see Fig. 2(a) for the simulated and Fig. 3(a), (b) for the WoL dataset. The distortion of hyperbolic distances is between half to two orders of magnitude lower for testing and training compared to the Euclidean distances. As the theory suggests, fewer dimensions in the hyperbolic spaces are enough to achieve the same distortion as higher dimensional Euclidean spaces. For example, for high ILS simulated data, the testing distortion is lower with

4 hyperbolic dimensions compared to 128 Euclidean dimensions; or, with low ILS, the training distance obtained by $\mathbb{R}^{128}$ is approximately obtained by $\mathbb{H}^{16}$. WoL dataset also offers stark examples: the training distortion of $\mathbb{H}^4$ is similar to $\mathbb{R}^{16}$, and the $\mathbb{H}^{32}$ space results in the same distortion of *queries* as $\mathbb{R}^{512}$. Similar patterns are observed when we examine the weighted loss (5) instead of the simple (unweighted) distortion. These results show that far fewer hyperbolic dimensions can match the Euclidean distortion in high dimensions.

Visual inspection of distances shows several interesting patterns; see Fig. 2(c) for simulated and Fig. 3(d) for biological data. With $d = 4$, distortions are already low for the 201-taxon simulated dataset using hyperbolic but not Euclidean distances. However, on the larger WoL dataset, despite theoretical guarantees with infinite-precision computation, hyperbolic embedding shows not only considerable levels of distortion but also signs of a bias towards under-estimating distances. The problem is far worse at the testing time. Euclidean spaces, however, lead to far greater distortion and bias with $d = 4$ for both the small and the large dataset. Particularly, the predicted distances from the query seem divorced from true distances with $\mathbb{R}^4$ for the WoL dataset. At $d = 16$, no bias is detectable for hyperbolic distances in either dataset, but Euclidean distances remain biased *and* noisy for both datasets, especially the larger WoL. With $d = 64$ on WoL data, only a small amount of distortion is left in hyperbolic spaces, compared to high levels of noise (and some outliers) for Euclidean. Training and testing fidelities are high for Euclidean distance given the highest numbers of dimensions and are nearly perfect for hyperbolic.

The impact of lowered distortion on improved placement accuracy depends on the dataset. On the simulated data, hyperbolic distances have far lower placement error with small d; see Fig. 2(b). The placement error reduces for both methods with more dimensions, but in the high ILS case, hyperbolic continues to have an advantage in placement accuracy even when d is increased to 128. In particular, in this condition, Euclidean distances are never able to match the accuracy of EPA-ng, whereas hyperbolic distances do with 16 to 64 dimensions. On the low ILS dataset, where error levels are generally much lower, all methods converge to the same error with enough dimensions.

On the WoL data, the impact of reducing loss on placement error is somewhat muted (Fig. 3(c)). The placement error quickly drops for both spaces from 2 to 16 dimensions. In this range, hyperbolic embedding has lower placement error, and the reductions in error are substantial (e.g., 1.4 edge with $d = 4$). As the number of dimensions increases to 32 and beyond, the reduced distortion of backbone and query sequences stops translating into substantially reduced placement errors. With $d \geq 64$, the two methods are essentially tied in their placement errors and are also tied with ML placement using EPA-ng. Furthermore, when given $d > 16$, APPLES2 placements with DEPP distances are more accurate than APPLES2 placements using the simple JC model.

The slower pace or lack of improvements in placement accuracy for higher dimensions on the WoL and low ILS datasets can be explained in several ways. First, as the number of dimensions increases (e.g., beyond 32 for WoL), the

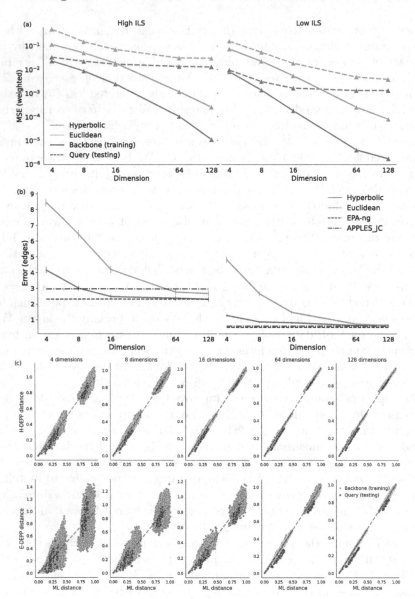

Fig. 2. Comparison of embeddings in Euclidean space and Hyperbolic space using the 'Loid model (exp. map) on simulated data. (a) Training and testing distortion versus dimensions. The distortion is measured by the median of mean square errors of predicted distances with normalization (b) The mean and standard error of the placement error versus dimensions. The horizontal lines show error with EPA-ng and APPLES2+JC model. (c) The comparison between ML distances on the reference tree used to train the model (i.e., ground truth) versus the distances obtained from the trained E-DEPP (Euclidean) or H-DEPP (hyperbolic) models. Each dot represents one pairwise distance between a pair of backbone species or a particular query selected arbitrarily and all backbones. The unit line is indicated on the graph as a dashed line.

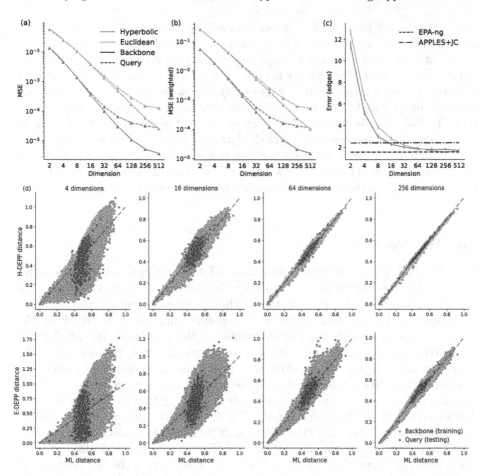

Fig. 3. Comparison of embeddings in Euclidean space and Hyperbolic space using the 'Loid model (exp. map) on WoL 16S data. (a, b) Training and testing distortion measured using the median of mean square errors of predicted distances without (a) or with (b) normalization. (c) The mean and standard error of the placement error versus the number of dimensions. The horizontal lines show error with EPA-ng and APPLES2+JC model. (d) The comparison between ML distances on the reference tree used to train the model (i.e., ground truth) versus the distances obtained from the trained E-DEPP (Euclidean) or H-DEPP (hyperbolic) models. Each dot represents one pairwise distance between a pair of backbone species (10% selected randomly) or a particular query selected arbitrarily and all backbones. The unit line is indicated on the graph as a dashed line.

reduction in testing distortion slows down compared to the training distortion and the placement error is a function of the testing distortion. However, there must be more to this pattern because testing distortion does continue to diminish, albeit slower. Another potential explanation is that APPLES-2 has some tolerance for error in its input distances. Such tolerance (or safety radius) is a key feature of all phylogenetic distance-based methods [1,11], and APPLES-2 may further increase the tolerance by focusing on small distances. We see evidence for such tolerance. For example, improving unweighted testing from an order of 10^{-2} to 10^{-3} as we go from $d = 2$ to $d = 8$ on WoL dramatically reduces placement error (from 12 edges to 3), but as we further reduce the testing error to 10^{-4} and closer to 10^{-5}, the placement error starts to plateau. Note that on WoL, all the methods have errors below 2 edges in a tree of roughly 7500 edges; thus, the errors are already low and have little room for improvement. Similarly, error is small on the low ILS simulated dataset. On the high ILS simulated dataset when errors are high (a random placement would have 14 edges of error), the improved accuracy of distortions do translate to better placements.

6.3   Tree Updates

Since H-DEPP learns better distances than E-DEPP, we next ask if it has better accuracy in updating trees with more species to get a fully resolved tree (as opposed to only the placements). Unlike the placement accuracy, the accuracy of the updated tree shows dramatic improvements compared to E-DEPP for the WoL datasets and consistent improvements for the simulated dataset (Fig. 4). The advantages are more clear when we measure the tree distance using the quartet score, which is not sensitive to rogue taxa; the presence of paralogous 16S copies and HGT can create unstable leaves. The quartet distance of the updated tree reduces by three (or two) folds for the full tree (or the tree induced down to queries) on the WoL datasets. The RF distance also improves, though less dramatically, likely due to the effects of rogue taxa. On the simulated dataset, clear improvements are obtained for the low ILS dataset. In particular, for most cases, the error is sufficiently low, which is remarkable given the fact that we are updating the species tree using a single gene. However, in the high ILS case, the error, which does reduce with hyperbolic compared to Euclidean embeddings, remains unacceptably high. Note that the task of updating a very high ILS species tree of only 190 species using data from only a single gene is likely futile, irrespective of the model used. Nevertheless, overall, the results indicate that the smaller distortion of hyperbolic distances has the potential to enable distance-based phylogenetic inference to update species trees using data from a single gene for conditions with sufficiently low gene tree discordance.

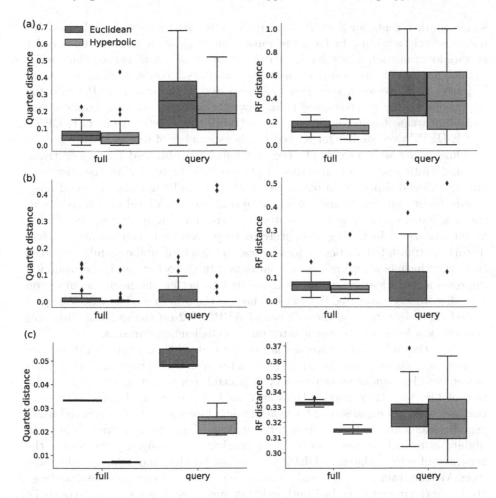

Fig. 4. Tree update error. Normalized quartet (left column) and RF (right column) distances between (a, b) true species trees and FastME trees inferred using DEPP distances on simulated data under High ILS (a) and Low ILS (b) model conditions, (c) the ASTRAL tree and a FastME tree inferred using DEPP distances on WoL 16S data. Here we show the results of trees including all the species (200 species for simulated data per tree and 7797 species for WoL tree) or the full tree restricted to query species (10 species for simulated data per tree and 319 species species for WoL tree), labeled as full and query respectively on the x-axis. For the WoL 16S data, results are obtained by sampling a single copy for multi-copy queries, repeating the process 100 times.

7 Conclusions and Future Work

The DEPP framework has shown promise in marrying deep learning with the phylogenetic placement problem in cases where one wants to update a species tree given limited data coming from a single gene. This application can become far more common than it may appear because most of the researchers using single marker genes such as 16S are really interested in species relationships not gene relationships. In fact, the most common use of marker genes is for species identification where gene tree discordance is only a nuisance. Thus, while updating a species tree using a single gene may seem counter-intuitive or even ill-posed, it addresses a very real biological need. By moving DEPP into the hyperbolic space, we attempted to improve its accuracy for such applications. This improvement comes at no cost as the training and testing times of H-DEPP and E-DEPP were similar for a fixed numbers number of dimensions.

Our results were mixed. On the one hand, we obtained far better training and testing accuracy in terms of the reduced distortion in the distances among embedded points. In particular, with hyperbolic models, orders of magnitude fewer dimensions are enough to obtain the same level of distortions as the Euclidean space. The lower distortions are not simply due to over-fitting as improvements in testing error were more pronounced than training. Lower distortion translated to better topological accuracy of updated fully resolved trees that include all queries on all datasets. On the other hand, the dramatic improvements in learned distances translated to better placement accuracy on high discordance simulated dataset but not on the other two datasets. We speculated the reason may be the reliance of APPLES-2 on the smallest distances and also low levels of placement error on less challenging datasets.

Given that hyperbolic distances are better trained than Euclidean distances, they open up the path for updating trees—a problem that is more ambitious than placement. Tree updating can be very impactful. For example, for the bacterial tree of life, very large species trees such as the one we used here and others that are similarly expansive [22] are available. Once such a tree is inferred using genome-wide data with much computational and human effort, we can hope to update it to include new species using marker genes only. By improving the accuracy of such updates, H-DEPP can enable building much larger references trees. We can train a model based on the existing trees and single-gene data (e.g., 16S), use the model to embed both existing species and new ones in hyperbolic spaces, compute distances from the embeddings, and infer updated trees using distance-based methods such as FastME. While our results are promising, we leave a full exploration of this approach to future work.

Our experiments were limited in scope, and future work should explore a wider range. For the WoL dataset, we only tested 16S, and only when the sequences are full length. In microbiome analyses, often fragmentary data are given. While the method can handle such data, we did not test the method with fragmentary queries. Also, our queries and reference sequences were aligned jointly, which is not how most analysis pipelines proceed (however, note that UPP aligns most of the sequences independently from others). Future work should further explore the impact of alignment errors.

Our experiments identified many challenges that had to be overcome to obtain high accuracy with hyperbolic embeddings. The tricks we used for training the curvature, data scaling, changing learning rates, and our results comparing exponential maps versus HNN layers can have downstream ramifications beyond training phylogenetic trees. Our techniques and findings may be useful more broadly for clustering applications using hyperbolic embedding, as attempted in various machine learning settings. Conversely, the rapidly evolving advances in building hyperbolic neural networks in the machine learning community may help us further improve the accuracy of our proposed H-DEPP framework.

References

1. Atteson, K.: The performance of neighbor-joining methods of phylogenetic reconstruction. Algorithmica **25**(2–3), 251–278 (1999). https://doi.org/10.1007/PL00008277
2. Bachmann, G., Bécigneul, G., Ganea, O.: Constant curvature graph convolutional networks. In: International Conference on Machine Learning, pp. 486–496. PMLR (2020)
3. Balaban, M., Jiang, Y., Roush, D., Zhu, Q., Mirarab, S.: Fast and accurate distance-based phylogenetic placement using divide and conquer. Mol. Ecol. Resour. (2021). https://doi.org/10.1111/1755-0998.13527. https://onlinelibrary.wiley.com/doi/10.1111/1755-0998.13527
4. Balaban, M., Sarmashghi, S., Mirarab, S.: APPLES: scalable distance-based phylogenetic placement with or without alignments. Syst. Biol. **69**(3), 566–578 (2020). https://doi.org/10.1093/sysbio/syz063. https://academic.oup.com/sysbio/advance-article/doi/10.1093/sysbio/syz063/5572672. https://academic.oup.com/sysbio/article/69/3/566/5572672
5. Billera, L.J., Holmes, S.P., Vogtmann, K.: Geometry of the space of phylogenetic trees. Adv. Appl. Math. **27**(4), 733–767 (2001). https://doi.org/10.1006/aama.2001.0759
6. Chami, I., Ying, Z., Ré, C., Leskovec, J.: Hyperbolic graph convolutional neural networks. In: Advances in Neural Information Processing Systems, vol. 32 (2019)
7. Chen, W., et al.: Fully hyperbolic neural networks. arXiv preprint arXiv:2105.14686 (2021)
8. Corso, G., Ying, Z., Pándy, M., Veličković, P., Leskovec, J., Liò, P.: Neural distance embeddings for biological sequences. In: Advances in Neural Information Processing Systems, vol. 34 (2021)
9. Fitch, W.M., Margoliash, E.: Construction of phylogenetic trees. Science **155**(3760), 279–284 (1967). https://doi.org/10.1126/science.155.3760.279. https://www.science.org/doi/10.1126/science.155.3760.279
10. Ganea, O., Bécigneul, G., Hofmann, T.: Hyperbolic entailment cones for learning hierarchical embeddings. In: International Conference on Machine Learning, pp. 1646–1655. PMLR (2018)
11. Gascuel, O., Steel, M.: A 'stochastic safety radius' for distance-based tree reconstruction. Algorithmica **74**(4), 1386–1403 (2016). https://doi.org/10.1007/s00453-015-0005-y. http://link.springer.com/10.1007/s00453-015-0005-y
12. Jiang, Y., Balaban, M., Zhu, Q., Mirarab, S.: DEPP: deep learning enables extending species trees using single genes. bioRxiv (abstract in RECOMB 2021) (2021). https://doi.org/10.1101/2021.01.22.427808. http://biorxiv.org/content/early/2021/01/24/2021.01.22.427808.abstract
13. Lagesen, K., Hallin, P., Rødland, E.A., Stærfeldt, H.H., Rognes, T., Ussery, D.W.: RNAmmer: consistent and rapid annotation of ribosomal RNA genes. Nucleic Acids Res. **35**(9), 3100–3108 (2007)

14. Layer, M., Rhodes, J.A.: Phylogenetic trees and Euclidean embeddings. J. Math. Biol. **74**(1–2), 99–111 (2017). https://doi.org/10.1007/s00285-016-1018-0

15. Lefort, V., Desper, R., Gascuel, O.: FastME 2.0: a comprehensive, accurate, and fast distance-based phylogeny inference program. Mol. Biol. Evol. **32**(10), 2798–2800 (2015). https://doi.org/10.1093/molbev/msv150. http://mbe.oxfordjournals.org/lookup/doi/10.1093/molbev/msv150

16. Linial, N., London, E., Rabinovich, Y.: The geometry of graphs and some of its algorithmic applications. Combinatorica **15**(2), 215–245 (1995). https://doi.org/10.1007/BF01200757

17. Liu, Q., Nickel, M., Kiela, D.: Hyperbolic graph neural networks. In: Advances in Neural Information Processing Systems, vol. 32 (2019)

18. Mallo, D., De Oliveira Martins, L., Posada, D.: SimPhy: phylogenomic simulation of gene, locus, and species trees. Syst. Biol. **65**(2), 334–344 (2016). https://doi.org/10.1093/sysbio/syv082. http://sysbio.oxfordjournals.org/content/early/2015/12/04/sysbio.syv082.short?rss=1. https://academic.oup.com/sysbio/article-lookup/doi/10.1093/sysbio/syv082. http://www.ncbi.nlm.nih.gov/pubmed/265

19. Matsumoto, H., Mimori, T., Fukunaga, T.: Novel metric for hyperbolic phylogenetic tree embeddings. Biol. Methods Protoc. **6**(1), bpab006 (2021)

20. Mirarab, S., Warnow, T.: ASTRAL-II: coalescent-based species tree estimation with many hundreds of taxa and thousands of genes. Bioinformatics **31**(12), i44–i52 (2015). https://doi.org/10.1093/bioinformatics/btv234. http://bioinformatics.oxfordjournals.org/cgi/content/long/31/12/i44. http://bioinformatics.oxfordjournals.org/lookup/doi/10.1093/bioinformatics/btv234

21. Nguyen, N.P.D., Mirarab, S., Kumar, K., Warnow, T.: Ultra-large alignments using phylogeny-aware profiles. Genome Biol. **16**(1), 124 (2015). https://doi.org/10.1186/s13059-015-0688-z. http://genomebiology.com/2015/16/1/124. https://genomebiology.biomedcentral.com/articles/10.1186/s13059-015-0688-z

22. Parks, D.H., et al.: A standardized bacterial taxonomy based on genome phylogeny substantially revises the tree of life. Nat. Biotechnol. **36**(10), 996–1004 (2018). https://doi.org/10.1038/nbt.4229. http://www.nature.com/articles/nbt.4229

23. Robinson, D., Foulds, L.: Comparison of phylogenetic trees. Math. Biosci. **53**(1–2), 131–147 (1981). http://www.sciencedirect.com/science/article/pii/0025556481900432

24. Sala, F., De Sa, C., Gu, A., Ré, C.: Representation tradeoffs for hyperbolic embeddings. In: International Conference on Machine Learning, pp. 4460–4469. PMLR (2018)

25. Sarkar, R.: Low distortion delaunay embedding of trees in hyperbolic plane. In: van Kreveld, M., Speckmann, B. (eds.) GD 2011. LNCS, vol. 7034, pp. 355–366. Springer, Heidelberg (2012). https://doi.org/10.1007/978-3-642-25878-7_34

26. Shimizu, R., Mukuta, Y., Harada, T.: Hyperbolic neural networks++. arXiv preprint arXiv:2006.08210 (2020)

27. Skopek, O., Ganea, O.E., Bécigneul, G.: Mixed-curvature variational autoencoders (2020)

28. Tabaghi, P., Dokmanić, I.: Hyperbolic distance matrices. In: Proceedings of the 26th ACM SIGKDD International Conference on Knowledge Discovery & Data Mining, pp. 1728–1738 (2020)

29. Tabaghi, P., Peng, J., Milenkovic, O., Dokmanić, I.: Geometry of similarity comparisons. arXiv preprint arXiv:2006.09858 (2020)

30. de Vienne, D.M., Ollier, S., Aguileta, G.: Phylo-MCOA: a fast and efficient method to detect outlier genes and species in phylogenomics using multiple co-inertia analysis. Mol. Biol. Evol. **29**(6), 1587–1598 (2012). https://doi.org/10.1093/molbev/msr317. https://academic.oup.com/mbe/article-lookup/doi/10.1093/molbev/msr317

31. Zhu, Q., et al.: Phylogenomics of 10,575 genomes reveals evolutionary proximity between domains Bacteria and Archaea. Nat. Commun. **10**(1), 5477 (2019). https://doi.org/10.1038/s41467-019-13443-4. http://www.nature.com/articles/s41467-019-13443-4

Phylogenetic Network Dissimilarity Measures that Take Branch Lengths into Account

Berk A. Yakici⬤, Huw A. Ogilvie⬤, and Luay Nakhleh[(✉)]⬤

Rice University, Houston, TX 77005, USA
{bay,huw.a.ogilvie,nakhleh}@rice.edu

Abstract. Dissimilarity measures for phylogenetic trees have long been used for analyzing inferred trees and understanding the performance of phylogenetic methods. Given their importance, a wide array of such measures have been developed, some of which are based on the tree topologies alone, and others that also take branch lengths into account. Similarly, a number of dissimilarity measures of phylogenetic networks have been developed in the last two decades. However, to the best of our knowledge, all these measures are based solely on the topologies of phylogenetic networks and ignore branch lengths. In this paper, we propose two phylogenetic network dissimilarity measures that take both topology and branch lengths into account. We demonstrate the behavior of these two measures on pairs of related networks. Furthermore, we show how these measures can be used to cluster a set of phylogenetic networks obtained by an inference method, illustrating this application on the posterior sample of phylogenetic networks. Both measures are implemented in the publicly available software package PhyloNet.

Keywords: Phylogenetic networks · Dissimilarity · Topology · Branch lengths

1 Introduction

Phylogenetic trees and networks are widely used to model the evolution of genes, species and languages. In the case of genomes and species, a deluge of data is available in the form of whole genomes being assembled and made available for phylogenetic inference [15,20,25]. However, the space of phylogenetic trees is notoriously complex due to a mix of discrete and continuous parameters. Therefore, this complexity must be confronted, as probability and likelihood distributions over phylogenetic trees often lack closed form solutions, requiring exploration of this space using algorithms such as hill climbing or Markov-chain Monte Carlo. Worse still, the discrete and continuous parameters are highly dependant, making this exploration fiendishly difficult. Phylogenetic network distributions are more complex again, among other reasons because dimensionality of the problem is no longer fixed.

© The Author(s), under exclusive license to Springer Nature Switzerland AG 2022
L. Jin and D. Durand (Eds.): RECOMB-CG 2022, LNBI 13234, pp. 86–102, 2022.
https://doi.org/10.1007/978-3-031-06220-9_6

Nonetheless, phylogenetic trees and networks remain successful models due to their strong and direct relationship to actual evolution. Evolving units such as species split and diverge over time, and patterns of splits are encoded by the nodes of phylogenetic trees. Each node represents a species (or other kind of evolving unit), and has one parent and a certain number of children (typically zero or two) to connect with its immediate ancestor and descendants respectively. Units which we have data for are typically represented as external nodes, often called taxa for trees of species or populations. The degree of divergence can be encoded as either the branch lengths or, for ultrametric phylogenies, as node ages. The pattern of splits and reticulations without continuous parameters is known as the tree topology and is an unordered discrete variable that grows hyperexponentially with the number of taxa [10].

Because phylogenetic trees do not account for reticulate evolution, such as introgression, hybridization, or horizontal gene transfer between species, or loanwords borrowed between languages [13], phylogenetic networks were developed in order to model splitting and reticulation. These networks may contain reticulation nodes, with two parents and one child, in addition to the nodes used to encode splits. The probability of inheriting evolutionary subunits, such as genes or words, may be encoded as additional continuous parameters associated with the reticulation nodes. By permitting two immediate ancestors with inheritance probabilities, phylogenetic networks can effectively encode the aforementioned examples of reticulate evolution [9,21]. The addition of reticulation nodes makes the number of possible phylogenetic network topologies far greater than the number of trees. Deriving these numbers are non-trivial, and so far have been restricted to specific classes of networks. For level-2 networks, there are 1,143 network topologies for three external nodes, compared with only three tree topologies [5].

The complexity of phylogenetic tree and network space has some immediate implications. For example, to quantify the dissimilarity between sets of continuous parameter values we can choose among the L^p-norms (e.g. the Euclidean norm), but this is not directly applicable to phylogenies due to their complex mixture of discrete and continuous parameters that cannot naturally be embedded in L^p spaces. However, we often wish to measure this dissimilarity to compare different inference methods with each other, to compare them with ground truths, or to study how different sets of evolutionary units co-evolve [2]. Furthermore, novel efficient proposal kernels such as Zig-Zag cannot be applied to phylogenetic trees without thorough preliminary theoretical work [17]. The Zig-Zag traverses tree space, randomly reversing direction along a given dimension at intervals. It is not obvious how a particle can sample multiple tree topologies by proceeding along or reversing the direction of travel, although some preliminary work has been done in this area [8].

These implications have motivated the development of myriad measures and metrics of dissimilarity and distances to compare phylogenetic trees, which have enabled the comparison of different phylogenies and improved our understanding of the algorithms used in phylogenetic inference [34]. We can classify many of these measures and metrics into three broad categories: clade-based, move-based, and geometric.

Clade-based measures are based on the presence or absence of clades (or, for unrooted phylogenies, splits). These may be limited to the difference in topology as in the Robinson-Foulds (RF) distance [26], or incorporate branch lengths as in weighted RF and branch score (BS) distances [18]. Move-based measures are based on the number of random-walk moves, such as Nearest-Neighbor Interchange (NNI), needed to modify one topology to be identical with another topology [1]. Geometric measures rely on the embedding of phylogenies in geometric spaces, such as the Billera-Holmes-Voghtmann (BHV), τ and t spaces [3,11]. The distance between phylogenies is then the shortest path between the corresponding points.

The additional complexity of phylogenetic networks means that available measures and metrics are far less developed. However, the need to accurately model evolution with reticulation demands much greater development and motivates our present study. Presently, beyond simple identity, there are several measures of topological dissimilarity with proofs of constituting metrics on subspaces of phylogenetic networks, e.g., [6,7,22].

None of these existing measures considers branch lengths or node ages, despite the importance of these distances in evolutionary biology. To address this absence, in this paper we propose rooted network branch score (rNBS) and average path distance (APD), two novel measures to compute dissimilarity between two rooted phylogenetic networks Ψ_1 and Ψ_2 that are sensitive to branch lengths in addition to the topology. When comparing pairs of simulated networks that undergo reticulation elimination and branch-length scaling, we observe an increase in dissimilarity value from both measures with respect to the amount of distortion applied to one of the pairs. Additionally, when we use rNBS, APD, and the topological distance of [22] to cluster networks obtained from a Bayesian Markov chain Monte Carlo (MCMC) sample based on their dissimilarity, we find that both rNBS and APD can highlight structure within searches of tree space that may not be obvious from other parameters and statistics. Thus, we believe both rNBS and APD are suitable in evaluating, prototyping and refining network inference. Both measures are implemented in PhyloNet [31,33] and are publicly available to download and use. The source code is available at https://github.com/NakhlehLab/PhyloNet/.

2   Methods

Our focus in this paper is binary evolutionary (or, explicit) phylogenetic networks. Furthermore, we assume all networks are leaf-labeled by the same set of taxa.

Definition 1. *The topology of a phylogenetic network Ψ is a rooted directed acyclic graph (V, E) such that V contains a unique node with in-degree 0 and out-degree 2 (the root) and every other node has either in-degree 1 and out-degree 2 (an internal tree node), in-degree 1 and out-degree 0 (an external tree node, or leaf), or in-degree 2 and out-degree 1 (a reticulation node). Edges incident into* *reticulation nodes are referred to as reticulation edges. The leaves are bijectively labeled by a set $\mathcal{X}$ of taxa, with $|\mathcal{X}| = n$. The phylogenetic network Ψ has branch lengths $\lambda : E \to \mathbb{R}^+$, where λ_b denotes the branch length of branch b in Ψ.*

In this section, we propose two different methods of measuring the dissimilarity of a pair of phylogenetic networks Ψ_1 and Ψ_2 while taking their branch lengths into account.

2.1   Rooted Network Branch Score (rNBS)

In this subsection, we describe a dissimilarity measure based on viewing a network in terms of the trees it displays, similar to the tree-based measure for topological comparison of phylogenetic networks implemented in PhyloNet [31].

Definition 2. *Let Ψ be a phylogenetic network leaf-labeled by set $\mathcal{X}$ of taxa. A tree T is displayed by Ψ if it can be obtained from Ψ by removing for each reticulation node exactly one of the edges incident into it followed by repeatedly applying forced contractions until no nodes of in- and out-degree 1 or in-degree 0 and out-degree 1 remain. A forced contraction of a node u of in-degree 1 and out-degree 1 consists of (i) adding an edge from u's parent to u's child, and (ii) deleting node u and the two edges that connect it to its parent and child. A forced contraction of a node u of in-degree 0 and out-degree 1 consists of removing the node u and its incident edge. The resulting tree has a unique root, whose in-degree is 0 and out-degree is 2, leaf nodes, whose in-degrees are 1 and out-degrees are 0, and other internal nodes, whose in-degrees are 1 and out-degrees are 2. We denote by $\mathcal{U}(\Psi)$ the set of all trees displayed by Ψ, where each tree in $\mathcal{U}(\Psi)$ is leaf-labeled by set $\mathcal{X}$ of taxa; that is, no tree whose leaves are not bijectively labeled by $\mathcal{X}$ is included in the set $\mathcal{U}(\Psi)$.*

It is important to note here that since branch lengths are taken into account, the set $\mathcal{U}(\Psi)$ can have trees with identical topologies but different branch lengths. This is illustrated in Fig. 1.

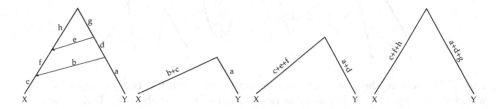

Fig. 1. A phylogenetic network (left) and its displayed trees. Since branch lengths are taken into account, the network displays three different trees. In the case of topology alone, the network displays a single tree.

We build a weighted complete bipartite graph $g = (V = \mathcal{U}(\Psi_1) \cup \mathcal{U}(\Psi_2), E)$ where the weight of edge $(t_1, t_2) \in E$ equals the rooted branch score of t_1 and

t_2 [12], which is the rooted equivalent of the branch score of [18]. The rNBS of Ψ_1 and Ψ_2 is then computed as the minimum-weight edge cover of g normalized by the number of edges in the edge cover. An edge cover of g is a subset $E' \subseteq E$ of its edges so that every node in V is the endpoint of at least one edge in E'. The weight of an edge cover E' is the sum of the weights of the edges in E'. A minimum-weight edge cover is an edge cover of g whose weight is smallest among all possible edge covers of g.

In our implementation, we use the Hungarian method to compute edge cover, which runs in $\mathcal{O}(|V|^3)$. However, in its current implementation, rNBS is not scalable with respect to the number of reticulations, since the size of V grows exponentially in the number of reticulation nodes. Exploring whether the rNBS value between two networks can be computed more efficiently without explicitly building the bipartite graph is a direction for future research. Figure 2 illustrates the rNBS of two phylogenetic networks.

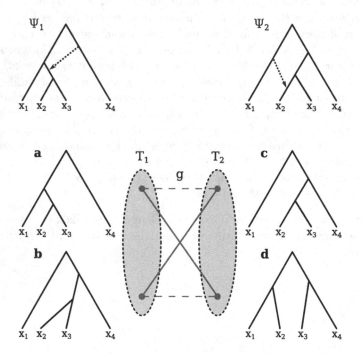

Fig. 2. Illustrating the rooted network branch score (rNBS) of two phylogenetic networks Ψ_1 and Ψ_2. $T_1 = \mathcal{U}(\Psi_1)$ and consists of the two trees (**a–b**) with distinct branch lengths and $T_2 = \mathcal{U}(\Psi_2)$ and consists of the two trees (**c–d**) with distinct branch lengths. The complete bipartite graph $g = (T_1 \cup T_2, E)$ is shown. Assuming the minimum-weight edge cover consists of the two edges depicted by the blue solid lines, then rNBS$(\Psi_1, \Psi_2) = w(\mathbf{a}, \mathbf{d}) \cdot 0.5 + w(\mathbf{b}, \mathbf{c}) \cdot 0.5$. (Color figure online)

Additionally, it is important to note that the rNBS is not a metric, in particular failing to satisfy the condition that rNBS(Ψ_1, Ψ_2) = 0 if and only if Ψ_1 and Ψ_2 are isomorphic (while respecting the leaf labeling). Figure 3 shows two networks that display the same set of trees even when branch lengths are included. One setting of the network branch lengths that would lead to this scenario is given by the following Rich Newick strings of the two networks:

Psi1 = (((((X2:7.0)#H2:4.0)#H1:6.0,X3:8.0):4.0,(#H1:4.0,
X4:5.0):2.0):3.0,(#H2:2.0,X1:2.0):1.0)Root;
Psi2 = ((((X2:4.0)#H1:13.0,X3:8.0):4.0,((#H1:3.0)#H2:8.0,
X4:5.0):2.0):3.0,(#H2:2.0,X1:2.0):1.0)Root;

In this case, rNBS(Ψ_1, Ψ_2) = 0 even though the two networks are different.

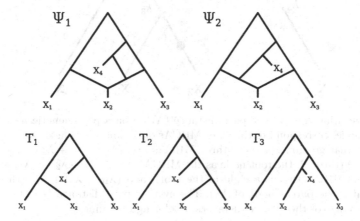

Fig. 3. Two networks Ψ_1 and Ψ_2 that display the same set of trees $\{T_1, T_2, T_2\}$. (adapted from [23])

2.2 Average Path Distance (APD)

Here we present a dissimilarity measure that is based directly on the networks, rather than the trees they display. We view phylogenetic networks Ψ_1 and Ψ_2 as two $n \times n$ matrices M_1 and M_2, respectively, where entries $[i, j]$ in M_1 and M_2 correspond to the average path distance between the two leaves labeled i and j in Ψ_1 and Ψ_2, respectively. Thus, it can be viewed as an extension of path distance by [28] to networks.

In a phylogenetic tree, the path distance between two leaves is the sum of weights of edges on the unique (simple) path between those two leaves. In a phylogenetic network, there could be more than one path between two leaves. Let $\mathcal{M}(\Psi, i, j)$ be the set of all most recent common ancestors (MRCAs) of i and j in network Ψ. Here, an MRCA is a node from which there is a path to i and a path to j and these two paths do not share any edge. The average path distance (APD) between two leaves i and j is the average of all such paths between the two leaves. For example, in the network shown in Fig. 4, there are three MRCAs of X_2 and X_3.

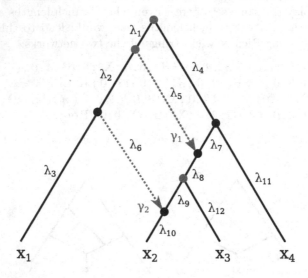

Fig. 4. Illustrating the average path distance (APD) on a phylogenetic network. The blue solid circles correspond to the three MRCAs of X_2 and X_3. The distances between X_2 and X_3 that go through these three MRCAs are: $(\lambda_{10} + \lambda_6 + \lambda_2 + \lambda_1) + (\lambda_{12} + \lambda_8 + \lambda_7 + \lambda_4)$ (through the root node as the MRCA), $(\lambda_{10} + \lambda_6 + \lambda_2) + (\lambda_{12} + \lambda_8 + \lambda_5)$ (through the MRCA that is a child of the root), and $(\lambda_{10} + \lambda_9) + \lambda_{12}$ (through the MRCA that is the parent node of X_3). The average path distance (APD) of X_2 and X_3 is the average of these three distances. (Color figure online)

To compute matrices M_1 and M_2 we utilize a BFS-like approach to traverse networks Ψ_1 and Ψ_2, starting from their leaves and finishing at the root, visiting each internal node if and only if we have explored all of its children first. If we are currently at an MRCA of leave pairs i and j, we add their path distance to matrix entries $[i, j]$ and $[j, i]$. After traversing the graph, we divide each entry in the matrices by the number of MRCAs per pair of leaves to obtain the average path distance. After building matrices M_1 and M_2 for networks Ψ_1 and Ψ_2, respectively, APD can be computed by taking the Frobenius norm of the matrix difference:

$$\mathrm{APD}(\Psi_1, \Psi_2) = \|M_1 - M_2\|_F. \tag{1}$$

Overall, the graph traversal runs in $\mathcal{O}(m)$, where m is the number of edges in the network. Building and summarizing each matrix takes $\mathcal{O}(n^2)$. The computation in Eq. 1 also takes $\mathcal{O}(n^2)$ operations. Thus, APD is computable in polynomial time with respect to the number of nodes in the phylogenetic network.

However, it is important to note that the APD of the two different networks of Fig. 3 is also 0 under the branch length settings above. Therefore, APD is not a metric either.

Normalized Average Path Distance (NormAPD). We can obtain a normalized APD measure (NormAPD) as follows, which assumes Ψ_1 is a reference network and is useful in settings where an inferred network is compared to a reference one.

$$\text{NormAPD}(\Psi_1, \Psi_2) = \frac{\|M_1 - M_2\|_F}{\|M_1\|_F} = \frac{\text{APD}(\Psi_1, \Psi_2)}{\|M_1\|_F}. \tag{2}$$

3 Results and Discussion

3.1 Dissimilarity Under Various Network Perturbations

To study the behavior of our dissimilarity measures on pairs of related networks, we simulated rooted phylogenetic networks that are leaf-labeled by the same set of 8 taxa using the SpeciesNetwork [35] add-on in BEAST 2.5 [4]. For the simulation, we set the origin to 0.1, birth rate to 30, and hybridization rate to 5. We then filtered out networks that contain more than 7 reticulations to limit the number of trees displayed by the networks. Our final data set contained 500 rooted phylogenetic networks. For each network Ψ_0 in the set of simulated networks, we generated perturbed versions of Ψ_0 and calculated the dissimilarity between the perturbed networks and Ψ_0.

Uniform Scaling. Here we obtain Ψ_i by scaling each branch of Ψ_0 by a factor of 1.5 for 10 iterations. At the end of iteration i, we have network Ψ_i whose branch lengths are $(1.5)^i \lambda_0$, where λ_0 is the branch length of the particular branch on the original network Ψ_0. We computed $\text{rNBS}(\Psi_0, \Psi_i)$ and $\text{APD}(\Psi_0, \Psi_i)$, for $i = 1, 2, \ldots, 10$, and plotted the results as a function of the iteration number. The results for rNBS and APD are shown in Figs. 5a and 5b, respectively. Both figures show an exponential relationship between the number of iterations and the dissimilarity between the original and perturbed networks measured by rNBS and APD. Furthermore, we observe that neither of the measures have a consistent rate of increase in the dissimilarity values across pairs of networks. In fact, without normalization, both measures report higher values of dissimilarity when the network contains more edges.

Non-uniform Scaling. Here, we obtain Ψ_i, for $i = 1, 2, \ldots, 10$ from Ψ_0 in a similar fashion to what we did in the case of uniform scaling, except that in each of the 10 iterations, we scaled each branch by a value drawn from $\mathcal{N}(1.5, 0.1^2)$. The results for rNBS and APD are shown in Figs. 5c and 5d, respectively. Even though scale factors are sampled from a distribution, the rates of increase of dissimilarity values computed by both rNBs and APD are consistent with the results from uniform scaling.

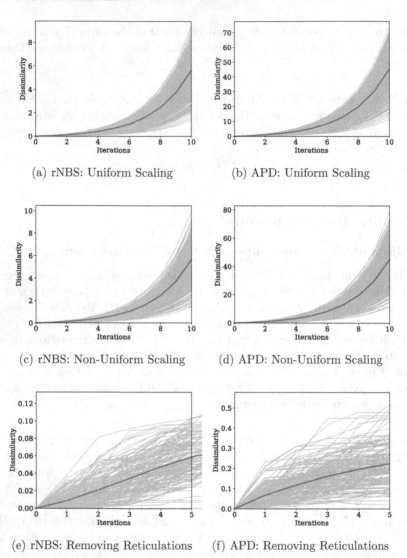

Fig. 5. Dissimilarity measures on pairs of networks using three different perturbations. (a–b) Perturbed networks are obtained by uniform scaling of all branches and comparing perturbed networks to a reference network. (c–d) Perturbed networks are obtained by non-uniform scaling of all branches and comparing perturbed networks to a reference network. (e–f) Perturbed networks are obtained by removing reticulation edges and comparing perturbed networks to a reference network. Each gray line shows the relationship between number of iterations and the amount of dissimilarity between the original and perturbed networks. Blue lines show the average dissimilarity value at each iteration. (Color figure online)

Reticulation Elimination. Here, we further filter out networks that have fewer than 5 reticulation nodes, thus limiting our dataset to networks with at least 5 and at most 7 reticulations, which accounts for 157 networks. We produce Ψ_i at each iteration as follows: Ψ_1 is obtained by removing a random reticulation edge from Ψ_0, and then Ψ_i is obtained by removing a random reticulation edge from Ψ_{i-1}, for $i = 2, \ldots, 5$. The results for rNBS and APD are shown in Figs. 5e and 5f, respectively.

As the figure shows, the dissimilarity values increase as more reticulations are removed, but that increase slows down for the APD values with the number of removed reticulations, which is apparent in Fig. 5f. For the rNBS values (Fig. 5e), observe that the change in values is slow as the average value goes from 0 to around 0.06 when 5 reticulations are removed. The reason for this is that while reticulations are removed, the rest of the topologies and branch length are unperturbed. Therefore, it is natural that the removal of reticulations would have much less of an effect than, say, having all branches differ in lengths between the reference and perturbed networks.

In summary, perturbation experiments show that the measures we introduce here are sensitive to both changes in branch lengths and reticulation edges and behave as expected from dissimilarity measures.

3.2 Analyzing Posterior Samples Using the Dissimilarity Measures

Phylogenetic analyses oftentimes produce thousands of candidates. To summarize the tree candidates, a consensus tree is often computed, e.g., [16,30]. As an alternative to a single consensus tree, clustering was offered as early as 1991 by W. Maddison [19]. Subsequently, it was shown that clustering indeed provides a more powerful and informative summary than single consensus trees [29]. Here, we explore the use of our dissimilarity measures for clustering networks in the posterior sample of a Bayesian inference method. We also compare them to clustering based on the topological dissimilarity measure of [22], referred to hereafter as TD.

MCMC_SEQ [32] is a method in PhyloNet [31,33] for Bayesian inference of phylogenetic network topology, divergence times, and inheritance probabilities, along with various other parameters from sequence alignment data. The method samples from the posterior distribution of these parameters. We analyzed the posterior samples obtained by MCMC_SEQ on one simulated data set and one empirical data set.

Analysis of a Simulated Data Set. We analyze the posterior sample obtained on a simulated data set of 5 species and a single individual per species. To generate the data set, we first simulated the homogeneous gene trees with MS [14], obtaining 100 gene trees with 500 sites per gene tree. Finally, we use Seq-Gen [24] to generate the simulated sequences. For this, we set the base frequencies $\{A, C, G, T\} = \{0.2112, 0.2888, 0.2896, 0.2104\}$ and theta to 0.018. We also varied the substitution rates to follow a flat Dirichlet distribution. The sequences are then inputted to MCMC_SEQ, setting the Markov chain Monte Carlo (MCMC) chain length to 50, 000, 000, the burn-in period to 10, 000, 000, and the sample frequency to 5000.

The networks in the posterior sample vary in their number of reticulation nodes between 0 (i.e., trees) and 4. However, if we only consider the samples after burn-in period is completed (that is, ignoring the first 2,000 samples), the number of reticulations alternate between 3 and 4. Since clustering would require computing all pairwise distances between the networks in the sample, we further reduced the sample by keeping only every 5th sample, i.e., the 2000th, 2005th, ..., and 9995th samples (the first 2, 000 samples were discarded as part of the burn-in period). We then computed pairwise dissimilarity matrices using rNBS, APD, and TD of [22]. Afterwards, we clustered the posterior networks in two different ways:

1. Clustering I: We partitioned the samples based on the number of reticulation nodes in the networks, thus resulting in two clusters, one consisting of all networks with 3 reticulations and another consisting of all networks with 4 reticulations.
2. Clustering II: We applied agglomerative clustering to the pairwise distance matrices, and set the linkage criterion to "complete," which makes the clustering algorithm use the maximum distances between all observations of the two sets when merging pairs of clusters. To determine the number of clusters, we looked at the dendrogram from each figure and manually curated the number to split. Note that cluster labels are arbitrarily assigned independently for each plot.

As the number of reticulations in networks is a major distinguishing factor when contrasting networks inferred on biological data, the rationale of the two ways of clustering is to understand (i) how the different iterations of MCMC_SEQ correlated in terms of the clustering, and (ii) whether clustering II is a refinement of clustering I.

For visualization, we plot the log-posterior density per iteration, number of reticulations per iteration (which corresponds to clustering I), the assigned cluster per iteration (which corresponds to clustering II), and the pairwise dissimilarity matrix computed by rNBS, APD, and TD of [22]. Here, we did not visualize the clustered heatmaps, but rather kept them ordered according to the sampling order since we are focused on understanding the correlation, in terms of dissimilarity, between adjacent samples. The results are shown in Fig. 6.

From Fig. 6 we can observe that rNBS, APD, and TD all cluster the networks with respect to the number of reticulations, shown by the purple vs blue shades in Fig. 6(a) for rNBS, purple vs green shades in Fig. 6(b) for APD, and green vs yellow shades in Fig. 6(c) for TD. While all heat-maps support the clustering by the number of reticulations, the clusters are best defined based on APD pairwise distances, followed by rNBS pairwise distances, and then least defined based on TD pairwise distances.

Additionally, agglomerative clustering obtained based on rNBS and APD pairwise distances shows distinct sub-clusters within the two clusters based on

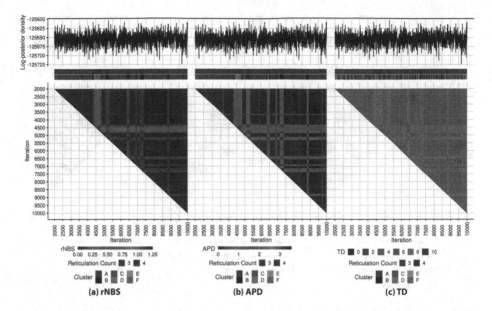

Fig. 6. Clustering of samples (networks and their branch lengths) from the posterior distribution obtained by MCMC_SEQ on a simulated data set. Each panel shows the log-posterior density, the number of reticulations in each sampled network, the assigned clustering label obtained by agglomerative clustering, and the heatmap of the dissimilarity values arranged by the number of MCMC iterations. Panels (a)–(c) correspond to the three dissimilarity measures rNBS, APD, and TD, respectively. (Color figure online)

the number of reticulations alone. For example samples in two distinct clusters B and C for both rNBS and APD contain samples with 4 reticulations, suggesting cluster analyses can help discern patterns of similarity in terms of the posterior values and obtain a refined clustering of the posterior sample. The fact that these structures were only sampled in one segment of the posterior chain also suggests it is less converged than the log-posterior probability trace suggests, and that APD- and rNBS-based clustering are more powerful diagnostic tools for convergence.

Analysis of an Empirical Data Set. For the empirical results, we analyzed the yeast data set of [27] with seven *Saccharomyces* species. We utilized the same methods as in the previous section, with the exception of setting the MCMC chain length to $35,000,000$, obtaining $5,000$ samples after discarding the first $2,000$. The results are shown in Fig. 7.

From Fig. 7, we see that both rNBS and APD highlight similar pairwise dissimilarity characteristics between posterior samples. While TD mostly agrees with both rNBS and APD, there exists an anomalous region with high dissimilarity values based on TD at around iteration $6,250$. Upon inspection, we

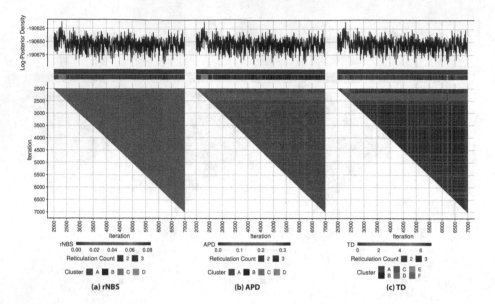

Fig. 7. Clustering of samples (networks and their branch lengths) from the posterior distribution obtained by MCMC_SEQ on an empirical yeast data set. Each panel shows the log-posterior density, the number of reticulations in each sampled network, the assigned clustering label obtained by agglomerative clustering, and the heatmap of the dissimilarity values arranged by the number of MCMC iterations. Panels (a)–(c) correspond to the three dissimilarity measures rNBS, APD, and TD, respectively.

observe a minor topological change close to a pair of leaves, causing the topological dissimilarity value to peak at 7.0. While both rNBS and APD report a slight increase in dissimilarity in that region as well, the reported dissimilarity from neither measure is high enough to affect the dissimilarity scales of rNBS and APD. This highlights topological dissimilarity's over-sensitivity to minor topological changes, which is not the case for either rNBS or APD.

3.3   Runtime Comparison

We report on the runtimes of the current implementations of rNBS, APD, and TD of [22] with respect to the number of reticulations in the pairs of networks. All experiments were run on a desktop running Linux Mint 20.3 on a single AMD Ryzen 9 5900X 12-Core Processor and 31.3 GiB available memory. Figure 8 summarizes the results.

We find that APD is one fold slower than TD. Additionally, while the runtime increases as the number of reticulations increases, the rate of increase is very small. When there are no reticulations, APD can compute approximately 2,082 dissimilarities per second, and when there are 10 reticulations, APD can compute approximately 649 dissimilarities per second. In contrast, rNBS slows down significantly as the number of reticulations increases, which supports the

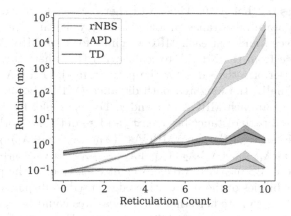

Fig. 8. Runtimes of rNBS, APD, and TD of [22] with respect to number of reticulation nodes in the networks. Time is measured in milliseconds.

asymptotic time complexity of the edge covering algorithms used for bipartite graphs formulated for rNBS (11, 278 per second when there are no reticulations vs. 0.03 per second when there are 10 reticulations). It should be noted that our implementation of APD is written in Java and does not currently use any libraries or hardware acceleration to transform matrices.

4 Conclusions and Future Work

By taking branch lengths into account, our two novel dissimilarity measures of phylogenetic networks will address an important deficit in the ability to analyse phylogenetic network space and its exploration. By taking different approaches to measuring dissimilarity, and through our analysis of scaled, perturbed, and MCMC sampled phylogenetic networks, we have shown that the path distance approach has more immediate promise than edge-covering of displayed trees. We implemented both measures in PhyloNet [31, 33] and studied their properties on pairs of perturbed networks. Furthermore, we illustrated the use of the dissimilarity measures for clustering and summarizing the phylogenetic networks in a posterior sample of the Bayesian inference method MCMC_SEQ. There are many directions for future work, two of which we will discuss here.

Incorporating Inheritance Probabilities in the Dissimilarity Measures. Statistical inference of phylogenetic networks results not only in estimates of the topologies and branch lengths, but also of inheritance probabilities, which annotate the reticulation edges. The measures we presented above ignore inheritance probabilities. In particular, the APD measure treats all paths equally. One possible extension that accounts for inheritance probabilities is to weight each path that is counted by a combination of the inheritance probabilities used by the reticulation edges on that path. Thus, a further extension to average path distance can incorporate inheritance probabilities by weighting the path distance between two leaves i and j at each MRCA. Thus, for each MRCA, we multiply the path distance from the MRCA to leaf i with the inheritance probabilities along the path, perform the same for the path from the MRCA to leaf j, and sum them up. Finally, the weighted path distance (WPD) becomes the sum of weighted distances at each MRCA of i and j. For example, for the network of Fig. 4, the weighted path distance between x_2 and x_3 in the network is the sum of weighted path distances at the three MRCAs: $[(\lambda_{10}+\lambda_6+\lambda_2+\lambda_1)\cdot\gamma_2+(\lambda_{12}+\lambda_8+\lambda_7+\lambda_4)\cdot(1-\gamma_1)]+[(\lambda_{10}+\lambda_6+\lambda_2)\cdot\lambda_2+(\lambda_{12}+\lambda_8+\lambda_5)\cdot\lambda_1]+[(\lambda_{10}+\lambda_9)\cdot(1-\lambda_2)+\lambda_{12}]$. Similarly, for the rNBS measure, we can weight each tree by the product of the inheritance probabilities of the reticulation edges used to display the tree. However, it is important to note here that these measures could be very sensitive to inaccuracies of the inheritance probability estimates. For example, two networks that are identical in terms of topologies and branch lengths but vary significantly in the inheritance probabilities of even a small number of reticulation edges could result in a very large dissimilarity value, unless the inheritance probabilities are weighted carefully. We will explore these directions.

Tree-Based Dissimilarity in the Presence of Incomplete Lineage Sorting. Zhu et al. [36] showed that when incomplete lineage sorting occurs, inference and analysis of phylogenetic networks are more adequately done with respect to the set of parental trees of the network, rather than the set of displayed trees. Thus, a different approach to defining the rooted network branch score could involve building a complete bipartite graph on the sets of parental trees of the two networks. A major challenge here could be computational. As we demonstrated in our analysis of rNBS, the runtime of the current implementation increases exponentially with the number of reticulations k. Worse, while a phylogenetic network has up to 2^k displayed trees, the number of parental trees could be significantly much larger to the extent that 2^k is feasible in the case of a small k whereas computing the set of all parental trees explicitly for the same value of k can be infeasible. Therefore, faster computations and/or heuristics for computing such a dissimilarity measure would be needed. As we discussed above, we will study whether constructing the bipartite graph explicitly is necessary for computing rNBS, as avoiding such construction could result in significant improvement to the computational requirements.

Acknowledgements. We thank Zhen Cao for contributing the MCMC posterior sample files for the simulated data set. This work was supported in part by NSF grants CCF-1514177, CCF-1800723 and DBI-2030604 to L.N.

References

1. Allen, B.L., Steel, M.: Subtree transfer operations and their induced metrics on evolutionary trees. Ann. Comb. 5(1), 1–15 (2001). https://doi.org/10.1007/s00026-001-8006-8

2. Avino, M., Ng, G.T., He, Y., Renaud, M.S., Jones, B.R., Poon, A.F.Y.: Tree shape-based approaches for the comparative study of cophylogeny. Ecol. Evol. **9**(12), 6756–6771 (2019). https://doi.org/10.1002/ece3.5185

3. Billera, L.J., Holmes, S.P., Vogtmann, K.: Geometry of the space of phylogenetic trees. Adv. Appl. Math. **27**(4), 733–767 (2001)

4. Bouckaert, R., et al.: BEAST 2.5: an advanced software platform for Bayesian evolutionary analysis. PLoS Comput. Biol. **15**(4), 1–28 (2019). https://doi.org/10. 1371/journal.pcbi.1006650

5. Bouvel, M., Gambette, P., Mansouri, M.: Counting phylogenetic networks of level 1 and 2. J. Math. Biol. **81**(6), 1357–1395 (2020). https://doi.org/10.1007/s00285-020-01543-5

6. Cardona, G., Llabrés, M., Rosselló, F., Valiente, G.: Metrics for phylogenetic networks I: generalizations of the Robinson-Foulds metric. IEEE/ACM Trans. Comput. Biol. Bioinform. **6**(1), 46–61 (2008)

7. Cardona, G., Llabrés, M., Rosselló, F., Valiente, G.: Metrics for phylogenetic networks II: nodal and triplets metrics. IEEE/ACM Trans. Comput. Biol. Bioinform. **6**(3), 454–469 (2008)

8. Dinh, V., Bilge, A., Zhang, C., Matsen IV, F.A.: Probabilistic path hamiltonian Monte Carlo. In: Precup, D., Teh, Y.W. (eds.) Proceedings of the 34th International Conference on Machine Learning. Proceedings of Machine Learning Research, vol. 70, pp. 1009–1018. PMLR (2017)

9. Elworth, R.A.L., Ogilvie, H.A., Zhu, J., Nakhleh, L.: Advances in computational methods for phylogenetic networks in the presence of hybridization. In: Warnow, T. (ed.) Bioinformatics and Phylogenetics. CB, vol. 29, pp. 317–360. Springer, Cham (2019). https://doi.org/10.1007/978-3-030-10837-3_13

10. Felsenstein, J.: The number of evolutionary trees. Syst. Zool. **27**(1), 27–33 (1978)

11. Gavryushkin, A., Drummond, A.J.: The space of ultrametric phylogenetic trees. J. Theor. Biol. **403**, 197–208 (2016). https://doi.org/10.1016/j.jtbi.2016.05.001

12. Heled, J., Drummond, A.J.: Bayesian inference of species trees from multilocus data. Mol. Biol. Evol. **27**(3), 570–580 (2009). https://doi.org/10.1093/molbev/msp274

13. Hoffer, B.L.: Language borrowing and the indices of adaptability and receptivity. Intercult. Commun. Stud. **14**(2), 53–72 (2005)

14. Hudson, R.R.: Generating samples under a Wright-Fisher neutral model of genetic variation. Bioinformatics **18**(2), 337–338 (2002). https://doi.org/10.1093/bioinformatics/18.2.337

15. Jain, C., Rodriguez-R, L.M., Phillippy, A.M., Konstantinidis, K.T., Aluru, S.: High throughput ANI analysis of 90K prokaryotic genomes reveals clear species boundaries. Nat. Commun. **9**(1), 5114 (2018). https://doi.org/10.1038/s41467-018-07641-9

16. Kannan, L., Wheeler, W.: Maximum parsimony on phylogenetic networks. Algorithms Mol. Biol. **7**, 9 (2012). https://doi.org/10.1186/1748-7188-7-9

17. Koskela, J.: Zig-zag sampling for discrete structures and non-reversible phylogenetic MCMC. J. Comput. Graph. Stat. (2022). https://doi.org/10.1080/10618600. 2022.2032722. Accepted for publication

18. Kuhner, M.K., Felsenstein, J.: A simulation comparison of phylogeny algorithms under equal and unequal evolutionary rates. Mol. Biol. Evol. **11**(3), 459–468 (1994)

19. Maddison, D.R.: The discovery and importance of multiple islands of most-parsimonious trees. Syst. Biol. **40**(3), 315–328 (1991)

20. Michael, T.P., VanBuren, R.: Building near-complete plant genomes. Curr. Opin. Plant Biol. **54**, 26–33 (2020). https://doi.org/10.1016/j.pbi.2019.12.009. Genome studies and molecular genetics
21. Nakhleh, L.: Evolutionary phylogenetic networks: models and issues. In: Heath, L., Ramakrishnan, N. (eds.) Problem Solving Handbook in Computational Biology and Bioinformatics, pp. 125–158. Springer, Boston (2010). https://doi.org/10.1007/978-0-387-09760-2_7
22. Nakhleh, L.: A metric on the space of reduced phylogenetic networks. IEEE/ACM Trans. Comput. Biol. Bioinform. **7**(2), 218–222 (2010). https://doi.org/10.1109/TCBB.2009.2
23. Pardi, F., Scornavacca, C.: Reconstructible phylogenetic networks: do not distinguish the indistinguishable. PLoS Comput. Biol. **11**(4), 1–23 (2015). https://doi.org/10.1371/journal.pcbi.1004135
24. Rambaut, A., Grass, N.C.: Seq-Gen: an application for the Monte Carlo simulation of DNA sequence evolution along phylogenetic trees. Bioinformatics **13**(3), 235–238 (1997). https://doi.org/10.1093/bioinformatics/13.3.235
25. Rhie, A., et al.: Towards complete and error-free genome assemblies of all vertebrate species. Nature **592**(7856), 737–746 (2021). https://doi.org/10.1038/s41586-021-03451-0
26. Robinson, D., Foulds, L.: Comparison of phylogenetic trees. Math. Biosci. **53**(1), 131–147 (1981). https://doi.org/10.1016/0025-5564(81)90043-2
27. Rokas, A., Williams, B.L., King, N., Carroll, S.B.: Genome-scale approaches to resolving incongruence in molecular phylogenies. Nature **425**(6960), 798–804 (2003)
28. Steel, M.A., Penny, D.: Distributions of tree comparison metrics-some new results. Syst. Biol. **42**(2), 126–141 (1993)
29. Stockham, C., Wang, L.S., Warnow, T.: Statistically based postprocessing of phylogenetic analysis by clustering. Bioinformatics **18**(suppl_1), S285–S293 (2002)
30. Swofford, D., Olson, G., Waddell, P., Hillis, D.: Phylogenetic inference. In: Hillis, D., Moritz, C., Mable, B. (eds.) Molecular Systematics, pp. 407–514. Sinauer Associates, Sunderland (2004). Chap. 11
31. Than, C., Ruths, D., Nakhleh, L.: PhyloNet: a software package for analyzing and reconstructing reticulate evolutionary relationships. BMC Bioinform. **9**(1), 1–16 (2008). https://doi.org/10.1186/1471-2105-9-322
32. Wen, D., Nakhleh, L.: Coestimating reticulate phylogenies and gene trees from multilocus sequence data. Syst. Biol. **67**(3), 439–457 (2017). https://doi.org/10.1093/sysbio/syx085
33. Wen, D., Yu, Y., Zhu, J., Nakhleh, L.: Inferring phylogenetic networks using PhyloNet. Syst. Biol. **67**(4), 735–740 (2018)
34. Whidden, C., Matsen, F.A., IV.: Quantifying MCMC exploration of phylogenetic tree space. Syst. Biol. **64**(3), 472–491 (2015). https://doi.org/10.1093/sysbio/syv006
35. Zhang, C., Ogilvie, H.A., Drummond, A.J., Stadler, T.: Bayesian inference of species networks from multilocus sequence data. Mol. Biol. Evol. **35**(2), 504–517 (2017). https://doi.org/10.1093/molbev/msx307
36. Zhu, J., Yu, Y., Nakhleh, L.: In the light of deep coalescence: revisiting trees within networks. BMC Bioinform. **17**(14), 415 (2016). https://doi.org/10.1186/s12859-016-1269-1

Homology and Reconciliation

The Complexity of Finding Common Partitions of Genomes with Predefined Block Sizes

Manuel Lafond[1(✉)], Adiesha Liyanage[2], Binhai Zhu[2(✉)], and Peng Zou[2]

[1] Department of Computer Science, Université de Sherbrooke, Sherbrooke, QC J1K 2R1, Canada
manuel.lafond@usherbrooke.ca
[2] Gianforte School of Computing, Montana State University, Bozeman, MT 59717, USA
bhz@montana.edu, peng.zou@student.montana.edu

Abstract. Partitioning genomes into syntenic blocks has many uses in comparative genomics, such as inferring phylogenies or ancestral gene order. These blocks are usually required to contain enough genes to be qualified as syntenic. This leads to the problem of finding a common partition of the genomes in which the size of the blocks are above a certain threshold (usually at least two). When this is not feasible, one can ask to remove a minimum number of "noisy" genes so that such a partition exists. This is known as the Strip Recovery problem and is similar to the well-known Minimum Common String Partition problem, but also quite different since the latter has no restriction on the block sizes.

The algorithmic aspects of Strip Recovery are not well-understood, especially in the presence of duplicated genes. In this work, we present several new complexity results. First, we solve an open problem mentioned by Bulteau and Weller in 2019 who asked whether, in polynomial time, one can decide if a common partition with block sizes at least two can be achieved without deleting any genes. We show that the problem is actually NP-hard for any fixed list of allowed block sizes, unless blocks sizes are all multiples of the minimum allowed size. The problem is also hard on fixed alphabets if this list is part of the input. However, the minimum number of required gene deletions can be found in polynomial time if both the allowed blocks sizes and alphabet are fixed.

1 Introduction

In comparative genomics, a common task is to partition two genomes with duplicated genes into *syntenic blocks*, which are blocks that have the same gene content. Finding these blocks can have many applications, including the inference of ancestral gene order [2,3], the reconstruction of phylogenies [15,20], gene tree/species tree reconciliation [14], the study of whole genome duplications [16], or the prediction of orthologous genes [27,29]. Synteny is a somewhat loosely defined concept, but almost every definition requires that syntenic blocks have

© The Author(s) under exclusive license to Springer Nature Switzerland AG 2022
L. Jin and D. Durand (Eds.): RECOMB-CG 2022, LNBI 13234, pp. 105–123, 2022.
https://doi.org/10.1007/978-3-031-06220-9_7

a minimum number of genes in common. By representing the two genomes G_1 and G_2 as strings, finding syntenic blocks can be modeled as a common string partition problem, where the goal is to split the genomes into two equal sets of substrings. A popular formulation is the *Minimum Common String Partition* (MCSP), where one asks for the minimum number of blocks in a common partition. This problem was initially proposed for the purpose of orthology assignment [9]. However, it has been argued that MCSP does not necessarily find the "best" partition in terms of syntenic blocks. This is because MCSP might infer blocks that are too small for syntenies, possibly of size 1, in order to attain its optimization criterion.

In [35], the authors argued that these small blocks can be treated as noise in the context of genome comparison, and propose the alternate *Strip Recovery* formulation. In this problem, the goal is to find a common string partition in which every block has length at least k, where k is a given parameter (the simplest version puts $k = 2$). Of course, such a partition does not always exist, which led to the *Maximal Strip Recovery* (MSR) problem. Here, the objective is to find maximum-length subsequences G_1' and G_2' of G_1 and G_2, respectively, such that G_1' and G_2' do admit a common string partition with the desired block sizes. The minimization version of the problem is called *Complementary* MSR, and asks to delete a minimum number of genes from the two strings so that they admit the desired block partition. Either way, this formulation has the advantage over MCSP of placing more genes into syntenic blocks of interesting sizes.

Despite this, MSR has received much less attention than its MCSP sister problem. In particular, an open problem from [8] states that it is not even known whether, in polynomial time, one can find a common partition into blocks of size two or more *without deleting any gene*. Moreover, previous work on MSR mainly focused on the permutation model, i.e., genomes without duplicates, whereas most of the aforementioned applications deal with paralogs. In this paper, we are interested in gaining a better understanding of the theoretical and algorithmic aspects of MSR with duplicated genes. In fact, we generalize the problem in several ways (see Fig. 1). First, we allow the input to specify an arbitrary list F of allowed block sizes in the desired partition, with F possibly infinite. The Maximal Strip Recovery problem becomes a special case that asks for a common partition with block sizes in $F = \{2, 3, \ldots\}$. More generally, blocks of size at least k can be specified with $F = \{k, k+1, \ldots\}$. Our framework does handle block sizes of non-contiguous integer sets, for instance requiring blocks of size 7 or 11 with $F = \{7, 11\}$. Although this does not appear to be biologically relevant, this level of genericity is actually crucial for our purposes. As we will show in Section 3, any infinite set F of block sizes can be reduced to an equivalent finite set F' that is not necessarily contiguous but that is easier to handle algorithmically. For instance, the case $F = \{4, 5, \ldots\}$ is equivalent to the case $F' = \{4, 5, 6, 7, 9\}$, motivating the need to handle arbitrary F. Second, we allow the input to consist of two *sets* of strings $\mathcal{W}_1$ and $\mathcal{W}_2$ instead of individual strings. The two sets may differ in size and in total string length. This can be useful for genome analyses that require the comparison of multiple chromosomes or unrelated segments between two species. Moreover, this can be useful to devise simple NP-hardness reductions. We provide several algorithmic results for these generalizations.

$$sequence\ A\ =\ abccabcbab$$
$$sequence\ B\ =\ cbabcababc$$
$$partition\ of\ A\ =\ \boxed{abc}\cdot\boxed{cab}\cdot\boxed{cbab}$$
$$partition\ of\ B\ =\ \boxed{cbab}\cdot\boxed{cab}\cdot\boxed{abc}$$

Fig. 1. An example of partition of A and B without deleting any letter, with $F = \{3, 4\}$ and the alphabet $\Sigma = \{a, b, c\}$. Note that when we change F to $F = \{4\}$, then no such partition exists. In this case, it seems the best one can do is to delete the first 6 letters of A and the last 6 letters of B.

Related Work. Wang and Zhu first proved that both (the decision versions of) MSR and Complementary MSR are NP-complete [34] in the permutation model. Almost at the same time, MSR was proved to be APX-hard [6,26], and CMSR was also proved to be APX-hard by Jiang [26]. On the positive side, some heuristic approaches based on Maximum Independent Set and Maximum Clique were applied [11,35] and they were shown to be effective. For MSR, Chen et al. proposed a factor-4 (polynomial-time) approximation algorithm for MSR [10]. In [23], a factor-3 approximation algorithm was proposed for CMSR and an $O^*(3^k)$ fixed-paramter tractable (FPT) algorithm, where k is the parameter representing the minimum number of deleted genes, was also proposed for CMSR (the O^* notation suppresses polynomial factors). The approximation factor for CMSR was improved to 2.33 [32], then more recently to 2 [22]. The best FPT algorithmic bound for CMSR is $O(2.36^k n^2)$ [4]. In 2014, Jiang and Zhu obtained a linear kernel of size $78k$ for CMSR [24]. This kernel was improved to $58k$ [21] and more recently to $42k$ [31]. Combined with these kernel bounds, CMSR can be solved in $O(n^2 + 2.36^k k^2)$ time. Of course, the MCSP problem also received quite some algorithmic attention [12,19]. The best known approximation (to minimize the number of blocks) has a factor of $O(\log n)$ [17]. For these reasons, the fixed-parameter tractability of MCSP was considered in 2008 [13]. For d-occurrence strings, the problem was first shown to be FPT in [5,25]. In 2014, it was shown that MCSP parameterized by the solution size ℓ is FPT [7]. In 2021, Lafond and Zhu also studied the important variant when a permutation (on the final blocks) is given in advance [28].

Our Contributions. In addition to generalizing the MSR problem to arbitrary block sizes F and multiple sets of strings, we solve the open problem stated in [8] mentioned above. Recall that the problem asks whether it is possible to decide in polynomial time if a common partition with block sizes in $F = \{2, 3, \ldots\}$ can be found without deleting *any* marker. We actually obtain a dichotomy result in terms of F: the problem is in P if each member of F is a multiple of its minimum, and is NP-hard otherwise. This holds whether the input consists of two strings or two sets of strings, and even on instances in which each symbol of the alphabet occurs at most 6 times. The open problem of [8] is solved as a special case of this result.

On the positive side, we show that if F and the alphabet Σ are both fixed finite sets, then the generalized strip recovery problem can be solved in polynomial time. More precisely, we show that the problem is in the XP complexity class with respect to the parameters $\max(F)$ and $|\Sigma|$ by proposing an algorithm that runs in time $O(n^{|F|}|\Sigma|^{\max(F)+3})$. Although this may be impractical, we show that unfortunately, fixing both Σ and F is necessary and in some sense best possible, since the problem is NP-hard if Σ is fixed but not F.

2   Preliminary Notions

We use the notation $[n] = \{1, 2, \ldots, n\}$. Let S be a string. We write $|S|$ for the size of S and $S[i..j]$ for the substring of S containing positions from i to j, inclusively. Each individual character of S is called a *marker*. Thus, S has $|S|$ distinct markers, and two markers may represent the same symbol but are considered distinct if they are located at different positions of S. We say that S' is a *subsequence* of S if S' can be obtained by deleting some set of markers form S. A *permutation string* is a string in which each symbol occurs exactly once.

A *partition* of S is a multiset of strings that can be concatenated to obtain S. The strings in the partition are called *blocks*. If W is a set of strings, a partition of W is the union (in the multiset sense) of a partition of each of its strings. For a set of integers F, we say that a partition M is an *F-partition* if the length of each block of M is in F. A *common partition* of two sets of strings W_1 and W_2 is a multiset of strings M such that M is a partition of both W_1 and W_2. Also, M is a *common F-partition* if M is also an F-partition.

The General and Exact F-Strip Recovery Problems
Assume that F is a fixed set of integers, possibly infinite. The general version of the original Strip Recovery problem that we propose is the following.

The General F-Strip Recovery Problem (GSR-F)
Input : two sets of strings W_1 and W_2;
Question : delete a minimum number of markers from the strings in W_1 and W_2 so that the resulting strings admit a common F-partition.

We will also study the following variant.

The Exact F-Strip Recovery Problem (XSR-F)
Input : two sets of strings W_1 and W_2;
Question : does there exist a common F-partition of W_1 and W_2?

The original Strip Recovery problem is the special case of GSR-F where $|W_1| = |W_2| = 1$ and $F = \{2, 3, \ldots\}$. As we mentioned, the problem of [8] asks whether XSR-F is polynomial-time solvable with $F = \{2, 3, \ldots\}$. Notice that F is considered fixed, not part of the input. We will consider the case where F is in the input later.

3 The Exact F-Strip Recovery Problem with Fixed F

We now obtain a dichotomy result on XSR-F. If all elements of F are multiples of its minimum, then XSR-F can easily be solved in polynomial time, and otherwise XSR-F is NP-hard (assuming that F is enumerable, as we describe below). Before we proceed, let us discuss some aspects of fixed but infinite F sets. The fact that F can be infinite introduces ambiguity as to how an algorithm can access its elements, since it depends on how F is specified. As it turns out, infinite sets can always be replaced by a finite, but equivalent set. This requires knowledge of the greatest common divisor $gcd(F)$ of F, but it always exists.

We need a few more definitions beforehand. For two sets F and F', we say that XSR-F and XSR-F' are *equivalent* if W_1, W_2 is a Yes-instance of XSR-F if and only if W_1, W_2 is a Yes-instance of XSR-F'. For a set F of integers and an integer t, we say that t is *expressible by* F if there exist $\{g_1, \ldots, g_k\} \subseteq F$ and positive integer coefficients $a_1, \ldots, a_k$ such that $t = \sum_{i=1}^{k} a_i g_i$.

For instance, when $F = \{2, 3, \ldots\}$ as in the MSR problem, every integer equal to or greater than 2 is expressible by F.

Lemma 1. *XSR-F and XSR-F' are equivalent if and only if the set of integers expressible by F is equal to the set of integers expressible by F'.*

Proof. Assume that XSR-F and XSR-F' are equivalent. Consider $k \in F$. Take an instance of XSR-F that consists of $W_1 = W_2 = \{a^k\}$, where a^k is the string with k a's. This is clearly a Yes-instance of XSR-F. By equivalence, this is also a Yes-instance of XSR-F', which is only possible if k is expressible by F'. Since this holds for every $k \in F$, all elements of F are expressible with F'. By the same argument, all elements of F' are expressible by F.

Conversely, suppose that F and F' express the same set of elements. Let W_1, W_2 be a Yes-instance of XSR-F and let M be a common F-partition. Let S be any block in M. Then $|S|$ is expressible by F, and thus also by F'. This means that S can be partitioned into substrings of lengths in F'. We can apply this to every S of M, resulting in a common F'-partition, and thus W_1, W_2 is also a Yes-instance of XSR-F'. By the same argument, a Yes-instance of XSR-F' is also a Yes-instance of XSR-F, and thus they are equivalent.   □

Again relating to the $F = \{2, 3, \ldots\}$ case, the above implies that MSR is equivalent to the case $F' = \{2, 3\}$, since they suffice to express every integer in F. We show that this behavior works for every infinite F.

Proposition 1. *Let F be an infinite set of positive integers. Then there exists a finite set $F' \subseteq F$ such that XSR-F and XSR-F' are equivalent.*

Proof. First observe that there must exist a finite set $G \subseteq F$ such that $gcd(G) = gcd(F)$. One way to see this is to start with $G = \{\min(F)\}$. If $gcd(G) = gcd(F)$, we are done. Otherwise, there is some $k \in F$ such that $gcd(G)$ does not divide k. Add k to G. The gcd of $\{\min(F), k\}$ is strictly smaller than the gcd of $\{\min(F)\}$. We can repeat this process of adding elements of F to G until we obtain a set G that satisfies $gcd(G) = gcd(F)$. Because $gcd(F) \geq 1$ and because the gcd of G starts at $\min(F)$ and decreases monotonically, the process finishes in at most $\min(F)$ step, a finite number.

Let us denote $g := gcd(G) = gcd(F)$ for the rest of the proof. Now, since every $k \in F$ is a multiple of g, every integer expressible by F (and by G) is a multiple of g. Let us now argue that there exists an integer $t(G)$ such that *every* multiple of g greater than $t(G)$ is expressible by G. If $g = 1$, this is saying that every integer above some $t(G)$ is a linear combination of elements of G. In this case, $t(G)$ is known to exist by one of Schur's theorems (note that finding the smallest such $t(G)$ is known as the coin problem, see [1]). Note that the best known upper bound for $t(G)$ is in the order of $O(\max(G)^2)$ [30]. Now assume that $g \geq 2$. Let $G' = \{x/g : x \in G\}$. Then $gcd(G') = 1$, as otherwise G would have the common divisor $g \cdot gcd(G')$, which is larger than g. Thus there is some $t(G')$ such that every integer greater than $t(G')$ is expressible by G'. This implies that every multiple of g greater than $t(G) := g \cdot t(G')$ is expressible by G, as desired.

Since $G \subseteq F$, every multiple of g greater than $t(G)$ can be expressed by F, and these are the only expressible integers greater than $t(G)$. It follows the integers expressible by F and by $F' = G \cup \{x \in F : x \leq t(G)\}$ are equal. By Lemma 1, XSR-F and XSR-F' are equivalent. □

Note that if F' is the finite subset of F generated by the above proof, then $\max(F')$ (and thus $|F'|$) can be upper bounded by $\max(G)^2$, where G satisfies $G \subset F$ and $gcd(G) = gcd(F)$. It seems difficult to make a more precise analysis of these values for generic F. We now obtain our dichotomy result.

Proposition 2. *Let F be a set of integers and let $\ell = \min(F)$. If every member of F is a multiple of ℓ, then XSR-F can be solved in polynomial time.*

Proof. Let $\mathcal{W}_1, \mathcal{W}_2$ be an instance of XSR-F. By Lemma 1, XSR-F and XSR-$\{\ell\}$ are equivalent. Thus, we may assume that $F = \{\ell\}$. In this case, all strings of the partition must have the same length and there is only one way to split $\mathcal{W}_1$ and $\mathcal{W}_2$. It is then easy to see that $\mathcal{W}_1$ and $\mathcal{W}_2$ admit a common F-partition if and only if each substring in the forced partitions appears with the same number of occurrences. □

Theorem 1. *Let F be any set of integers in which some element is not a multiple of $\min F$. Then XSR-F is NP-hard, even if each symbol occurs at most 6 times in the input strings.*

Proof. Our reduction is from the Positive Cubic XSAT problem. In this variant of SAT, we are given a boolean formula ϕ in CNF form in which each variable occurs exactly three times, and in which each clause has exactly three literals that are all positive. The goal is to decide whether there exists an assignment of the variables such that for each clause, exactly one of its variables is true (and two are false). When this is possible, we say that ϕ is *1-in-3 satisfiable*. Note that this can also be seen as an exact set cover problem in which each set has three elements and each element has frequency three. The Positive Cubic XSAT problem was shown NP-hard in [33].

Let ϕ be an instance of this problem. Let $x_1, \ldots, x_n$ be its variables and let $C_1, \ldots, C_m$ be its clauses. Let $\ell = \min F$ and let h be the smallest element of F such that h is not a multiple of ℓ. By the previous proposition, we may assume that h exists. Moreover by Proposition 1, we may assume that F is finite. Indeed, if F is infinite, it is equivalent to some finite F'. Since F is fixed, F' is also fixed and we may assume that an algorithm has knowledge of F' for the purposes of the reduction. We will thus treat F as a finite set. Hence, we can find ℓ and h in constant time (this allows the reduction to run in polynomial time). Note that every element between ℓ and h (exclusively) in F is a multiple of ℓ and, by Lemma 1, we may ignore them and still have an equivalent instance. We will therefore assume that h is the second smallest element of F.

Let $r = \lfloor h/\ell \rfloor$ be the number of times we can fit ℓ into h. Let $s = \ell(r+1) - h$, which is the excess we get if we try to fit one more ℓ. Since ℓ is not a divisor of h, we know that $0 < s < \ell$. Also note for later reference that $h = r\ell + \ell - s$ (obtained by rearranging).

For our reduction, we will define several short strings that serve as building blocks for our instance. Unless stated otherwise, each string we define is a permutation string, and no symbol occurs in two of those strings. Thus, only the length of these strings matters, and the exact nature of their symbols is arbitrary.

To each clause C_a in ϕ corresponds four strings $\mathbb{L}_a, \mathbb{M}_a, \mathbb{M}'_a$ and $\mathbb{R}_a$ (here, $\mathbb{L}, \mathbb{M}$ and $\mathbb{R}$ stand for left, middle and right, respectively), where

- $\mathbb{L}_a$ and $\mathbb{R}_a$ are permutation strings of length s;
- $\mathbb{M}_a$ and $\mathbb{M}'_a$ are permutation strings of length $h - s$;
- none of the $\mathbb{L}_a, \mathbb{M}_a, \mathbb{M}'_a, \mathbb{R}_a$ have a symbol in common;

Intuitively, the strings $\mathbb{L}_a\mathbb{M}_a$ and $\mathbb{M}'_a\mathbb{R}_a$ will represent the fact that clause C_a is "chosen" by a variable, whereas the strings $\mathbb{M}_a\mathbb{R}_a$ and $\mathbb{L}_a\mathbb{M}'_a$ will represent the fact that clause C_a is "not chosen" by a variable.

Now let x_i be a variable. Let C_a, C_b, C_c be the three clauses containing x_i. The x_i variable has two corresponding strings X_i and X'_i built from the clause strings from above, and from other substrings $P_i^j, P_i^*, Q_i, S_i, T_i^*$ and T_i^j, where $j \in [r]$. They are defined as follows:

- for $j \in [r]$, P_i^j is a permutation string of length ℓ;
- P_i^*, Q_i, S_i, T_i^* are permutation strings of length $\ell - s$;
- for $j \in [r]$, T_i^j is a permutation string of length ℓ.

The above strings do not share any symbol, neither between themselves or with strings from other variables or from clauses.

Define X_i and X_i' as

$$X_i = P_i^1 \, P_i^2 \, \cdots \, P_i^r \, P_i^* \, \mathbb{L}_a \, \mathbb{M}_a \, \mathbb{R}_a \, Q_i \, \mathbb{L}_b \, \mathbb{M}_b \, \mathbb{R}_b \, S_i \, \mathbb{L}_c \, \mathbb{M}_c \, \mathbb{R}_c \, T_i^* \, T_i^1 \, T_i^2 \, \cdots \, T_i^r$$

$$X_i' = P_i^1 \, P_i^2 \, \cdots \, P_i^r \, P_i^* \, \mathbb{L}_a \, \mathbb{M}_a' \, \mathbb{R}_a \, Q_i \, \mathbb{L}_b \, \mathbb{M}_b' \, \mathbb{R}_b \, S_i \, \mathbb{L}_c \, \mathbb{M}_c' \, \mathbb{R}_c \, T_i^* \, T_i^1 \, T_i^2 \, \cdots \, T_i^r$$

Notice that each X_i and each X_i' is a permutation string, and that X_i and X_i' do have symbols in common (in fact, they only differ in terms of the $\mathbb{M}_a, \mathbb{M}_a', \mathbb{M}_b, \mathbb{M}_b', \mathbb{M}_c, \mathbb{M}_c'$ substrings).

We put $\mathcal{W}_1 = \{X_1, X_1', X_2, X_2', \ldots, X_n, X_n'\}$.

As for $\mathcal{W}_2$, each of its strings has length ℓ or h. First, for each variable x_i, with C_a, C_b, C_c the clauses containing x_i, add the following strings to $\mathcal{W}_2$:

$$
\begin{array}{ll}
-\ P_i^1, \ P_i^2, \ \ldots, \ P_i^r & (r \text{ strings of length } \ell) \\[4pt]
-\ P_i^1 P_i^2 \ldots P_i^r P_i^* & (\text{one string of length } r\ell + \ell - s = h\} \\[4pt]
-\ P_i^* \mathbb{L}_a & (\text{one string of length } \ell - s + s = \ell) \\[4pt]
-\ \mathbb{R}_a Q_i, \ Q_i \mathbb{L}_b, \ \mathbb{R}_b S_i, \ S_i \mathbb{L}_c & (\text{four strings of length } s + \ell - s = \ell) \\[4pt]
-\ \mathbb{R}_c T_i^* & (\text{one string of length } s + \ell - s = \ell) \\[4pt]
-\ T_i^* T_i^1 T_i^2 \ldots T_i^r & (\text{one string of length } r\ell + \ell - s = h\} \\[4pt]
-\ T_i^1, \ T_i^2, \ \ldots, \ T_i^r & (r \text{ strings of length } \ell)
\end{array}
$$

Then, for each clause C_a, add the following strings to $\mathcal{W}_2$:

$$
\begin{array}{ll}
-\ \mathbb{L}_a \mathbb{M}_a, \ \mathbb{M}_a' \mathbb{R}_a & (\text{two strings of length } h) \\[4pt]
-\ \mathbb{M}_a \mathbb{R}_a, \ \mathbb{L}_a \mathbb{M}_a', \ \mathbb{M}_a \mathbb{R}_a, \ \mathbb{L}_a \mathbb{M}_a' & (\text{four strings of length } h)
\end{array}
$$

Note that the latter includes the same string $\mathbb{M}_a \mathbb{R}_a$ twice, and the same string $\mathbb{L}_a \mathbb{M}_a'$ twice. However, $\mathbb{L}_a \mathbb{M}_a$ and $\mathbb{M}_a' \mathbb{R}_a$ is included once.

This concludes the construction of $\mathcal{W}_1$ and $\mathcal{W}_2$. Note that because ℓ and h are constants, we can easily build $\mathcal{W}_1$ and $\mathcal{W}_2$ in polynomial time. Let us consider the maximum number of occurrences of a symbol. Note that for each $i \in [n]$, each variable substring $P_i^j, P_i^*, Q_i, S_i, T_i^j, T_i^*$ occurs only twice in $\mathcal{W}_1$, namely in X_i and X_i'. One can also see from the construction of $\mathcal{W}_2$ that they occur twice. As for the clause substrings $\mathbb{L}_a, \mathbb{M}_a, \mathbb{M}_a'$ and $\mathbb{R}_a$, we note that $\mathbb{L}_a$ and $\mathbb{R}_a$ occur six times in $\mathcal{W}_1$, twice for each variable that C_a contains. Also, each of $\mathbb{M}_a$ and $\mathbb{M}_a'$ occurs three times. The same holds for $\mathcal{W}_2$. It follows that the maximum number of occurrences of a symbol is six.

It might be useful to describe how the reasoning before proceeding further. As we will show, for each x_i, there are only two ways to partition X_i and X_i'. Putting $x_i = true$ will correspond to the partition in Fig. 2:

$$X_i = \underline{P_i^1 \, P_i^2} \, \cdots \, \underline{P_i^r \, P_i^*} \, \underline{\mathbb{L}_a \, \mathbb{M}_a} \, \mathbb{R}_a \, \underline{Q_i \, \mathbb{L}_b} \, \underline{\mathbb{M}_b \, \mathbb{R}_b \, S_i} \, \underline{\mathbb{L}_c \, \mathbb{M}_c} \, \mathbb{R}_c \, \underline{T_i^* \, T_i^1} \, \underline{T_i^2} \, \cdots \, T_i^r$$

$$X_i' = \underline{P_i^1 \, P_i^2} \, \cdots \, \underline{P_i^r \, P_i^*} \, \mathbb{L}_a \, \underline{\mathbb{M}_a' \, \mathbb{R}_a} \, \underline{Q_i \, \mathbb{L}_b} \, \underline{\mathbb{M}_b' \, \mathbb{R}_b} \, \underline{S_i \, \mathbb{L}_c} \, \underline{\mathbb{M}_c' \, \mathbb{R}_c} \, \underline{T_i^* \, T_i^1} \, \underline{T_i^2} \, \cdots \, T_i^r$$

Fig. 2. Partition of X_i and X_i' corresponding to $x_i = true$.

where a double-underline indicates a substring of length h, and a single underline a substring of length ℓ. The $\mathbb{L}_a \mathbb{M}_a$ and $\mathbb{M}'_a \mathbb{R}_a$ substrings are included as blocks here, which correspond to choosing x_i to satisfy clause C_a. The same holds for C_b and C_c. Since $\mathbb{L}_a \mathbb{M}_a$ and $\mathbb{M}'_a \mathbb{R}_a$ are only present once in $\mathcal{W}_2$, no other x_j can choose these substrings again. Putting $x_i = false$ corresponds to the partition in Fig. 3.

$$X_i = \underline{P_i^1}\ \underline{P_i^2}\ \cdots\ \underline{P_i^r}\ \underline{P_i^*}\ \underline{\mathbb{L}_a\ \mathbb{M}_a\ \mathbb{R}_a}\ \underline{Q_i}\ \underline{L_b}\ \underline{\mathbb{M}_b\ \mathbb{R}_b}\ \underline{S_i}\ \underline{L_c}\ \underline{\mathbb{M}_c\ \mathbb{R}_c}\ \underline{T_i^*}\ \underline{T_i^1}\ \underline{T_i^2}\ \cdots\ \underline{T_i^r}$$

$$X_i' = \underline{\underline{P_i^1}}\ \underline{\underline{P_i^2}}\ \cdots\ \underline{\underline{P_i^r}}\ \underline{\underline{P_i^*\ \mathbb{L}_a}}\ \underline{\underline{\mathbb{M}'_a\ \mathbb{R}_a}}\ \underline{\underline{Q_i\ L_b}}\ \underline{\underline{\mathbb{M}'_b\ \mathbb{R}_b}}\ \underline{\underline{S_i\ L_c}}\ \underline{\underline{\mathbb{M}'_c\ \mathbb{R}_c}}\ \underline{\underline{T_i^*}}\ \underline{\underline{T_i^1}}\ \underline{\underline{T_i^2}}\ \cdots\ \underline{\underline{T_i^r}}$$

Fig. 3. Partition of X_i and X_i' corresponding to $x_i = false$.

where here, the $\mathbb{M}_a \mathbb{R}_a$ and $\mathbb{L}_a \mathbb{M}'_a$ substrings are included. They correspond to not choosing x_i to satisfy clause C_a. These substrings are present twice in $\mathcal{W}_2$, and so two variables of C_a will need to choose not to satisfy it.

Let us now proceed with the details. We show that ϕ is 1-in-3 satisfiable if and only if there exists a common F-partition of $\mathcal{W}_1$ and $\mathcal{W}_2$.

($\Rightarrow$) Suppose that ϕ is 1-in-3 satisfiable by some assignment A of the x_i's. We describe how to partition the strings of $\mathcal{W}_1$. For each x_i such that $x_i = true$ under assignment A, partition X_i and X_i' as in Fig. 2. If instead $x_i = false$, partition X_i and X_i' as in Fig. 3. One can verify through manual inspection that each substring is of length either ℓ or h.

As for $\mathcal{W}_2$, none of its strings is partitioned further in our solution. Thus each substring is of length ℓ or h.

Let us argue that there is a bijection between the strings of our partition of $\mathcal{W}_1$ and the strings of $\mathcal{W}_2$. Consider a clause C_a and let x_i, x_j, x_k be its variables. Assume without loss of generality that $x_i = true$ and $x_j = x_k = false$. Hence we partitioned X_i as in Fig. 2 and X_j, X_k as in Fig. 3. Note that in $\mathcal{W}_1$, the substrings $\mathbb{L}_a \mathbb{M}_a, \mathbb{M}_a \mathbb{R}_a, \mathbb{M}_a \mathbb{R}_a, \mathbb{M}'_a \mathbb{R}_a$ occur only in $X_i, X_i', X_j, X_j', X_k, X_k'$. Moreover, $\mathbb{L}_a \mathbb{M}_a$ occurs exactly once in our partition of $\mathcal{W}_1$, namely in the partition of X_i. This is correct since $\mathbb{L}_a \mathbb{M}_a$ is also present once in $\mathcal{W}_2$. The string $\mathbb{M}_a \mathbb{R}_a$ is present exactly twice in our partition of $\mathcal{W}_1$ because of how we partition X_j and X_k, and $\mathbb{M}_a \mathbb{R}_a$ is also twice in $\mathcal{W}_2$. Similarly, $\mathbb{M}'_a \mathbb{R}_a$ is chosen once in X_i', and $\mathbb{L}_a \mathbb{M}'_a$ is chosen twice, namely in X_j' and X_k'. These correspond to the number of times these substrings occur in $\mathcal{W}_2$, and we may put those in a one-to-one correspondence.

Consider now the strings of $\mathcal{W}_2$ added for a variable x_i. If $x_i = true$, then $P_i^1 \ldots P_i^r P_i^*, \mathbb{R}_a Q_i, \mathbb{R}_b S_i, \mathbb{R}_c T_i^*$ and the T_i^j strings are covered by X_i. Moreover, the $P_i^j, P_i^* L_a, Q_i L_b, S_i L_c$ and $T_i^* T_i^1 \ldots T_i^r$ strings are covered by X_i'. These are distinct for each X_i and can be put in a one-to-one correspondence. If $x_i = false$, one can easily check that the same strings are covered, but with the roles of X_i and X_i' reversed. Therefore, a common partition exists.

($\Leftarrow$) Suppose that $\mathcal{W}_1$ and $\mathcal{W}_2$ admit a common partition. We first argue that none of the strings of $\mathcal{W}_2$ is partitioned into smaller substrings. Since ℓ is the minimum of F, none of its strings of length ℓ can be partitioned. Consider a string of length h of $\mathcal{W}_2$. Recall that by assumption h is the second smallest value of F. Thus, it cannot be partitioned further either, since the only way would be to split it into substrings of length ℓ, which is not possible since ℓ is not a divisor of h.

Now, consider a string X_i of $\mathcal{W}_1$ and let C_a, C_b, C_c be the clauses that x_i occurs in. We argue that X_i can only be split in two ways, those shown in Figs. 2 and 3. To see this, let $W_1, W_2, \ldots, W_k$ be the blocks in the partition of X_i, such that the concatenation $W_1 W_2 \ldots W_k$ is X_i. Since all strings of $\mathcal{W}_2$ are of length ℓ or h, there are only two options for W_1: it must either be the prefix of length h or the prefix of length ℓ of X_i. That is, W_1 must be $P_i^1 \ldots P_i^r P_i^*$, or it must be P_i^1.

First assume that $W_1 = P_i^1 \ldots P_i^r P_i^*$. It is easily seen that the partition of Fig. 2 is forced from that point. This is because in X_i, W_1 is followed by $\mathbb{L}_a$, and the only strings of $\mathcal{W}_2$ that starts with the first symbol of $\mathbb{L}_a$ are $\mathbb{L}_a \mathbb{M}_a$ and $\mathbb{L}_a \mathbb{M}'_a$, so here we are forced to include $\mathbb{L}_a \mathbb{M}_a$ from X_i. Then, the only substring of X_i that starts with a prefix of $\mathbb{R}_a$ and that is in $\mathcal{W}_2$ is $\mathbb{R}_a Q_i$. Then, in a similar fashion, we are forced to choose $\mathbb{L}_b \mathbb{M}_b$ then $\mathbb{R}_b S_i$, then $\mathbb{L}_c \mathbb{M}_c$ and then $\mathbb{R}_c T_i^*$, followed by the T_i^j's individually. Thus X_i is partitioned as in Fig. 2. Now consider X_i'. The partition of $\mathcal{W}_1$ must include $P_i^1, \ldots, P_i^r$, $P_i^* \mathbb{L}_a$, $Q_i \mathbb{L}_b$, $S_i \mathbb{L}_c$, and $T_i^* T_i^1 \ldots, T_i^r$. All of these include symbols that are unique to X_i and X_i', and since they were not included in the X_i partition above, they must be substrings of the X_i' partition. This implies that X_i' must be partitioned as in Fig. 2.

The other possibility is that $W_1 = P_i^1$. In this case, the first string of X_i' in the partition must be $P_i^1 \ldots P_i^r P_i^*$. We can apply the arguments from above symmetrically by interchanging the roles of X_i and X_i' to deduce that in this case, X_i and X_i' are partitioned as in Fig. 3.

Now assign $x_i = true$ if X_i and X_i' are partitioned as in Fig. 2, and assign $x_i = false$ if instead X_i and X_i' are partitioned as in Figure 3. Let C_a be a clause with variables x_i, x_j, x_k. Since the $\mathbb{L}_a \mathbb{M}_a$ and $\mathbb{M}'_a \mathbb{R}_a$ strings are in $\mathcal{W}_2$ exactly once, it follows that exactly one of X_i, X_j or X_k is partitioned as in Fig. 2. This means that exactly one of x_i, x_j or x_k is $true$. Thus ϕ is 1-in-3-satisfiable.   □

The above reduction made heavy use of the fact that we allow the input to be sets of strings instead of individual strings. The problem might not be NP-hard if we receive only two input strings. Unfortunately, we show that it does not make the problem easier.

Theorem 2. *Let F be any set of integers in which at least one element is not a multiple of $\min F$. The XSR-F problem is NP-hard, even when restricted to instances in which $\mathcal{W}_1, \mathcal{W}_2$ both contain only one string, and even if each symbol occurs at most 6 times.*

Proof. We reduce from the XSR-F problem in which the input sets may have more than one string but occurrences at most 6, which is NP-hard by Theorem 1. Let $\mathcal{W}_1, \mathcal{W}_2$ be an instance of this problem. Denote $\mathcal{W}_1 = \{A_1, \ldots, A_n\}$ and $\mathcal{W}_2 = \{B_1, \ldots, B_m\}$, and assume without loss of generality that $n \geq m$. Let ℓ be the minimum of F. Let $\Lambda = \{X_1, Y_1, Z_1, X_2, Y_2, Z_2, \ldots, X_n, Y_n, Z_n\}$ be new permutation strings, each of length ℓ and each containing symbols that occur in no other string. Form the new instance of XSR-F with string sets $\mathcal{W}'_1 = \mathcal{W}_1 \cup \Lambda$ and $\mathcal{W}'_2 = \mathcal{W}_2 \cup \Lambda$. It is clear that $\mathcal{W}_1, \mathcal{W}_2$ is a Yes-instance if and only if $\mathcal{W}'_1, \mathcal{W}'_2$ is a Yes-instance, since the Λ strings must be matched. We shall therefore use the $\mathcal{W}'_1, \mathcal{W}'_2$ instance for our reduction.

Now build single strings S and T as follows:

$$S = A_1 \; X_1 \; Y_1 \; Z_1 \; A_2 \; X_2 \; Y_2 \; Z_2 \; \ldots \; A_n \; X_n \; Y_n \; Z_n$$

$$T = B_1 \; Z_1 \; Y_1 \; X_1 \; B_2 \; Z_2 \; Y_2 \; X_2 \; \ldots \; B_m \; Z_m \; Y_m \; X_m \; Z_{m+1} \; Y_{m+1} \; X_{m+1} \; \ldots \; Z_n \; Y_n \; X_n$$

Note that the order of the Λ triples is flipped from S to T, i.e. $X_i Y_i Z_i$ occurs in S but $Z_i Y_i X_i$ occurs in T. Also note that A_n and B_m are visually aligned above, which is slightly misleading since n is possibly much larger than m. We now show that $\mathcal{W}'_1, \mathcal{W}'_2$ admit a common F-partition if and only if S, T admit a common F-partition.

Suppose that $\mathcal{W}'_1, \mathcal{W}'_2$ admit a common F-partition. Observe that S is a concatenation of the elements of $\mathcal{W}'_1$ and T of the elements of $\mathcal{W}'_2$. For this reason, this partition is also a common F-partition of S and T.

Conversely, assume that S, T admit a common F-partition. Consider some X_i substring. Notice that X_i is preceded and succeeded by a different symbol in S and T (i.e. preceded by some symbol of A_i in S and preceded by Y_i in T, and succeeded by Y_i in S and succeeded by some B_i symbol or by Z_{i+1} in T). Because of this, and because X_i has length ℓ, any common F-partition must contain X_i as a block. By the same reasoning, each Y_i and Z_i substring must also be a block. It follows that the other blocks must form a partition of $\{A_1, \ldots, A_n\}$ and $\{B_1, \ldots, B_m\}$, and thus there is a common F-partition of $\mathcal{W}'_1$ and $\mathcal{W}'_2$. $\square$

4   GSR-F in Polynomial Time for Fixed F and Alphabet

We now return to the optimization version GSR-F, where our goal is to delete a minimum number of characters from the input strings so that they admit a common F-partition. Note that this problem is NP-hard, by the previous section. However, the hardness reduction requires strings with alphabets of arbitrary size. We show that if Σ is a fixed alphabet, and if F is a fixed set of integers, then GSR-F can be solved in polynomial time. More specifically, we show that GSR-F can be solved in time $O(n^{|F||\Sigma|^k+3})$, where $k = \max(F)$.

Note that this can be used to solve the XSR-F problem under the same conditions. Moreover, owing to Proposition 1, requiring F to be finite is not a strong constraint for XSR-F, since most infinite sets of interests reduce to the finite case (including $F = \{2, 3, \ldots\}$, which can be replaced with $F = \{2, 3\}$).

Let B be the set of strings on alphabet Σ whose length is in F. Note that $|B| = \sum_{c \in F} |\Sigma|^c \leq \sum_{c=1}^{k} |\Sigma|^c \leq |F||\Sigma|^k$, which is a constant. Moreover, in any F-partition, each block must be a string of B.

To obtain a polynomial time algorithm, it suffices to compute the set of possible ways to split the input into blocks that are in B, and to check whether there is a common split between the input sets of strings. This is not too hard to achieve using dynamic programming, as we describe.

A *block count table* is a map $C : B \to \mathbb{N}$ that assigns to each $X \in B$ an entry $C(X)$ that, for our purposes, represents a number of desired occurrences for X in a common F-partition. For a set of strings $\mathcal{W}$ and a block count table C, we say that $\mathcal{W}$ *is C-splittable* if there is an F-partition of $\mathcal{W}$ in which each string $X \in B$ occurs exactly $C(X)$ times.

In our dynamic programming, we order the strings of $\mathcal{W}_1$ and $\mathcal{W}_2$ arbitrarily, and process each string in order one marker at a time, from left to right. We will rely on the table described as follows.

Definition 1. *Let $\mathcal{W}$ be a set of strings and let C be a block count table. Then we denote by $D(\mathcal{W}, C)$ the minimum number of markers to delete from the strings in $\mathcal{W}$ so that the resulting strings are C-splittable (with $D(\mathcal{W}, C) = \infty$ if this is not possible).*

Our goal is to find $\min_C(D(\mathcal{W}_1, C) + D(\mathcal{W}_2, C))$. Note that this is always less than infinity, since in the worst case we can take C with all zeroes, i.e. delete everything from $\mathcal{W}_1$ and $\mathcal{W}_2$. However, the number of C's to consider is possibly infinite. We will bound it to polynomial size later.

We need to add a few definitions. For two strings X and Y, we write $X \prec Y$ if X is a subsequence of Y. Furthermore, for a block count table C and $X \in B$, let $C - X$ denote the block count table in which $C(X)$ is reduced by one. Precisely $C - X = C'$, where $C'(X) = C(X) - 1$ and $C'(Y) = C(Y)$ for all $Y \in B \setminus \{X\}$. Note that $C'(X) = -1$ is possible under this definition. By a slight abuse of notation, we still call C' a block count table, and allow negative values (observing that in this case, no string is C'-splittable).

We can now define a recurrence, stated in Lemma 2, for $D(\mathcal{W}, C)$ where $\mathcal{W}$ is a general set of strings. Aside from a few trivial cases, the main idea is that to find a partition of $\mathcal{W}$ with counts C, we can fix some $W_h \in \mathcal{W}$ and focus on the last block of W_h in this partition. For each suffix $W_h[j + 1..|W_h|]$ and each possible $X \in B$ that is a subsequence of this suffix, we count $|W_h| - j - |X|$ deletions from that suffix to obtain X. We then remove this suffix, resulting in a set of strings of shorter total length. We can then use a recurrence on this shorter set of strings, but with the target count table $C - X$.

Lemma 2. *Let $\mathcal{W} = \{W_1, \dots, W_h\}$ be a set of strings and let C be a block count table. Then the following holds:*

1. *if $C(X) = 0$ for all $X \in B$, then $D(\mathcal{W}, C) = \sum_{i=1}^{h} |W_i|$;*
2. *otherwise, if $C(X) < 0$ for some X, then $D(\mathcal{W}, C) = \infty$;*
3. *otherwise, if $\mathcal{W} = \emptyset$, then $D(\mathcal{W}, C) = \infty$;*
4. *otherwise, if $|W_h| < \min(F)$, then $D(\mathcal{W}, C) = D(\mathcal{W} \setminus \{W_h\}, C) + |W_h|$ (note that this includes the case where W_h is the empty string);*
5. *otherwise, for $j \in \{0, 1, \dots, |W_h| - \min(F)\}$, define $\mathcal{W}^{(j)} = (\mathcal{W} \setminus \{W_h\}) \cup \{W_h[1..j]\}$. Then $D(\mathcal{W}, C)$ is the minimum between $D(\mathcal{W} \setminus \{W_h\}, C) + |W_h|$ and the expression*

$$\min_{0 \le j \le |W_h| - \min(F)} \left(\min_{X \in B : X \prec W_h[j+1..|W_h|]} \left(D(\mathcal{W}^{(j)}, C - X) + |W_h| - j - |X| \right) \right)$$

Proof. We begin with the first four cases. If $C(X) = 0$ for each X, then the only way to achieve a count table with all zeroes is to delete every marker from every $\mathcal{W}$ string, which justifies Case 1. Otherwise, if $C(X) < 0$ for some X, then this is impossible to achieve, which justifies Case 2. At this point, we may assume that $C(X) \geq 0$ for all X, and that $C(X) > 0$ for at least one $X \in B$. If $\mathcal{W} = \emptyset$, then we must achieve count table C with an empty set of strings, but this is not possible for those X with $C(X) > 0$, and so $D(\mathcal{W}, C) = \infty$ as in Case 3. If $|W_h| < \min(F)$, we must delete all of its markers since it cannot contribute to a block, which justifies Case 4.

We now proceed which Case 5. Let W_h and $\mathcal{W}^{(j)}$ be defined as in the statement. We shall refer to the longer, double-minimization expression of Case 5 as *the long expression*. First note that the long expression is well-defined, i.e. the two minimizations are not over empty sets. Indeed, at this point we may assume that $|W_h| \geq \min(F)$, which means that for any $j \leq |W_h| - \min(F)$, there is some $X \in B$ that is a subsequence of $W_h[j + 1..|W_h|]$.

Now, let $j^* \in \{0, 1, \ldots, |W_h| - \min(F)\}$ and $X^* \in B$, with $X^* \prec W_h[j^* + 1..|W_h|]$, such that j^* and X^* minimize the long expression. Let us define

$$d = \min \begin{cases} D(\mathcal{W} \setminus \{W_h\}, C) + |W_h| \\ D(\mathcal{W}^{(j^*)}, C - X^*) + |W_h| - j^* - |X^*| \end{cases}$$

which is the value chosen in Case 5. Let M be a minimum-size set of markers to delete from $\mathcal{W}$ so that after deleting these markers, we obtain a C-splittable set of string. By definition, we have $|M| = D(\mathcal{W}, C)$, and we need to prove that $|M|$ is both an upper bound and a lower bound on d.

Let us start by showing that $|M| \leq d$. If $d = \infty$, then this is clear. Otherwise, we can find a set of d markers to delete from $\mathcal{W}$ to get a C-splittable string as follows. First if $d = D(\mathcal{W} \setminus \{W_h\}, C) + |W_h|$, we delete all markers from W_h, plus $D(\mathcal{W} \setminus \{W_h\}, C)$ markers from the other strings to obtain a C-splittable set of strings. Second if $d = D(\mathcal{W}^{(j^*)}, C - X^*) + |W_h| - j^* - |X^*|$, we delete $|W_h| - j^* - |X^*|$ markers from $W_h[j^* + 1..|W_h|]$ to obtain X^*, thereby using the suffix of W_h to make the X^* block (which is in B). Then, we delete $D(\mathcal{W}^{(j^*)}, C - X^*)$ markers in $\mathcal{W}^{(j^*)}$ to obtain a $(C - X^*)$-splittable string. The result is a C-splittable string obtained by the deletion of d markers. In either case, the optimality of M implies that $|M| \leq d$.

We next show that $|M| \geq d$. Let $\mathcal{W}'$ be the set of strings obtained after deleting each marker of M from $\mathcal{W}$. Consider an F-partition P' of $\mathcal{W}'$ that attains count table C. One possibility is that all markers of W_h get deleted by M. If this is the case, P' is obtained only with $\mathcal{W} \setminus \{W_h\}$ and obviously, $|M| = |W_h| + D(\mathcal{W} \setminus \{W_h\}, C) \geq d$.

We thus assume that W_h contributes to at least one block. Then, there is a suffix $W_h[j' + 1..|W_h|]$ of W_h of length at least $\min(F)$ such that some markers of M are deleted from that suffix, resulting in a block X' that is part of P'. Note that j' is between 0 and $|W_h| - \min(F)$, and that $|W_h| - j' - |X'|$ markers must be deleted from that suffix. Also, this means that the block count table

$C - X'$ is attained by deleting markers from $(\mathcal{W} \setminus \{W_h\}) \cup \{W_h[1..j']\} = \mathcal{W}^{(j')}$. Thus $D(\mathcal{W}, C) = |M| = D(\mathcal{W}^{(j')}, C - X') + |W_h| - j' - |X'|$. The value of d is smaller of equal to that value, since d is obtained by minimizing over all possible values of j and X, including j' and X'. Therefore, $|M| \geq d$. This concludes the proof.   □

It only remains to describe the computational aspects of our approach. The recurrence needs to be computed on both $\mathcal{W}_1$ and $\mathcal{W}_2$. Let $n = \max(\sum_{W \in \mathcal{W}_1} |W|, \sum_{W \in \mathcal{W}_2} |W|)$. Notice that if there is a common F-partition P of $\mathcal{W}_1$ and $\mathcal{W}_2$, then no string of B occurs more than n times in P. Let $\mathbb{C}(n)$ be the set of all possible count tables whose values are all at most n. That is, $C \in \mathbb{C}(n)$ if and only if $C(X) \leq n$ for every $X \in B$. Then $|\mathbb{C}(n)| = (n+1)^{|B|} \in \Theta(n^{|B|})$, which is polynomial in n since $|B|$ is constant[1].

To solve the strip recovery problem for $\mathcal{W}_1$ and $\mathcal{W}_2$, we must find C such that $\mathcal{W}_1$ and $\mathcal{W}_2$ are both C-splittable after a minimum number of deletions. It therefore suffices to compute

$$\min_{C \in \mathbb{C}(n)} (D(\mathcal{W}_1, C) + D(\mathcal{W}_2, C))$$

Also, note that this idea generalizes trivially to the case where more than two sets of strings are given. The main algorithm is given below.

```
1  function getMinStripRecovery(W₁, W₂)
2      computeD(W₁)
3      computeD(W₂)
4      return min_{C∈ℂ(n)}(D(W₁, C) + D(W₂, C))
5  function computeD(Wₐ = {W₁, ..., Wₗ})
6      Let D be a map, initially empty
7      for h = 1..l do
8          for i = 1..|Wₕ| do
9              Let W = {W₁, W₂, ..., Wₕ₋₁, Wₕ[1..i]}
10             for C ∈ ℂ(n) do
11                 │ Compute D(W, C) using Lemma 2
12             end
13         end
14     end
```

Algorithm 1: Pseudocode for the computation of the dynamic programming.

Theorem 3. *Algorithm 1 solves the GSR-F problem in time* $O(n^{|F|}|\Sigma|^{\max(F)+3})$

---

[1] Let us note that in practice, the same block can't actually occur more than $n/\min(F)$ times, where here n can be the minimum total length of $\mathcal{W}_1$ or $\mathcal{W}_2$. Hence, the smaller set $\mathbb{C}(n/\min(F))$ should be considered in practice, but this does not help the complexity analysis.

Proof. The correctness of the algorithm is easy to see if we assume that *computeD* computes the correct D values. To see that this is the case, notice that it computes $D(\mathcal{W}, C)$ in increasing order of number of strings and increasing order of the prefix of the last string. Focus on one particular iteration with some $h \in [l]$ and $\mathcal{W} = \{W_1, \ldots, W_{h-1}, W_h[1..i]\}$. As can be seen from the recurrence of Lemma 2, the computation of $D(\mathcal{W}, C)$ requires either access to $D(\mathcal{W} \setminus \{W_h\}, C)$, which has one less string, or requires some $D(\mathcal{W}^{(j)}, C - X)$, where $\mathcal{W}^{(j)}$ has the same number of strings as $\mathcal{W}$ but whose last string is shorter. In either case, the order of computation lets us assume that all values required for a specific $D(\mathcal{W}, C)$ value have been calculated on previous iterations. It follows that the algorithm is correct.

Let us now look at the time complexity, starting with the time spent in $computeD(\mathcal{W}_1)$. One can see from the structure of the loops that this procedure iterates over $n \cdot |\mathbb{C}(n)|$ combinations of $\mathcal{W}$ and C. We can thus focus on the time required to calculate one $D(\mathcal{W}, C)$ entry. Note that Case 5 is the worst case of the recurrence. It needs to enumerate $O(n)$ values of j and $|B|$ values of X. For each such j and X combination, we must spend time $O(n)$ to check $X \prec W_h[j+1..|W_h|]$. Thus, one $D(\mathcal{W}, C)$ computation takes time $O(n^2 \cdot |B|) = O(n^2)$. The total time spent in $computeD(\mathcal{W}_1)$ is therefore $O(n^3 \cdot |\mathbb{C}(n)|) = O(n^{|B|+3})$. Similarly, the time spent in $computeD(\mathcal{W}_2)$ is $O(n^{|B|+3})$. Finding the best C after these computations does not add complexity, and so the running time is $O(n^{|B|+3}) = O(n^{|F||\Sigma|^{\max(F)}+3})$. $\square$

5   Fixed Alphabet with Unbounded F is NP-Hard

So far, we have assumed that F was a fixed set of integers. We now show that if F is part of the input, then XSR-F is NP-hard, even if the alphabet is constant. This shows that unless P = NP, polynomial-time feasibility requires F and Σ to be fixed.

We use a reduction from 3-Partition which is known to be strongly NP-complete [18]. In 3-Partition, we are given an integer D and a set of integers $S = \{a_i | i = 1..3m\}$ with $D/4 < a_i < D/2$ for all i, and the question is whether we can partition S into m subsets, each with three integers such that the sum of each subset is exactly D. Since 3-Partition is strongly NP-complete, each $a_i \in S$ is bounded by a polynomial in m.

For our XSR-F instance, we construct the input strings A, B and the input set F as follows. The alphabet of A and B is $\Sigma = \{0, 1, \#, \$\}$, and we put $F = \{a_1, a_2, ..., a_{3m}, D+1\}$. The strings A and B are defined as

$$A = \overbrace{\$...\$}^{D+1} \overbrace{1...1}^{a_1} \overbrace{\$...\$}^{D+1} \overbrace{1...1}^{a_2} \cdots \overbrace{\$...\$}^{D+1} \overbrace{1..1}^{a_{3m}} \overbrace{\#...\#}^{D+1} \overbrace{0...0}^{m(D+1)} \ ,$$

$$B = \overbrace{\#...\#}^{D+1} \overbrace{\$...\$}^{3m(D+1)} \overbrace{(0...0}^{D+1} \overbrace{1...1)^m}^{D} .$$

The idea is that because F only contains the a_is and $D + 1$, any common F-partition must have the 1...1 blocks of A. To handle the 1...1 blocks of B, there must be a way to partition them into groups of three that achieve length D. We now proceed with the details.

Theorem 4. *The XSR problem is NP-hard, even if the input consists of two strings on an alphabet of size four.*

Proof. First note that the above construction can easily be carried out in polynomial time, since the values of the a_i's (and therefore of D) are polynomial. We show that a 3-Partition instance (S, D) admits a 3-Partition if and only if the constructed A and B strings admit a common F-partition.

First assume that (S, D) admit a 3-Partition $\{S_1, \ldots, S_m\}$. Create a common F-partition M of A and B as follows. In M, we add $3m$ copies of \$...\$ of length $D + 1$, one copy of #...# of length $D + 1$, m copies of 0...0 of length $D + 1$, and for each $a_i \in S$, we add a 1...1 substring of length a_i. It is clear that each string of M has its length in F, and it remains to argue that they can be concatenated to obtain A or B. For A, this is easy to see, as M is partitioned as the braces indicate in the above Figure (plus the m copies of 0...0 that result in the last 0 block of A). For B, the #-block and \$-blocks are easily covered. We must form m copies of 0...01...1 blocks. This can be done by taking, for each $S_t = \{a_i, a_j, a_k\}$ in the 3-Partition, one 0...0 block of length $D + 1$ followed by the 1...1 blocks of length a_i, a_j, and a_k. Since these sum to D, this can be used to create the desired m blocks of the form 0...01...1.

Conversely, assume that A and B admit a common F-partition M. Note that the unique #-block in A is adjacent to the unique 0-block and to the last 1-block, but in B the #-block is not adjacent to a 0 or a 1. This means that in M, the substrings that contain a # can only contain # symbols. Also, the \$-blocks in A are adjacent to the 1-blocks in A but no \$ is adjacent to a 1 in B, and so the substrings of M with a \$ only contain \$ symbols. Moreover, the 0-block in A is only adjacent to the #-block but not in B, and thus each substring of M with a 0 only contains 0s. It follows that the substrings of M that contain a 1 only contain 1s. Now, consider the 1...1 block of A of length a_i. Because $D/4 < a_i < D/2$, that block must be in M (otherwise, that block would be obtained by concatenating two or more blocks of 1s from M with a smaller length in F, which is not possible because of the bounds on a_i). Thus, M contains a 1-block of length a_i, for each $a_i \in S$. It is then easy to see that each of the 1-blocks in B must be partitioned into 3 blocks, of length a_i, a_j and a_k respectively such that $a_i + a_j + a_k = D$. This corresponds to a solution for the 3-Partition instance.   □

6   Conclusion

We have provided several negative results on the complexity of the MSR problem in the presence of duplicated genes. Notably, it is hard to even partition two genomes into blocks of sizes of any fixed list, unless we restrict ourselves to one size. Limiting the number of duplicates does not seem to help, although the problem is still open if less than 6 copies per gene are allowed. Our polynomial-time algorithm shows that there may be hope for cases where block sizes are fixed and the number of distinct genes is bounded, but the algorithm presented here needs improvement to be usable in practice. In fact, heuristics are currently used to partition genomes, and our results lead us to believe that it will be difficult to aim for practical algorithms that also offer theoretical guarantees. This is not the end of the road though, as future work may focus on approximation algorithms with a bounded ratio on the number of deleted genes, or on novel parameters not considered here that can be used for efficient parameterized algorithms.

References

1. Ramírez Alfonsín, J.L.: The Diophantine Frobenius Problem. OUP, Oxford (2005)
2. Anselmetti, Y., Berry, V., Chauve, C., Chateau, A., Tannier, E., Bérard, S.: Ancestral gene synteny reconstruction improves extant species scaffolding. BMC Genomics **16**(10), 1–13 (2015)
3. Bourque, G., Pevzner, P.A., Tesler. G.: Reconstructing the genomic architecture of ancestral mammals: lessons from human, mouse, and rat genomes. Genome Res. **14**(4):507–516 (2004)
4. Bulteau, L., Fertin, G., Jiang, M., Rusu, I.: Tractability and approximability of maximal strip recovery. Theor. Comput. Sci. **440–441**, 14–28 (2012). https://doi.org/10.1016/j.tcs.2012.04.034
5. Moret, B.M.E.: Extending the reach of phylogenetic inference. In: Darling, A., Stoye, J. (eds.) WABI 2013. LNCS, vol. 8126, pp. 1–2. Springer, Heidelberg (2013). https://doi.org/10.1007/978-3-642-40453-5_1
6. Bulteau, L., Fertin, G., Rusu, I.: Maximal strip recovery problem with gaps: hardness and approximation algorithms. J. Discrete Algorithms **19**, 1–22 (2013). https://doi.org/10.1016/j.jda.2012.12.006
7. Bulteau, L., Komusiewicz, C.: Minimum common string partition parameterized by partition size is fixed-parameter tractable. In: Proceedings of 25th ACM-SIAM Symposium on Discrete Algorithms, SODA2014, pp. 102–121. SIAM (2014)
8. Bulteau, L., Weller, M.: Parameterized algorithms in bioinformatics: an overview. Algorithms **12**(12), 256 (2019). https://doi.org/10.3390/a12120256
9. Chen, X., Zheng, J., Zheng, F., Nan, P., Zhong, Y., Lonardi, S., Jiang, T.: Assignment of orthologous genes via genome rearrangement. IEEE/ACM Trans. Computational Biology and Bioinformatics **2**(4), 302–315 (2005). https://doi.org/10.1109/TCBB.2005.48. https://doi.org/10.1109/TCBB.2005.48
10. Chen, Z., Fu, B., Jiang, M., Zhu, B.: On recovering syntenic blocks from comparative maps. J. Comb. Optim. **18**(3):307–318 (2009). https://doi.org/10.1007/s10878-009-9233-x
11. Choi, V., Zheng, C., Zhu, Q., Sankoff, D.: Algorithms for the extraction of synteny blocks from comparative maps. In: Giancarlo, R., Hannenhalli, S. (eds.) WABI 2007. LNCS, vol. 4645, pp. 277–288. Springer, Heidelberg (2007). https://doi.org/10.1007/978-3-540-74126-8_26
12. Chrobak, M., Kolman, P., Sgall, J.: The greedy algorithm for the minimum common string partition problem. ACM Trans. Algorithms **1**(2), 350–366 (2005). https://doi.org/10.1145/1103963.1103971

13. Damaschke, P.: Minimum common string partition parameterized. In: Crandall, K.A., Lagergren, J. (eds.) WABI 2008. LNCS, vol. 5251, pp. 87–98. Springer, Heidelberg (2008). https://doi.org/10.1007/978-3-540-87361-7_8

14. Delabre, M., et al.: Evolution through segmental duplications and losses: a super-reconciliation approach. Algorithms Mol. Biol. **15**, 1–15 (2020)

15. Drillon, G., Champeimont, R., Oteri, F., Fischer, G., Carbone, A.: Phylogenetic reconstruction based on synteny block and gene adjacencies. Mol. biol. Evol. **37**(9), 2747–2762 (2020)

16. Eichler, E.E., Sankoff, D.: Structural dynamics of eukaryotic chromosome evolution. Science **301**(5634), 793–797 (2003)

17. Ganczorz, M., Gawrychowski, P., Jez, A., Kociumaka, T.: Edit distance with block operations. In: Proceedings of ESA'2018, LIPIcs, vol. 112, pp. 33:1–33:14. Schloss Dagstuhl-Leibniz-Zentrum fuer Informatik (2018)

18. Garey, M.R., Johnson, D.S.: Computers and Intractability, vol. 174. Freeman, San Francisco (1979)

19. Goldstein, A., Kolman, P., Zheng. J.: Minimum common string partition problem: hardness and approximations. Eur. J. Comb. **12** (2005)

20. Goodstadt, L., Ponting, C.P.: Phylogenetic reconstruction of orthology, paralogy, and conserved synteny for dog and human. PLoS Comput. Biol. **2**(9), e133 (2006)

21. Hu, S., Li, W., Wang, J.: An improved kernel for the complementary maximal strip recovery problem. In: Xu, D., Du, D., Du, D. (eds.) COCOON 2015. LNCS, vol. 9198, pp. 601–608. Springer, Cham (2015). https://doi.org/10.1007/978-3-319-21398-9_47

22. Jiang, H., Guo, J., Zhu, D., Zhu, B.: A 2-approximation algorithm for the complementary maximal strip recovery problem. In: Pisanti, N., Pissis, S.P. (eds.) 30th Annual Symposium on Combinatorial Pattern Matching, CPM 2019, 18–20 June 2019, Pisa, Italy, vol. 128 LIPIcs, pp. 5:1–5:13. Schloss Dagstuhl - Leibniz-Zentrum für Informatik (2019). https://doi.org/10.4230/LIPIcs.CPM.2019.5

23. Jiang, H., Li, Z., Lin, G., Wang, L., Zhu, B.: Exact and approximation algorithms for the complementary maximal strip recovery problem. J. Comb. Optim. **23**(4), 493–506 (2012). https://doi.org/10.1007/s10878-010-9366-y

24. Jiang, H., Zhu, B.: A linear kernel for the complementary maximal strip recovery problem. J. Comput. Syst. Sci. **80**(7), 1350–1358 (2014). https://doi.org/10.1016/j.jcss.2014.03.005

25. Jiang, H., Zhu, B., Zhu, D., Zhu, H.: Minimum common string partition revisited. J. Comb. Optim. **23**(4), 519–527 (2012). https://doi.org/10.1007/s10878-010-9370-2

26. Jiang, M.: Inapproximability of maximal strip recovery. Theor. Comput. Sci., **412**(29), 3759–3774 (2011). https://doi.org/10.1016/j.tcs.2011.04.021

27. Lafond, M., Semeria, M., Swenson, K.M., Tannier, E., El-Mabrouk, N.: Gene tree correction guided by orthology. BMC Bioinform. **14**, 1–9 (2013)

28. Lafond, M., Zhu, B.: Permutation-constrained common string partitions with applications. In: Lecroq, T., Touzet, H. (eds.) SPIRE 2021. LNCS, vol. 12944, pp. 47–60. Springer, Cham (2021). https://doi.org/10.1007/978-3-030-86692-1_5

29. Lechner, M., et al.: Orthology detection combining clustering and synteny for very large datasets. PLoS ONE **9**(8), e105015 (2014)

30. Lewin, M.: A bound for a solution of a linear diophantine problem. J. Lond. Math. Soc. **2**(1), 61–69 (1972)

31. Li, W., Liu, H., Wang, J., Xiang, L., Yang, Y.: An improved linear kernel for complementary maximal strip recovery: simpler and smaller. Theor. Comput. Sci., **786**, 55–66 (2019). https://doi.org/10.1016/j.tcs.2018.04.020

32. Lin, G., Goebel, R., Li, Z., Wang, L.: An improved approximation algorithm for the complementary maximal strip recovery problem. J. Comput. Syst. Sci. **78**(3), 720–730 (2012). https://doi.org/10.1016/j.jcss.2011.10.014
33. Cristopher Moore and John Michael Robson: Hard tiling problems with simple tiles. Discrete Comput. Geom. **26**(4), 573–590 (2001)
34. Wang, L., Zhu, B.: On the tractability of maximal strip recovery. J. Comput. Biol. **17**(7), 907–914 (2010). https://doi.org/10.1089/cmb.2009.0084
35. Zheng, C., Zhu, Q., Sankoff, D.: Removing noise and ambiguities from comparative maps in rearrangement analysis. IEEE ACM Trans. Comput. Biol. Bioinform. **4**(4), 515–522 (2007). https://doi.org/10.1145/1322075.1322077

Reconciliation with Segmental Duplication, Transfer, Loss and Gain

Yoann Anselmetti[1,2] , Mattéo Delabre[1] , and Nadia El-Mabrouk[1(✉)]

[1] Département d'informatique (DIRO), Université de Montréal, Montreal, Canada
mabrouk@iro.umontreal.ca
[2] Laboratoire CoBIUS, Département d'informatique, Université de Sherbrooke,
Sherbrooke, Canada

Abstract. We generalize the reconciliation approach, used for inferring the evolution of a single gene family given a species tree, to groups of co-localized genes, also called syntenies. More precisely, given a set $\mathcal{X}$ of syntenies in a set Σ of genomes, a tree T for $\mathcal{X}$ and a tree S for Σ, the problem is to find a most parsimonious history for $\mathcal{X}$ with respect to a given evolutionary model. We extend a previous model involving segmental duplications and losses, to also include segmental horizontal gene transfers (HGTs) and gene gains. We present a polynomial-time dynamic programming algorithm to solve the problem. We apply it to CRISPR-associated (Cas) gene syntenies. These genes are part of CRISPR-Cas systems, one of its members (CRISPR-Cas9) well-known as currently the most reliable and accurate molecular scissor technology for genome editing. The inferred evolutionary scenario is a plausible explanation of the diversification of this system into its different types. An implementation of the algorithm presented in this paper is available at: https://github.com/UdeM-LBIT/superrec2/releases/tag/rcg2022.

Keywords: Reconciliation · Synteny · CRISPR-Cas · Horizontal gene transfer

1 Introduction

The incongruence between the tree of a given gene family and the phylogenetic tree of the corresponding species can be explained through *reconciliation* (an embedding of the gene tree into the species tree) by the fact that genes have been subject to events changing their occurrence in genomes, typically gene duplications (D) and gene losses (L) [12]. The standard parsimony criteria used to choose among all possible reconciliations is to minimize the number of duplications (D distance) and losses (DL distance) induced by the reconciliation. This can be computed in linear time by LCA-mapping [12,31,32].

Horizontal gene transfer (HGT), largely involved in shaping bacterial gene content, has also been considered in the analysis of gene families through reconciliation. In this case, the parsimony problem consists in finding a minimum scenario of duplication, loss and transfer events (DTL distance) explaining a gene tree with respect to a given species tree. The problem of finding a most parsimonious acyclic DTL scenario has been shown NP-hard, becoming polynomial when the acyclicity requirement is relaxed [1,27].

Although used successfully for many years, one of the major drawbacks of the reconciliation model is that gene families are considered to evolve independently from one another. While some work has been done on inferring the evolution of co-localized genes (such as operons in bacteria or paralogons), also called *syntenic groups of genes* (or simply *syntenies*) [2,9], adjusting the computation of the evolutionary cost to favour co-evolution events—hence grouping individual events into single segmental ones [7]—or inferring the minimum number of "duplication episodes" defined as sets of single duplications mapped to the same node in the species tree [6,23], none of these methods explicitly seek for an evolutionary scenario minimizing segmental duplication, loss and HGT events (see a recent review [8]).

The first attempt to generalize the reconciliation approach to a set of gene trees was described in [5]. Given a set of gene families grouped into ordered syntenies (i.e., ordered groups of genes), a gene tree for each gene family and a species tree, the *DL Super-reconciliation* problem was defined as finding an evolutionary scenario for the syntenies agreeing with the individual gene trees, whilst minimizing the number of segmental duplications and losses. The problem admits a solution only in the case of "consistent" gene trees and gene orders. It was shown that the associated decision problem is NP-hard, and that a two-steps method on the synteny tree (obtained as a supertree of the gene trees), first assigning an event labeling from the LCA-mapping and then inferring ancestral syntenies and losses, leads to an optimal solution. Moreover, ignoring gene orders, a polynomial-time algorithm exists for the second step.

In this paper, we describe DTL Super-reconciliation, generalizing the model to handle HGT and gene gain events. We restrict the problem to the unordered case, where syntenies are defined as unordered groups of genes. We introduce the evolutionary model in Sect. 3 and the formal optimization problem in Sect. 4. We show in Sect. 5 that the two-steps method for solving the DL Super-reconciliation problem does not apply in this case, then present a polynomial-time dynamic programming algorithm for DTL Super-reconciliation in Sect. 6. Finally, in Sects. 7 and 8, we apply our algorithm to CRISPR-associated (Cas) gene syntenies. These genes are part of CRISPR-Cas systems, one of its members (CRISPR-Cas9) well-known as currently the most reliable and accurate molecular scissor technology for genome editing. The inferred evolutionary scenario leads to an interesting explanation of the diversification of this system into its different types, which opens the door to further investigations.

2   Preliminary Definitions

All trees are considered rooted. Given a tree T, we denote by $r(T)$ its root, $V(T)$ its node set and by $L(T) \subseteq V(T)$ its leaf set. A node v' is an *ancestor* of v if

v' is on the path from $r(T)$ to v; the *parent* $\mathrm{p}(v)$ of v, of which v is a *child*, directly precedes v on this path. Conversely, v is a *descendant* of v'. Notice that a node v is both an ancestor and a descendant of itself; where this case needs to be excluded, we will talk about strict ancestors and descendants. Two nodes are *separated* in T if neither is an ancestor of the other. We denote by $\mathrm{E}(T)$ the set of edges of T, where each edge is represented by a pair of nodes $(\mathrm{p}(v), v)$. For any two nodes v_1 and v_2 of T, the node distance $\mathrm{D}_T^{\mathrm{node}}(v_1, v_2)$ is defined as the number of edges on the unique path from v_1 to v_2.

Given a node v of T, $T[v]$ is the *subtree* of T rooted at v (i.e., containing only the descendants of v). The *lowest common ancestor* (LCA) of a subset V of nodes, denoted $\mathrm{lca}_T(V)$, is the ancestor of all nodes in V that is the most distant from the root. A node is said to be *unary* if it has exactly one child and *binary* if it has exactly two. A *binary tree* is a tree where all internal (non-leaf) nodes are binary. If all internal nodes are unary or binary, then the tree is called *partially binary*. The two children of a binary node v are denoted v_l and v_r for the "left" and "right" child. Notice that the considered trees are unordered, and thus left and right are set arbitrarily.

If A is a set of labels (on a given finite alphabet), then any tree T such that there exists a one-to-one relation between A and $\mathrm{L}(T)$ is said to *be a tree for* A. In particular, a *species tree* S for a set Σ of species represents an ordered set of speciation events that led to Σ (i.e., each internal node of S represents an ancestral species preceding a speciation event). Similarly, a *gene tree* T for a gene family Γ is a branching history encoding each gene divergence that led to the gene family Γ. For a gene g of Γ, we denote by $s(g)$ the species of Σ the gene g belongs to.

Let $\mathcal{F}$ be a set of gene families. In this paper, a *syntenic group* or *synteny* X is a non-empty subset of $\mathcal{F}$ representing a group of co-localized genes, where the relative order of genes in the genomic region is ignored. The genes of a synteny are considered to all belong to different gene families (i.e., duplications inside a syntenic group are not allowed), therefore the genes are simply identified by their family Γ of $\mathcal{F}$. Given two syntenies X and Y, we say that there is a loss between X and Y if at least one of the gene families from X is absent in Y. Notice that, due to the possibility of gene gains, Y may contain genes not found in X. The loss indicator function $\mathrm{D}^{\mathrm{sub}}(X, Y)$ (*sub* for "subset") is therefore defined as $\mathrm{D}^{\mathrm{sub}}(X, Y) = 0$ if $X \subseteq Y$, otherwise $\mathrm{D}^{\mathrm{sub}}(X, Y) = 1$.

A *synteny family* is a set $\mathcal{X}$ of syntenies. A *synteny tree* T is a tree for a synteny family $\mathcal{X}$, $x(l)$ being the synteny of $\mathcal{X}$ associated to each leaf l of T. In this paper, the species and synteny trees are considered binary. See the left part of Fig. 1 for an example of a synteny tree.

3   Evolutionary Histories for Syntenies

The *Super-reconciliation* framework introduced in [5] generalizes the reconciliation framework from a single gene family Γ to a family $\mathcal{X}$ of syntenies. More precisely, while an instance of a reconciliation problem is a tuple $\langle \Gamma, T, \Sigma, S \rangle$, T

being a gene tree for Γ, an instance of a Super-reconciliation problem is a tuple $\langle \mathcal{X}, \mathcal{F}, T, \Sigma, S \rangle$ where T is a synteny tree for the synteny family $\mathcal{X}$ on a set $\mathcal{F}$ of gene families. Notice that, while in [5] the synteny tree is inferred from a set of "consistent" gene trees, in this paper we start from the synteny tree itself, whatever the way used to infer it (e.g., from a set of gene trees, or alternatively from an alignment of the concatenated gene sequences).

The goal of the reconciliation approach is to infer a *correct* and *optimal* evolutionary history explaining T given S. Correctness depends on the considered evolutionary events, while optimality is stated as a parsimony criterion, given a cost function for the evolutionary events. As for Super-reconciliation, in addition to evolutionary events, ancestral syntenies should also be inferred.

The DL Super-reconciliation model [5], defined both on ordered and unordered syntenies, only involves speciations, duplications and losses. Here, we extend the unordered version to also allow for HGTs. We assume that evolution only takes place inside the S species tree, excluding speciations and transfers to and from extinct or unsampled lineages [26, 30]. Also notice that the constraints on HGTs do not exclude cyclic reconciliations. In addition, in order to avoid the unrealistic assumption of having all gene families present at the root, we allow for gene gains.

Definition 1 (Evolutionary history for syntenies). *Let $\mathcal{X}$ be a set of syntenies on a set $\mathcal{F}$ of gene families, Σ be the set of taxa to which these syntenies belong and S be the species tree for Σ. An $\mathcal{E}$ evolutionary history, with $\mathcal{E} \subseteq \{\text{Spe}, \text{Dup}, \text{HGT}, \text{Loss}, \text{Gain}\}$, is a partially binary tree with leaves mapped to $\mathcal{X}$ and where each internal node v corresponds to an event $e(X)$ of $\mathcal{E}$ with $X \subseteq \mathcal{F}$ being the synteny at v belonging to a genome $s(v)$ such that:*

- Spe *produces two syntenies Y and Z verifying $X = Y = Z$ and $s(Y)$, $s(Z)$ are the two children of $s(X)$ in S.*
- Dup (D) *produces two syntenies Y and Z verifying $Y = X$, $Z \subseteq X$, and $s(X) = s(Y) = s(Z)$.*
- HGT (T) *produces two syntenies Y and Z verifying $Y = X$, $Z \subseteq X$, $s(Y) = s(X)$ and $s(Z)$ is separated from $s(X)$.*
- Loss (L) *produces a single synteny Y verifying $Y \subset X$ and $s(Y) = s(X)$. A loss is* full *if Y is the empty synteny (i.e., $Y = \emptyset$) and* partial *otherwise.*
- Gain (G) *produces a single synteny Y verifying $X \subset Y$ and $s(Y) = s(X)$.*

We denote by $H = \langle H^{\text{tree}}, e, x, s \rangle$ such a history where H^{tree} is the supporting partially binary tree and for each of its nodes v, $e(v)$ is the event, $x(v)$ is the synteny, and $s(v)$ is the species to which $x(v)$ belongs. We have $\mathcal{X} = \{x(l) : l \in \text{L}(H^{\text{tree}})\}$ (Fig. 1, center). Note that for an internal node v of H^{tree}, $x(v)$ is not necessarily in $\mathcal{X}$. In general, x is not an onto function, as not all possible syntenies on $\mathcal{F}$ are required to be represented in the history.

We next define a Super-reconciliation from a history. In Definition 2, the $\mathcal{E}$-Super-reconciliation obtained from H is $\langle T_H, e, x, s \rangle$ where T_H is the binary tree obtained from H^{tree} by removing edges adjacent to empty syntenies (due to full losses) and then removing unary nodes, and e, x and s are restrictions of the event, synteny and species mappings to the nodes of T_H (Fig. 1, right).

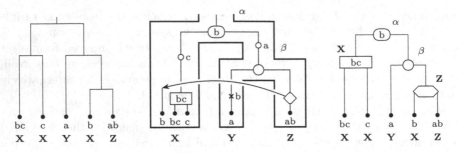

Fig. 1. (Left.) A synteny tree T for the family $\mathcal{X} = \{\{b\}, \{b,c\}, \{c\}, \{a\}, \{a,b\}\}$ on the set of species $\Sigma = \{\mathbf{X}, \mathbf{Y}, \mathbf{Z}\}$. Under each leaf l are shown its associated synteny $x(l)$ and species $s(l)$. (Center.) An evolutionary history H^{tree} for T, embedded in the species tree S. For binary nodes, rounded rectangles correspond to Spe, plain rectangles to Dup, and chamfered rectangles to HGT. For unary nodes, circles correspond to Gain, and crosses to Loss. For a binary node v, the value of $x(v)$ is shown inside the shape (omitted when unchanged from its parent). For unary nodes, the synteny difference is shown beside the node. (Right.) The Super-reconciliation $\langle T, x, s \rangle$ obtained from H^{tree}. For each internal node v, we exhibit $e(v)$ (shape of the node), $s(v)$ (letter beside the node) and $x(v)$ (content of the node).

Definition 2 (Super-reconciliation). *Let $I = \langle \mathcal{X}, \mathcal{F}, T, \Sigma, S \rangle$ be an instance of a Super-reconciliation problem. An $\mathcal{E}$-Super-reconciliation for I is a tuple $\mathcal{R} = \langle T_H, e, x, s \rangle$ obtained from an $\mathcal{E}$ history $H = \langle H^{\text{tree}}, e, x, s \rangle$. We say that H is a history for I leading to $\mathcal{R}$.*

Note that an $\mathcal{E}$-*Reconciliation* is defined as an $\mathcal{E}$-Super-reconciliation but on an instance $\langle \Gamma, T, \Sigma, S \rangle$. The next lemma, which will be required later for a simple inference of the event labeling from the species labeling, directly follows from Definition 1 and Definition 2.

Lemma 1 (Synteny and species trees coincide). *Let $I = \langle \mathcal{X}, \mathcal{F}, T, \Sigma, S \rangle$ and $\mathcal{R} = \langle T, e, x, s \rangle$ be a DTLG Super-reconciliation for I (i.e. a Super-reconciliation on the set of events $\mathcal{E} = \{\text{Spe}, D, T, L, G\}$). Then s satisfies the following conditions: (1) neither $s(v_l)$ nor $s(v_r)$ is a strict ancestor of $s(v)$ (i.e. each one can either be a descendant of $s(v)$ or be separated from it in the case of HGTs); (2) at least one of $s(v_l)$ or $s(v_r)$ is a descendant of $s(v)$.*

Restrictions on Gain Points: Allowing for segmental gains and losses may lead to unrealistic optimal scenarios where full syntenies are lost and gained again. In order to ensure more realistic scenarios, we will assume that a gene family can only appear once in the history (i.e., not allowing for convergent evolution).

Definition 3 (Gain point). *Let T be a synteny tree for a synteny family $\mathcal{X}$ on $\mathcal{F}$ and x be a synteny assignment on the internal nodes of T. A node $v \in V(T)$ is a gain point for $\Gamma \in \mathcal{F}$ in $\langle T, x \rangle$ iff $\Gamma \in x(v)$ and either $v = \mathrm{r}(T)$ or $\Gamma \notin x(\mathrm{p}(v))$. The set of gain points for Γ is denoted as $\text{Gain}_{\langle T,x \rangle}(\Gamma)$.*

Moreover, as gain of function affects genes individually, we will restrict a gain to an event inserting a single gene in a synteny. Formally, we add a restriction to the Gain event of Definition 1 specifying that the synteny Y produced from a synteny X by a gain is such that $|Y| = |X| + 1$.

Consequently, each gene family $\Gamma \in \mathcal{F}$ can only be gained once in the history, leading to exactly $|\mathcal{F}|$ gains. In other words, we account for gene gains only to avoid imposing all gene families to be present at the root of T, but without including them in the cost function for inferring a most parsimonious history. Consequently, we can define our problem as a DTL (for D, T, and L events) rather than a DTLG problem (D, T, L, and G events) as follows, considering the above restriction on the Gain event.

Definition 4 (DTL Super-reconciliation). *Let $I = \langle \mathcal{X}, \mathcal{F}, T, \Sigma, S \rangle$. A DTL Super-Reconciliation for I is a DTLG Super-reconciliation $\langle T, e, x, s \rangle$ such that, for each $\Gamma \in \mathcal{F}$, $|\mathrm{Gain}_{\langle T,x \rangle}(\Gamma)| = 1$.*

4   Most Parsimonious Super-Reconciliations

In this paper, we assume a null cost for speciations and, according to the restriction on Gain events described above, we can ignore the cost of gains. We then define $\delta = \langle c_{\mathrm{Dup}}, c_{\mathrm{HGT}}, c_{\mathrm{Loss}} \rangle \in (\mathbb{R}^+ \cup \{\infty\})^3$ the cost for, respectively, a duplication, an HGT and a loss event. For a history $H = \langle H^{\mathrm{tree}}, e, x, s \rangle$ and a node $v \in \mathrm{V}(H^{\mathrm{tree}})$, we define $c_\delta(H, v)$ to be the sum of costs of events in the $H^{\mathrm{tree}}[v]$ subtree, up to and including v itself. The history's overall cost $c_\delta(H)$ is equal to $c_\delta(H, \mathrm{r}(H^{\mathrm{tree}}))$.

The goal is to find a most parsimonious history (i.e., a history of minimum cost), explaining a given synteny tree T with respect to a species tree S. From Definition 2, a Super-reconciliation $\mathcal{R}$ for an instance I represents not a single, but rather a set of histories H from which $\mathcal{R}$ can be obtained. In the rest of this section, we give some results allowing to reduce the problem to the exploration of the Super-reconciliation space rather than the history space.

The next lemma states that gain points can be inferred from the synteny tree. For each family $\Gamma \in \mathcal{F}$, denote by $\mathrm{L}(T)_\Gamma = \{l \in \mathrm{L}(T) : \Gamma \in x(l)\}$ the set of leaves whose corresponding synteny contains Γ and $\mathrm{lca}(\Gamma, T) = \mathrm{lca}_T(\mathrm{L}(T)_\Gamma)$.

Lemma 2 (Optimal gain point position). *Let $I = \langle \mathcal{X}, \mathcal{F}, T, \Sigma, S \rangle$. There is an optimal history H for I such that, in the Super-reconciliation $\mathcal{R} = \langle T, e, x, s \rangle$ obtained from H, for each $\Gamma \in \mathcal{F}$, $\mathrm{lca}(\Gamma, T)$ is the gain point for Γ in T.*

Proof. This proof and all subsequent ones can be found in the appendix.

We denote by $x^{\mathrm{gain}}(v)$ the set of genes gained at node v of T. We next introduce a way to assign a cost to a Super-reconciliation which, as we show in the subsequent lemma, matches the minimum cost of any history leading to that Super-reconciliation.

Definition 5 (Super-reconciliation cost). *Let* $\mathcal{R} = \langle T, e, x, s \rangle$ *be a Super-reconciliation for* $I = \langle \mathcal{X}, \mathcal{F}, T, \Sigma, S \rangle$. *The cost* $C_\delta(\mathcal{R}, v)$ *of* $\mathcal{R}$ *for the* $T[v]$ *subtree (or simply* $C(\mathcal{R}, v)$ *or* $C(v)$ *if no ambiguity) is defined recursively as follows:*

- *if* v *is a leaf, then* $C_\delta(\mathcal{R}, v) = 0$;
- *if* $e(v) = \text{Spe}$ *and* v_l, v_r *are the two children of* v, *then*

$$C_\delta(\mathcal{R}, v) = C_\delta(\mathcal{R}, v_l) + C_\delta(\mathcal{R}, v_r) +$$
$$c_{\text{Loss}} \times (\text{D}^{\text{sub}}(x(v), x(v_l)) + \text{D}^{\text{sub}}(x(v), x(v_r))$$
$$+ \text{D}_S^{\text{node}}(s(v), s(v_l)) + \text{D}_S^{\text{node}}(s(v), s(v_r)) - 2);$$

- *if* $e(v) = \text{Dup}$ *and* v_l, v_r *are the two children of* v, *then*

$$C_\delta(\mathcal{R}, v) = c_{\text{Dup}} + C_\delta(\mathcal{R}, v_l) + C_\delta(\mathcal{R}, v_r) +$$
$$c_{\text{Loss}} \times (\min \left\{ \text{D}^{\text{sub}}(x(v), x(v_l)), \text{D}^{\text{sub}}(x(v), x(v_r)) \right\}$$
$$+ \text{D}_S^{\text{node}}(s(v), s(v_l)) + \text{D}_S^{\text{node}}(s(v), s(v_r)));$$

- *if* $e(v) = \text{HGT}$, v' *is the separated child, and* v'' *is the conserved one, then*

$$C_\delta(\mathcal{R}, v) = c_{\text{HGT}} + C_\delta(\mathcal{R}, v_l) + C_\delta(\mathcal{R}, v_r)$$
$$+ c_{\text{Loss}} \times (\text{D}^{\text{sub}}(x(v), x(v'')) + \text{D}_S^{\text{node}}(s(v), s(v''))).$$

The global cost of $\mathcal{R}$ *is defined as* $C_\delta(\mathcal{R}) = C_\delta(\mathcal{R}, r(T))$ *(or simply* $C(\mathcal{R})$ *if no ambiguity).*

Lemma 3 (Super-reconciliations minimize history cost). *Let* $\mathcal{R} = \langle T, e, x, s \rangle$ *be a Super-reconciliation for* $I = \langle \mathcal{X}, \mathcal{F}, T, \Sigma, S \rangle$ *and* $\mathcal{H}$ *be the set of histories leading to* $\mathcal{R}$. *Then, for any* $C_\delta(\mathcal{R}) = \min_{H \in \mathcal{H}} c_\delta(H)$.

Searching for a most parsimonious history is therefore equivalent to searching in the smaller space of super-reconciliations. Finally, the following definition makes the link between the values of the s mapping and the evolutionary events at the internal nodes of a Super-reconciliation.

Definition 6 (Min-event labeling). *Given* $I = \langle \mathcal{X}, \mathcal{F}, T, \Sigma, S \rangle$, *an internal node* v *of* T, *and three species* $\sigma, \sigma', \sigma''$ *of* S, *we define the* min-event *function* $e^{\min}(\sigma, \sigma', \sigma'')$, *used to label* v *if* $s(v) = \sigma$, $s(v_l) = \sigma'$, *and* $s(v_r) = \sigma''$, *as follows:*

- *If* σ *is an ancestor of both* σ' *and* σ'', *then*
  - *if* σ' *is separated from* σ'' *and* $\sigma = \text{lca}(\sigma', \sigma'')$, *then* $e^{\min}(\sigma, \sigma', \sigma'') = \text{Spe}$;
  - *if* σ' *and* σ'' *are not separated or* $\sigma \neq \text{lca}(\sigma', \sigma'')$, *then* $e^{\min}(\sigma, \sigma', \sigma'') = \text{Dup}$.
- *If either* σ' *or* σ'' *is separated from* σ, *then* $e^{\min}(\sigma, \sigma', \sigma'') = \text{HGT}$.
- *Species* σ' *and* σ'' *cannot be both separated from* σ *as per Lemma 1, therefore in such cases* $e^{\min}(\sigma, \sigma', \sigma'')$ *equals an error value not in* $\{\text{Spe}, \text{Dup}, \text{HGT}\}$.

Additionally, given a mapping s from $V(T)$ *to* $V(S)$, *we define* $e_s^{\min}$ *so that for any internal node* v *of* T, $e_s^{\min}(v) = e^{\min}(s(v), s(v_l), s(v_r))$.

The following lemma shows that the min-event labeling leads to the most parsimonious Super-reconciliation.

Lemma 4 (Min-event labeling is optimal). *Let* $\mathcal{R} = \langle T, e, x, s \rangle$ *and* $\mathcal{R}^{\min} = \langle T, e_s^{\min}, x, s \rangle$ *be two DTL Super-reconciliations for* $I = \langle \mathcal{X}, \mathcal{F}, T, \Sigma, S \rangle$. *Then,* $C_\delta(\mathcal{R}^{\min}) \leq C_\delta(\mathcal{R})$.

It follows that event labeling can be ignored, as it can be directly inferred from the species mapping. Therefore, from now on, a Super-reconciliation will be simply designed as a triplet $\langle T, x, s \rangle$. We are now ready to formally define the considered optimization problem.

δ-SUPER-RECONCILIATION Problem
Input: An input $I = \langle \mathcal{X}, \mathcal{F}, T, \Sigma, S \rangle$.
Output: A Super-reconciliation $\mathcal{R} = \langle T, x, s \rangle$ for I minimizing $C_\delta(\mathcal{R})$.

5   A Two-Steps Method

Finding a DL Reconciliation of a gene tree with a species tree is a classical problem [10,12,14,22]. Given an instance $I = \langle \Gamma, T, \Sigma, S \rangle$, define its LCA-mapping from T to S as $s(v) = \mathrm{lca}_S(\{s(l) \mid l \in L(T[v])\})$ for any $v \in V(T) \setminus L(T)$, and $s(l) = x$ for any $l \in L(T)$, where x is the extant species to which l belongs. This mapping leads to an optimal DL Reconciliation (with constant costs on operations) and can be computed in time $\mathcal{O}(|V(T)| + |V(S)|)$ [13].

As for DL Super-reconciliation, an approach using two steps to infer an optimal $\mathcal{R} = \langle T, x, s \rangle$ for an instance $I = \langle \mathcal{X}, \mathcal{F}, T, \Sigma, S \rangle$ was presented in [5]:

1. Compute s as the LCA-mapping of T to S;
2. Infer x in a way minimizing segmental losses and with the constraint that $x(r(T)) = \mathcal{F}$ (as gains were not allowed).

Using this approach, an optimal DL Super-reconciliation can be computed in time $\mathcal{O}(|\mathcal{X}| \times |\mathcal{F}|)$. Crucially, this approach works because any optimal DL Super-reconciliation $\langle T, x, s \rangle$ is such that s is the LCA-mapping of T to S. Otherwise, we would not be able to compute s separately from x.

Unfortunately, this two-steps method does not work for solving the DTL Super-reconciliation problem. In fact, for a given input $I = \langle \Gamma, T, \Sigma, S \rangle$, the mapping s^{DT} from T to S allowing to minizime the duplication and loss cost is not necessarily the mapping s of an optimal DTL Super-reconciliation $\mathcal{R} = \langle T, x, s \rangle$. In particular, changing a speciation node (inferred from s^{DT}) to a HGT event may lead to less losses, and thus to a better cost in total. Figure 2 shows a counter example of the two-steps method for solving the DTL problem.

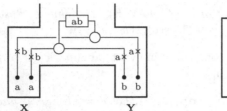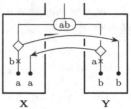

X   Y   X   Y

Fig. 2. Two solutions for the same DTL Super-reconciliation problem input. (Left.) Solution obtained by first computing s to minimize duplications and HGTs (yielding one such event) and then labeling internal nodes to minimize losses (yielding 4 losses). (Right.) A more parsimonious solution with two HGTs and only two losses.

6   A Dynamic Programming Algorithm for DTL Super-Reconciliation

In this section, we introduce SuperDTL, a polynomial-time algorithm for solving the DTL Super-Reconciliation problem. Let $I = \langle \mathcal{X}, \mathcal{F}, T, \Sigma, S \rangle$ be any input for this problem. For any node v of T, we define $x^{\text{lca}}(v)$ to be the leafset of $T[v]$ excluding genes gained below v, namely:

$$x^{\text{lca}}(v) = \begin{cases} x(v) & \text{if } v \in \text{L}(G) \\ x^{\text{lca}}(v_l) \cup x^{\text{lca}}(v_r) \setminus (x^{\text{gain}}(v_l) \cup x^{\text{gain}}(v_r)) & \text{otherwise.} \end{cases}$$

For a synteny X and a node v of T, denote by $C_X(v)$ the minimum cost of a Super-reconciliation $\langle T[v], x, s \rangle$ between $T[v]$ and S in which $x(v) = X$. Notice that $x^{\text{lca}}(v)$ should be a subset of X as otherwise there would be a gene family with two or more gain points (by Definition 3). In other words, if $x^{\text{lca}}(v) \nsubseteq X$, then $C_X(v) = \infty$. However, X may contain genes in $\mathcal{F} \setminus x^{\text{lca}}(v)$, which may allow grouping losses in the evolutionary history thus leading to a lower cost. Testing all possible subsets of $\mathcal{F} \setminus x^{\text{lca}}(v)$ would be costly, but as shown in [5], this can be avoided due to a property that still holds in our case: all that matters is to know whether $x^{\text{lca}}(v)$ is included in X in a strict or not strict way, and the nature and number of "extra" genes is irrelevant for the computation of the optimal cost. In other words, the following lemma from [5] holds.

Lemma 5. *Let v be an internal node of T and X, Y be two subsets of $\mathcal{F}$ such that $x^{\text{lca}}(v) \subseteq X, Y$. Then $C_X(v) = C_Y(v)$.*

We therefore only need to consider two possibilities for $x(v)$. We define $\text{C}(v, \sigma)$ (respec. $\text{C}^\star(v, \sigma)$) to be the minimum cost of a Super-reconciliation $\langle T[v], x, s \rangle$ between $T[v]$ and S in which $s(v) = \sigma$ and $x(v) = x^{\text{lca}}(v)$ (respec. $x(v) \supsetneq x^{\text{lca}}(v)$). Algorithms 1 and 2, described below, provide a method for computing those two functions.

Algorithm 1. Computing the value of $C(v, \sigma)$

> **function** $\mathcal{C}(v, \sigma)$
>> **if** $v \in L(T)$ **then**
>>> **return** 0 if $\sigma = s(v)$ **else** ∞
>> **else**
>>> $c \leftarrow \infty$
>>> **for** $\sigma', \sigma'' \in V(S)^2$ **do**
>>>> $star_l \leftarrow \infty$ if $x^{\mathrm{lca}}(v) \subseteq x^{\mathrm{lca}}(v_l)$ **else** 0
>>>> $star_r \leftarrow \infty$ if $x^{\mathrm{lca}}(v) \subseteq x^{\mathrm{lca}}(v_r)$ **else** 0
>>>> $partial_l \leftarrow \mathrm{D}^{\mathrm{sub}}(x^{\mathrm{lca}}(v), x^{\mathrm{lca}}(v_l))$
>>>> $partial_r \leftarrow \mathrm{D}^{\mathrm{sub}}(x^{\mathrm{lca}}(v), x^{\mathrm{lca}}(v_r))$
>>>> **if** $e^{\min}(\sigma, \sigma', \sigma'') = \mathrm{Spe}$ **then**
>>>>> $c \leftarrow \min\{c, \ \min\{\mathcal{C}^*(v_l, \sigma') + star_l, \mathcal{C}(v_l, \sigma') + c_{\mathrm{Loss}} \times partial_l\} \ +$
>>>>> $\qquad\qquad \min\{\mathcal{C}^*(v_r, \sigma'') + star_r, \mathcal{C}(v_r, \sigma'') + c_{\mathrm{Loss}} \times partial_r\} \ +$
>>>>> $\qquad\qquad c_{\mathrm{Loss}} \times (\mathrm{D}_S^{\mathrm{node}}(\sigma, \sigma') + \mathrm{D}_S^{\mathrm{node}}(\sigma, \sigma'') - 2)\}$
>>>> **else if** $e^{\min}(\sigma, \sigma', \sigma'') = \mathrm{Dup}$ **then**
>>>>> $c \leftarrow \min\{c, \ \min\{\mathcal{C}^*(v_l, \sigma') + \mathcal{C}^*(v_r, \sigma'') + star_l + star_r,$
>>>>> $\qquad\qquad \mathcal{C}(v_l, \sigma') + \mathcal{C}^*(v_r, \sigma'') + star_r,$
>>>>> $\qquad\qquad \mathcal{C}^*(v_l, \sigma') + \mathcal{C}(v_r, \sigma'') + star_l,$
>>>>> $\qquad\qquad \mathcal{C}(v_l, \sigma') + \mathcal{C}(v_r, \sigma'') \ +$
>>>>> $\qquad\qquad\qquad c_{\mathrm{Loss}} \times \min\{partial_l, partial_r\}\} \ +$
>>>>> $\qquad\qquad c_{\mathrm{Dup}} + c_{\mathrm{Loss}} \times (\mathrm{D}_S^{\mathrm{node}}(\sigma, \sigma') + \mathrm{D}_S^{\mathrm{node}}(\sigma, \sigma''))\}$
>>>> **else if** $e^{\min}(\sigma, \sigma', \sigma'') = \mathrm{HGT}$ **then**
>>>>> $full \leftarrow \mathrm{D}_S^{\mathrm{node}}(\sigma, \sigma'')$ if σ' is separated from σ **else** $\mathrm{D}_S^{\mathrm{node}}(\sigma, \sigma')$
>>>>> **if** σ' is separated from σ **then** $partial_l \leftarrow 0$
>>>>> **if** σ'' is separated from σ **then** $partial_r \leftarrow 0$
>>>>> $c \leftarrow \min\{c, \ \min\{\mathcal{C}^*(v_l, \sigma') + \mathcal{C}^*(v_r, \sigma''),$
>>>>> $\qquad\qquad \mathcal{C}(v_l, \sigma') + \mathcal{C}^*(v_r, \sigma'') + star_r + c_{\mathrm{Loss}} \times partial_l,$
>>>>> $\qquad\qquad \mathcal{C}^*(v_l, \sigma') + \mathcal{C}(v_r, \sigma'') + star_l + c_{\mathrm{Loss}} \times partial_r,$
>>>>> $\qquad\qquad \mathcal{C}(v_l, \sigma') + \mathcal{C}(v_r, \sigma'') + c_{\mathrm{Loss}}\} \ +$
>>>>> $\qquad\qquad c_{\mathrm{HGT}} + c_{\mathrm{Loss}} \times full\}$
>> **return** c

Lemma 6 (Termination and Correctness). *For any* $v \in V(T)$ *and* $\sigma \in V(S)$, $\mathcal{C}(v, \sigma) = C(v, \sigma)$ *and* $\mathcal{C}^*(v, \sigma) = C^*(v, \sigma)$ *(as computed by Algorithms 1 and 2 respectively).*

Algorithm SuperDTL computes the minimal cost of a Super-reconciliation for I by computing $\min_{\sigma \in V(S)} \mathcal{C}(\mathrm{r}(T), \sigma)$ using Algorithm 1, which recursively invokes itself and Algorithm 2. Additionally, an actual solution can be constructed by keeping track of which σ', σ'' pairs and which of C or C* yield the minimum values of the min expressions in both algorithms. To make SuperDTL efficient, it should not be implemented as a naive recursion, but rather C and C* should be considered as dynamic programming tables with $|V(T)| \times |V(S)|$

Algorithm 2. Computing the value of $C^*(v, \sigma)$

> **function** $C^*(v, \sigma)$
> > **if** $v \in L(T)$ **then return** ∞
> > **else**
> > > $c \leftarrow \infty$
> > > **for** $\sigma', \sigma'' \in V(S)^2$ **do**
> > > > **if** $e^{\min}(\sigma, \sigma', \sigma'') = \text{Spe}$ **then**
> > > > > $c \leftarrow \min\{c, \ \min\{C^*(v_l, \sigma'), C(v_l, \sigma') + c_{\text{Loss}}\} +$
> > > > > $\qquad \min\{C^*(v_r, \sigma''), C(v_r, \sigma'') + c_{\text{Loss}}\} +$
> > > > > $\qquad c_{\text{Loss}} \times (D_S^{\text{node}}(\sigma, \sigma') + D_S^{\text{node}}(\sigma, \sigma'') - 2)\}$
> > > > **else if** $e^{\min}(\sigma, \sigma', \sigma'') = \text{Dup}$ **then**
> > > > > $c \leftarrow \min\{c, \ \min\{C^*(v_l, \sigma') + C^*(v_r, \sigma''), C(v_l, \sigma') + C^*(v_r, \sigma''),$
> > > > > $\qquad C^*(v_l, \sigma') + C(v_r, \sigma''), C(v_l, \sigma') + C(v_r, \sigma'') + c_{\text{Loss}}\}$
> > > > > $\qquad + c_{\text{Dup}} + c_{\text{Loss}} \times (D_S^{\text{node}}(\sigma, \sigma') + D_S^{\text{node}}(\sigma, \sigma''))\}$
> > > > **else if** $e^{\min}(\sigma, \sigma', \sigma'') = \text{HGT}$ **then**
> > > > > $full \leftarrow D_S^{\text{node}}(\sigma, \sigma'')$ **if** σ' is separated from σ **else** $D_S^{\text{node}}(\sigma, \sigma')$
> > > > > $partial_l \leftarrow 0$ **if** σ' is separated from σ **else** 1
> > > > > $partial_r \leftarrow 0$ **if** σ'' is separated from σ **else** 1
> > > > > $c \leftarrow \min\{c, \ \min\{C^*(v_l, \sigma') + C^*(v_r, \sigma''),$
> > > > > $\qquad C(v_l, \sigma') + C^*(v_r, \sigma'') + c_{\text{Loss}} \times partial_l,$
> > > > > $\qquad C^*(v_l, \sigma') + C(v_r, \sigma'') + c_{\text{Loss}} \times partial_r,$
> > > > > $\qquad C(v_l, \sigma') + C(v_r, \sigma'') + c_{\text{Loss}}\} +$
> > > > > $\qquad c_{\text{HGT}} + c_{\text{Loss}} \times full\}$
> > **return** c

---

entries each. Using this implementation trick allows finding a minimal Super-reconciliation in polynomial time, as shown in the next theorem.

Theorem 1 (Time and space complexity). *Using SuperDTL, the* DTL SUPER-RECONCILIATION *problem can be solved in polynomial time* $\mathcal{O}(|\mathcal{X}| \times |\Sigma|^3)$ *and space* $\mathcal{O}(|\mathcal{X}| \times |\Sigma|)$.

7 Application to CRISPR-Associated (Cas) Gene Syntenies

7.1 Cas Gene Syntenies

Cas genes are part of CRISPR-Cas systems, one of its members, CRISPR-Cas9, being well-known as currently one of the most reliable and accurate "molecular scissor" biotechnology for genome editing. This technology, for which the Nobel Prize in Chemistry was awarded in 2020 [16], was derived from an adaptive bacterial immunity system targeting bacteriophages. The study and analysis of CRISPR-Cas systems over the past two decades has revealed their wide diversity and led to their categorization into two classes: Class 1, composed of multisubunit effector proteins, and Class 2, composed of a single large effector protein. Each class is further divided into several types themselves composed of several subtypes (for more details see Supplementary Table 2 in Makarova et al. [20]). Although the discovery of new CRISPR-Cas systems is an ongoing process [18], the classification is generally stable.

As the function of CRISPR-Cas systems highly depends on the syntenic organization of Cas genes, elucidating the evolution of these systems is crucial. Many studies have been dedicated to reconstructing the evolutionary histories of individual Cas gene families such as Cas1 [15], or to inferring the evolution of Cas gene syntenies to elucidate the syntenic events leading to the diversity inside the different subtypes [17,20,21]. From these multiple phylogenetic analyses, a global scenario has been predicted for the evolutionary formation of CRISPR-Cas systems, with the latest scenario described by Koonin et al. [18]. However, to the best of our knowledge, none of these studies take into account the species tree topology. Moreover, several studies point to evidence of HGTs involving Cas genes between prokaryote species [3,11,15,28], showing the need to take such events into consideration while inferring scenarios.

From this brief presentation, it appears that the DTL Super-reconciliation model is suitable for elucidating the evolution and radiation of CRISPR-Cas systems across prokaryotes.

7.2 Dataset

We used the Cas gene synteny subtypes from Class 1 of CRISPR-Cas systems described by Makarova et al. [20]. We limited the dataset to the 15 bacteria species and omitted archaea to avoid bias due to their underrepresentation in the dataset. Each of those species contains a Cas synteny. Taxonomical information of the 15 bacterial species has been recovered from the NCBI Taxonomy database [25] and the species tree topology is based on the phylogeny inferred by Coleman et al. [4].

We repurposed the phylogenetic tree given in Figure 1 of Makarova et al. [20] as our synteny tree. Some alterations of the syntenies were required to fit the constraints of the model. We considered Cas families Cas1–8 and Cas10–11. Since a part of Cas3, Cas3", can work as a standalone HD nuclease, we split Cas3 into Cas3' and Cas3". In [20], Cas10 and Cas8 share the same colour code as they provide similar functions in CRISPR-Cas complexes. Indeed, it was initially believed that Cas8 evolved from Cas10 [19]. Nevertheless, their sequence being extremely divergent, we decided to consider them as not homologous and to conserve separate Cas8 and Cas10 families. Finally, for syntenies with several copies of the same family, we conserved only one copy per synteny. The obtained Cas gene syntenies are illustrated in the left part of Fig. 3.

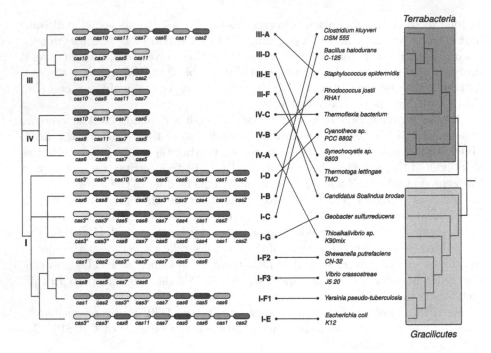

Fig. 3. Cas gene syntenies for the 15 considered bacterial species. (Left.) Phylogeny of Class 1 CRISPR-Cas systems with subtypes names, as presented in [20] with our preprocessing of syntenies as described in the text. (Right.) The species tree based on the topology inferred in [4] with representation of the two major groups of bacteria, Terrabacteria and Gracilicutes. Lines represent the correspondence between the Cas genes syntenies and the species they belong to, illustrating the high incongruence between the topology of the synteny tree and that of the species tree.

8   Results

8.1   DTL Super-Reconciliation Settings

We used SuperDTL to predict optimal DTL Super-reconciliations for the syntenies and trees depicted in Fig. 3. Notice that the synteny tree is non-binary and contains three multifurcations, one at the root and two in the subtree of Type I CRISPR-Cas systems. As our algorithm can only be applied to binary trees, we test all possible "binarizations" of the synteny tree and retain the overall minimal solutions.

We tested different values for the $\delta = \langle c_{\mathrm{Dup}}, c_{\mathrm{HGT}}, c_{\mathrm{Loss}} \rangle$ cost model, in agreement with a classical assumption that HGTs are less frequent than duplications, which are less frequent than losses. The number of solutions obtained for each setting is given in Table 1. We observe that the number of solutions decreases as the HGT cost increases, reaching a minimum for $c_{\mathrm{HGT}} = 4$, and then increases again. For a given c_{HGT}, the results are largely stable for different c_{Dup} values.

Table 1. Number of solutions obtained by Algorithm SuperDTL for different values of $\delta = \langle c_{\text{Dup}}, c_{\text{HGT}}, c_{\text{Loss}} \rangle$, with $c_{\text{Loss}} = 1$.

$c_{\text{HGT}} \backslash c_{\text{Dup}}$	1	1.5	2	2.5	3
1	1376	1376	1376	1376	1376
2	132	132	132	132	132
3	112	60	60	60	60
4	32	32	32	32	48
5	288	288	320	32	32
6	320	288	288	288	288

8.2   An Evolutionary Scenario

We analyzed the 32 solutions generated for $c_{\text{HGT}} = 4$ and $c_{\text{Dup}} \in \{1, 1.5, 2, 2.5\}$. All solutions lead to the same resolution of the three multifurcations in the synteny tree, and the incongruence observed between the species and synteny tree topologies is mainly resolved by inferring six HGTs and one duplication. The 32 solutions only differ from one another by minor variations, such as gene losses located either before or after a speciation or HGT event, which do not change the overall inferred evolutionary history for the Cas gene syntenies.

Figure 4 is a representation of one of these solutions. The evolutionary reconstruction is broadly consistent with the CRISPR-Cas evolution scenario established by Koonin et al. [18]. In both their scenario and ours, the CRISPR-Cas systems emerge from an initial adaptive immune system composed only of Cas genes involved in the effector complex (Cas 5–8 and Cas 10–11). We see from Fig. 4 that the CRISPR-Cas emergence is inferred at the root of Terrabacteria. The scenarios diverge on the fact that, in our scenario, Type IV emerges directly from this ancestral adaptive immune system before the acquisition of Cas1 and Cas2 genes, while in Koonin et al., Type IV emerges after the acquisition of Cas1 and Cas2 adaptation module genes and loss of several Cas genes. Aside from this difference, both scenarios are in agreement for Types I and III in terms of gene content in ancestral Cas syntenies. As it appears in Fig. 4, Types I and III Cas syntenies emerge from Cyanobacteria after the acquisition of Cas1 and Cas2 by an ancestor of Cyanobacteria. Type I emerges from the branch leading to *Cyanothece sp. PCC 8802* with acquisition of Cas3 and Cas4, while Type III emerges from the branch leading to *Synechocystis sp. 6803* without further gene gains.

Type I was spread to the other bacteria with a HGT to the ancestor of Proteobacteria. According to Wang et al. [29], the Cyanobacteria ancestor is estimated between 2,230 and 3,000 Mya, while the last common ancestor (LCA) of Alpha-, Beta-, and Gammaproteobacteria is estimated between 2,360 and 2,620 Mya. These calibrations make time-plausible the lateral transfer of Type I CRISPR-Cas from the *Cyanothece sp. PCC 8802* branch, close to the Cyanobacteria ancestor, to the LCA of Proteobacteria. A second HGT from *Geobacter sulfurreducens* brought Type I to the Firmicutes *Bacillus halodurans C-125* and

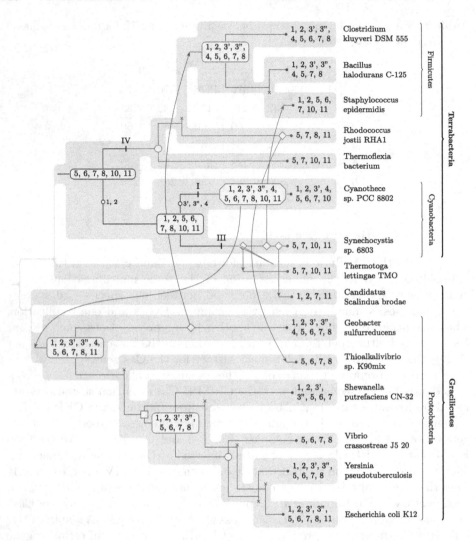

Fig. 4. Representation of one of the 32 solutions inferred by Algorithm SuperDTL reconstructing a DTL evolutionary history of the Class 1 Cas gene synteny dataset illustrated in Fig. 3. Event costs set to $c_{Loss} = 1$, $c_{HGT} = 4$ and $c_{Dup} \in \{1, 1.5, 2, 2.5\}$ yield equal solutions. The red, blue, and green parts of the tree reflect the evolution of Cas synteny types IV, I, and III respectively. Gains of Cas1+Cas2 and Cas3(Cas3'+Cas3")+Cas4 are illustrated. (Color figure online)

Clostridium kluyveri DSM 555. Type III was spread across bacteria with a succession of three HGTs, which also seem time-consistent.

The analysis of the synteny evolutionary history across the Proteobacteria subtree shows an unexpected scenario highlighting a limitation of our model. The SuperDTL algorithm inferred an ancestral synteny duplication before the LCA of *S. putrefaciens*, *V. crassostreae*, *Y. pseudotuberculosis* and *E. coli* resulting in a succession of three consecutive full synteny losses along the branch to the synteny in *E. coli* which is an unlikely evolutionary scenario. An alternative evolutionary scenario would consist in a speciation in place of the duplication, copying the ancestral synteny to an unsampled or extinct species which would later be transferred back to *Escherichia coli*. Such a model of HGT events as combinations of a speciation event outside the species tree (to an unsampled or dead lineage) followed by a transfer back inside the species tree has been described in [26]. This alternative scenario would replace one duplication and three losses with one speciation and one transfer, which yields the same cost under the $c_{Loss} = 1$, $c_{HGT} = 4$ and $c_{Dup} \in \{1, 1.5, 2, 2.5\}$ model but would be less costly for $c_{HGT} < 4$. Consideration of unsampled and dead lineages in the DTL Super-reconciliation will therefore be necessary to infer better synteny evolutionary scenarios.

9   Conclusion

We have developed SuperDTL, the first exact algorithm for inferring most parsimonious evolutionary histories for a set of syntenies, given a phylogenetic tree of the syntenies and a phylogenetic tree of the corresponding species, for an evolutionary model accounting for segmental duplications, gains, losses and HGT events. We only presented the unordered version of the problem in this paper, but the algorithm developed in [5] for the ordered version of the DL Super-reconciliation model can also be extended to the DTL Super-reconciliation model. However, as rearrangements are not considered in the model, a DT or DTL Super-reconciliation only exists if extant gene orders are pairwise consistent, which is a strong constraint, not verified, for example, in the case of the Cas gene syntenies of Fig. 3. Considering a unifying model accounting for both DTL and rearrangement events remains a challenge. The analysis of the synteny evolutionary scenario in Proteobacteria emphasized the need in DTL Super-reconciliation model to infer HGT to and from unsampled or extinct species to produce more realistic evolutionary scenarios. In addition, in order to avoid the unjustified constraint of having all genes at the root of the tree, we allowed for gene gains, which has been essential for the analysis of Cas gene syntenies. Although we defined the general evolutionary model in a way allowing for segmental gains (i.e. gains of a group of genes), we only considered single gene gains in our algorithm. An extension is however possible and may be considered for future work.

The reconstruction of an evolutionary history for Cas gene syntenies using SuperDTL provides a first attempt to reconcile the evolutionary scenario of Cas syntenies in the context of the evolution of bacterial species. However, several improvements could be brought to the considered Cas synteny dataset to better reconstruct its evolutionary history. First, we excluded archaea from our dataset while several studies show evidence for an emergence of CRISPR-Cas systems from archaeal species followed by horizontal transfers to bacteria. This is supported by the fact that most considered archaea have a complete CRISPR-Cas system while only part of bacteria have one. In fact, as of the latest update from January 21, 2021 of the CRISPRCasdb database[1] [24], 70.64% of analyzed archaeal species versus 36.27% of analyzed bacteria had a complete CRISPR-Cas system. A phylogenetic dating approach could also be used to produce a dated phylogeny of prokaryotes and to constrain HGT events. A larger dataset with species sampling representing the diversity of archaea and bacteria with dated phylogeny and Cas synteny tree based on synteny content and sequence divergence is required to further elucidate the evolutionary history of the CRISPR-Cas system in prokaryotes.

A   Additional Content for Sect. 4 ("Most Parsimonious Super-Reconciliations")

Lemma 2 (Optimal gain point position). *Let $I = \langle \mathcal{X}, \mathcal{F}, T, \Sigma, S \rangle$. There is an optimal history H for I such that, in the Super-reconciliation $\mathcal{R} = \langle T, e, x, s \rangle$ obtained from H, for each $\Gamma \in \mathcal{F}$, $\mathrm{lca}(\Gamma, T)$ is the gain point for Γ in T.*

Proof. Let $H = \langle H^{\mathrm{tree}}, e, x, s \rangle$. Let v be the node corresponding to $\mathrm{lca}(\Gamma, T)$ in H^{tree}. The fact that $\mathrm{lca}(\Gamma, T)$ is the gain point for Γ in T means that the gene family Γ is gained on the branch (u, v) where u is the root of H^{tree} or the node just preceding v in H^{tree}. The result then follows from the fact that: (1) Γ cannot be gained on a node of the left subtree $H^{\mathrm{tree}}[v_l]$ of $H^{\mathrm{tree}}[v]$ as in that case, Γ should also have been gained on another node on the right subtree $H^{\mathrm{tree}}[v_r]$ of $H^{\mathrm{tree}}[v]$ (or conversely); (2) thus Γ can only be gained on a node between the root r of H^{tree} and v; moving the gain point to an ancestor of v cannot decrease the number of losses.   □

Lemma 7. *Let $H = \langle H^{\mathrm{tree}}, e, x, s \rangle$ be a history, and $v \neq w$ be two nodes of H^{tree} such that none of the nodes on the path from v to w (excluding v and w themselves) is a HGT event and such that $s(v)$ and $s(w)$ are not separated. Then, there are at least $\mathrm{D}_S^{\mathrm{node}}(s(v), s(w))$ speciation nodes on the path from v to $\mathrm{p}(w)$.*

Proof. This follows from the constraints of Definition 1, which states that the only way to descend in the species tree, excluding HGT events, is through Spe events.   □

Lemma 8. *Let $H = \langle H^{\mathrm{tree}}, e, x, s \rangle$ be a history, and v, w_l, w_r be three nodes of H^{tree} such that v is a binary node, w_l descends from v_l and w_r from v_r, and none of the nodes on the paths from v to w_l and v to w_r (excluding v, w_l, and w_r themselves) are Dup or HGT events. Then:*

1. If $e(v) = \mathrm{Spe}$, there are at least $\mathrm{D}^{\mathrm{sub}}(x(v), x(w_l)) + \mathrm{D}^{\mathrm{sub}}(x(v), x(w_r))$ loss events on the path from w_l to w_r.

---

[1] Available at https://crisprcas.i2bc.paris-saclay.fr/MainDb/StrainList.

2. If $e(v) = $ Dup, there are at least $\min\{\mathrm{D}^{\mathrm{sub}}(x(v), x(w_l)), \mathrm{D}^{\mathrm{sub}}(x(v), x(w_r))\}$ loss events on the path from w_l to w_r.
3. If $e(v) = $ HGT, if $s(w_l)$ (resp. w_r) is not separated from $s(w)$ there is at least $\mathrm{D}^{\mathrm{sub}}(x(v), x(w_l))$ (resp. w_r) loss event on the path from w to w_l (resp. w_r).

Proof. This follows from the constraints of Definition 1, which states that the only way to loose part of a synteny, excluding Dup and HGT events, is through Loss events, and that Dup events allow loosing part of their synteny on either of their children, while HGT events allow loosing part of their synteny on their conserved child. □

Lemma 3 (Super-reconciliations minimize history cost). *Let $\mathcal{R} = \langle T, e, x, s \rangle$ be a Super-reconciliation for $I = \langle \mathcal{X}, \mathcal{F}, T, \Sigma, S \rangle$ and $\mathcal{H}$ be the set of histories leading to $\mathcal{R}$. Then, for any $C_\delta(\mathcal{R}) = \min_{H \in \mathcal{H}} c_\delta(H)$.*

Proof. Let $H = \langle H^{\mathrm{tree}}, e, x, s \rangle$ be a history leading to $\mathcal{R}$. Let us first prove that $c_\delta(H) \geq C_\delta(\mathcal{R})$, by structural induction on T. Let v be a node of T and v' be its corresponding node in H^{tree}. If v is a leaf, then $c_\delta(H, v') \geq C_\delta(\mathcal{R}, v) = 0$; otherwise, let v'_l and v'_r be the nodes in H^{tree} corresponding to v_l and v_r respectively, and assume $c_\delta(H, v'_l) \geq C_\delta(\mathcal{R}, v_l)$ and $c_\delta(H, v'_r) \geq C_\delta(\mathcal{R}, v_r)$.

- If $e(v) = $ Spe, by Lemma 7, there are at least $\mathrm{D}^{\mathrm{node}}_S(s(v), s(v_l)) - 1$ speciation nodes on the path from $(v')_l$ to v'_l and at least $\mathrm{D}^{\mathrm{node}}_S(s(v), s(v_r)) - 1$ speciation nodes on the path from $(v')_r$ to v'_r. Each of those speciation nodes must have at least a full loss child, for otherwise v'_l or v'_r would not correspond to v_l and v_r. By Lemma 8.1, there are at least $\mathrm{D}^{\mathrm{sub}}(x(v), x(v_l)) + \mathrm{D}^{\mathrm{sub}}(x(v), x(v_r))$ loss events on the path from v'_l to v'_r.
- If $e(v) = $ Dup, by Lemma 7, there are at least $\mathrm{D}^{\mathrm{node}}_S(s(v), s(v_l))$ speciation nodes on the path from $(v')_l$ to v'_l and at least $\mathrm{D}^{\mathrm{node}}_S(s(v), s(v_r))$ speciation nodes on the path from $(v')_r$ to v'_r, each of which must have a full loss child as per the same argument as above. By Lemma 8.2, there are at least $\min\{\mathrm{D}^{\mathrm{sub}}(x(v), x(v_l)), \mathrm{D}^{\mathrm{sub}}(x(v), x(v_r))\}$ loss events on the path from v'_l to v'_r.
- If $e(v) = $ HGT, assume w.l.o.g. that v_l is the conserved child. By Lemma 7, there are at least $\mathrm{D}^{\mathrm{node}}_S(s(v), s(v_l))$ speciation nodes on the path from $(v')_l$ to v'_l, each of which must have a full loss child as per the same argument as above. By Lemma 8.3, there are at least $\mathrm{D}^{\mathrm{sub}}(x(v), x(v_l)$ loss events on the path from v' to v'_l.

In all three cases, $c_\delta(H, v') \geq C_\delta(\mathcal{R}, v)$, concluding the first part of the proof. Additionally, it is easy to see that a history $H_\mathcal{R}$ can be constructed from $\mathcal{R}$ such that $H_\mathcal{R} \in \mathcal{H}$ and $c_\delta(H_\mathcal{R}) = C_\delta(\mathcal{R})$, by inserting speciation nodes and loss nodes in the locations described above. □

Lemma 4 (Min-event labeling is optimal). *Let $\mathcal{R} = \langle T, e, x, s \rangle$ and $\mathcal{R}^{\mathrm{min}} = \langle T, e^{\mathrm{min}}_s, x, s \rangle$ be two DTL Super-reconciliations for $I = \langle \mathcal{X}, \mathcal{F}, T, \Sigma, S \rangle$. Then, $C_\delta(\mathcal{R}^{\mathrm{min}}) \leq C_\delta(\mathcal{R})$.*

Proof. Given s, the choice for e is constrained by Definition 1, as evidenced by Lemma 1. For any internal node v of T, if either $s(v_l)$ or $s(v_r)$ is separated from $s(v)$, then it must be that $e(v) = \text{HGT}$. If $s(v_l)$ and $s(v_r)$ are not separated (i.e., either one is an ancestor of the other), then it must be that $e(v) = \text{Dup}$. Finally, if $s(v_l)$ is separated from $s(v_r)$, then $e_s^{\min}(v) = \text{Spe}$, but it would also be valid to set $e(v)$ to Dup. However, in that case, replacing Dup back with Spe would save two full losses at the cost of adding at most one partial loss, which cannot lead to a more costly Super-reconciliation since speciations have a null cost and $c_{\text{Dup}}, c_{\text{Loss}} \geq 0$. □

B   Additional Content for Sect. 6 ("A Dynamic Programming Algorithm for DTL Super-Reconciliation")

Lemma 6 (Termination and correctness). *For any $v \in \text{V}(T)$ and $\sigma \in \text{V}(S)$, $\mathcal{C}(v, \sigma) = \text{C}(v, \sigma)$ and $\mathcal{C}^\star(v, \sigma) = \text{C}^\star(v, \sigma)$ (as computed by Algorithms 1 and 2 respectively).*

Proof. First, note that both algorithms terminate even though they are mutually recursive, since any call to $\mathcal{C}(v, \sigma)$ or $\mathcal{C}^\star(v, \sigma)$ calls $\mathcal{C}$ and $\mathcal{C}^\star$ only on v_l and v_r. We proceed by structural induction to prove correctness. Let v be a leaf of T. Leaves cannot be labeled by any synteny or mapped to any species other than the ones specified in the problem input, therefore $\text{C}(v, s(v)) = 0 = \mathcal{C}(v, s(v))$, $\text{C}(v, \sigma) = \infty = \mathcal{C}(v, \sigma)$ for any $\sigma \neq s(v)$, and $\text{C}^\star(v, \sigma) = \infty = \mathcal{C}^\star(v, \sigma)$ for any $\sigma \in \text{V}(S)$.

Now, let v be an internal node and assume that for any $\sigma \in \text{V}(S)$, $\mathcal{C}^\star(v_l, \sigma) = \text{C}^\star(v_l, \sigma)$, $\mathcal{C}^\star(v_r, \sigma) = \text{C}^\star(v_r, \sigma)$, $\mathcal{C}(v_l, \sigma) = \text{C}(v_l, \sigma)$, and $\mathcal{C}(v_r; \sigma) = \text{C}(v_r, \sigma)$. Let σ be any node of the species tree. Both $\mathcal{C}^\star(v, \sigma)$ and $\mathcal{C}(v, \sigma)$ explore all possible pairs $\sigma', \sigma'' \in \text{V}(S)^2$ that can be mapped respectively to v_l and v_r, and return the minimum cost of all those options.

Consider the computation of $\mathcal{C}^\star(v, \sigma)$ by Algorithm 2. Since in that case $x(v) \supsetneq x^{\text{lca}}(v)$, then $x(v) \not\subseteq x^{\text{lca}}(v_l)$. Therefore, setting $x(v_l) = x^{\text{lca}}(v_l)$ (which costs $\mathcal{C}(v_l, \sigma') = \text{C}(v_l, \sigma')$) would imply that $\text{D}^{\text{sub}}(x(v), x(v_l)) = 1$. On the contrary, setting $x(v_l) \supsetneq x^{\text{lca}}(v_l)$ (which costs $\mathcal{C}^\star(v_r, \sigma') = \text{C}^\star(v_r, \sigma')$) would give us the freedom to choose $x(v_l)$ such that $x(v) \subseteq x(v_l)$, implying that $\text{D}^{\text{sub}}(x(v), x(v_l)) = 0$. The same logic holds for v, v_r, and σ''. Note that each of the three cases in the innermost loop uses the same cost computation formula as Definition 5, albeit adapted to test all four options of setting $x(v_l)$ and $x(v_r)$ to either x^{lca} or a superset of it. By Lemma 5, those are the only four options to consider, so $\mathcal{C}^\star(v, \sigma) = \text{C}^\star(v, \sigma)$.

Consider the computation of $\mathcal{C}(v, \sigma)$ by Algorithm 1. In that case, $x(v) = x^{\text{lca}}(v)$, therefore setting $x(v_l) \supsetneq x^{\text{lca}}(v_l)$ is only allowed if $x^{\text{lca}}(v) \not\subseteq x^{\text{lca}}(v_l)$, for otherwise there would be at least one gene family with two or more gain points. When setting $x(v_l) = x^{\text{lca}}(v_l)$ (which costs $\mathcal{C}(v_l, \sigma') = \text{C}(v_l, \sigma')$), the presence of a loss depends on the value of $\text{D}^{\text{sub}}(x^{\text{lca}}(v), x^{\text{lca}}(v_l))$. On the other hand, setting

$x(v_l) \supsetneq x^{\mathrm{lca}}(v_l)$ (which costs $\mathcal{C}^\star(v_r, \sigma') = \mathrm{C}^\star(v_r, \sigma'')$) would give us the freedom to choose $x(v_l)$ such that $x(v) \subseteq x(v_l)$, implying that $\mathrm{D}^{\mathrm{sub}}(x(v), x(v_l)) = 0$. The same logic holds for v, v_r, and σ''. Note that this algorithm also follows Definition 5, handling all four options mentioned previously, and excluding disallowed cases. So $\mathcal{C}(v, \sigma) = \mathrm{C}(v, \sigma)$. $\qquad\square$

Theorem 1 (Time and space complexity). *Using SuperDTL, the* DTL SUPER-RECONCILIATION *problem can be solved in polynomial time* $\mathcal{O}(|\mathcal{X}| \times |\Sigma|^3)$ *and space* $\mathcal{O}(|\mathcal{X}| \times |\Sigma|)$.

Proof. Both dynamic programming tables have exactly $|\mathrm{V}(T)| \times |\mathrm{V}(S)|$ entries each. Computing a single entry of one of the tables takes time $\mathcal{O}(|\mathrm{V}(S)|^2)$, provided all the other required entries are made available in constant time and set and tree operations are implemented in an efficient way. A single bottom-up traversal of the synteny tree is enough to fill in both tables using Algorithms 1 and 2, therefore, computing the tables takes time $\mathcal{O}(|\mathrm{V}(T)| \times |\mathrm{V}(S)|^3)$. To compute the overall minimal cost is to compute the minimum value of the C table in column $\mathrm{r}(T)$, which can be done in time $\mathcal{O}(\mathrm{V}(S))$. To construct an optimal solution, pointers to optimal σ', σ'' pairs and to which of C and $\mathrm{C}^\star$ is used must be tracked for each table entry, taking up a constant time and space for each entry. After the tables have been computed, tracing back those pointers allows constructing a solution by following $\mathcal{O}(|\mathrm{V}(T)|)$ pointers. The overall time and space complexities are therefore $\mathcal{O}(|\mathrm{V}(T)| \times |\mathrm{V}(S)|^3)$ and $\mathcal{O}(|\mathrm{V}(T)| \times |\mathrm{V}(S)|)$, or, equivalently, $\mathcal{O}(|\mathcal{X}| \times |\Sigma|^3)$ and $\mathcal{O}(|\mathcal{X}| \times |\Sigma|)$. $\qquad\square$

References

1. Bansal, M.S., Alm, E.J., Kellis, M.: Efficient algorithms for the reconciliation problem with gene duplication, horizontal transfer and loss. Bioinformatics **28**(12), i283–i291 (2012)
2. Bérard, S., Gallien, C., Boussau, B., Szollosi, G.J., Daubin, V., Tannier, E.: Evolution of gene neighborhoods within reconciled phylogenies. Bioinformatics **28**(18), i382i388 (2012)
3. Chakraborty, S., Snijders, A.P., Chakravorty, R., Ahmed, M., Tarek, A.M., Anwar Hossain, M.: Comparative network clustering of direct repeats (DRs) and cas genes confirms the possibility of the horizontal transfer of CRISPR locus among bacteria. Mol. Phylogenet. Evol. **56**(3), 878–887 (2010)
4. Coleman, G.A., et al.: A rooted phylogeny resolves early bacterial evolution. Science **372** (2021)
5. Delabre, M., et al.: Evolution through segmental duplications and losses: a super-reconciliation approach. Algorithms Mol. Biol. **15**(12) (2020)
6. Dondi, R., Lafond, M., Scornavacca. C.: Reconciling multiple genes trees via segmental duplications and losses. Algorithms Mol. Biol. **14** (2019)
7. Duchemin, W.: Phylogeny of dependencies and dependencies of phylogenies in genes and genomes. Université de Lyon, Theses (2017)
8. El-Mabrouk, N.: Predicting the evolution of syntenies-an algorithmic review. Algorithms **14**(5), 152 (2021)

9. Duchemin, W., et al.: DeCoSTAR: reconstructing the ancestral organization of genes or genomes using reconciled phylogenies. Genome Biol. Evol. **9**, 1312–1319 (2017)

10. Fitch, W.M.: Distinguishing homologous from analogous proteins. Syst. Zool. **19**(2), 99 (1970)

11. Godde, J.S., Bickerton, A.: The repetitive DNA elements called CRISPRs and their associated genes: evidence of horizontal transfer among prokaryotes. J. Mol. Evol. **62**(6), 718–729 (2006)

12. Goodman, M., Czelusniak, J., Moore, G.W., Romero-Herrera, A.E., Matsuda, G.: Fitting the gene lineage into its species lineage, a parsimony strategy illustrated by cladograms constructed from globin sequences. Syst. Biol. **28**(2), 132–163 (1979)

13. Górecki, P., Tiuryn, J.: DLS-trees: a model of evolutionary scenarios. Theor. Comput. Sci. **359**(1–3), 378–399 (2006)

14. Guigo, R., Muchnik, I., Smith, T.F.: Reconstruction of ancient molecular phylogeny. Mol. Phylogenet. Evol. **6**(2), 189–213 (1996)

15. Horvath, P., Coûté-Monvoisin, A.-C., Romero, D.A., Boyaval, P., Fremaux, C., Barrangou, R.: Comparative analysis of CRISPR loci in lactic acid bacteria genomes. Int. J. Food Microbiol. **131**(1), 62–70 (2009)

16. Jinek, M., Chylinski, K., Fonfara, I., Hauer, M., Doudna, J.A., Charpentier, E.: A programmable dual-RNA-guided DNA endonuclease in adaptive bacterial immunity. Science **337**(6096), 816–821 (2012)

17. Koonin, E.V., Makarova, K.S.: Origins and evolution of CRISPR-Cas systems. Philos. Trans. R. Soc. B Biol. Sci. **374**(1772), 20180087 (2019)

18. Koonin, E.V., Makarova, K.S.: Evolutionary plasticity and functional versatility of CRISPR systems. PLOS Biol. **20**(1), e3001481 (2022)

19. Makarova, K.S., Aravind, L., Wolf, Y.I., Koonin, E.V.: Unification of Cas protein families and a simple scenario for the origin and evolution of CRISPR-Cas systems. Biol. Dir. **6**(1), 38 (2011)

20. Makarova, K.S., et al.: Evolutionary classification of CRISPR-Cas systems: a burst of class 2 and derived variants. Nat. Rev. Microbiol. **18**(2), 67–83 (2020)

21. Moya-Beltrán, A., et al.: Evolution of type IV CRISPR-Cas systems: insights from CRISPR loci in integrative conjugative elements of acidithiobacillia. CRISPR J. **4**(5), 656–672 (2021)

22. Roderic, D.M.: Page. Maps between trees and cladistic analysis of historical associations among genes, organisms, and areas. Syst. Biol. **43**(1), 58–77 (1994)

23. Paszek, J., Gorecki. P.: Efficient algorithms for genomic duplication models. IEEE/ACM Trans. Comput. Biol. Bioinform. **15**, 1515–1524 (2017)

24. Pourcel, C., et al.: CRISPRCasdb a successor of CRISPRdb containing CRISPR arrays and cas genes from complete genome sequences, and tools to download and query lists of repeats and spacers. Nucleic Acids Res. **48**(D1), D535–D544 (2020)

25. Schoch, C.L., et al.: NCBI taxonomy: a comprehensive update on curation, resources and tools. Database **2020** (2020)

26. Szöllősi, G.J., Tannier, E., Lartillot, N., Daubin, V.: Lateral gene transfer from the dead. Syst. Biol. **62**(3), 386–397 (2013)

27. Tofigh, A., Hallett, M., Lagergren, J.: Simultaneous identification of duplications and lateral gene transfers. IEEE/ACM Tran. Comput. Biol. Bioinform. **8**(2), 517–535 (2011)

28. Tyson, G.W., Banfield, J.F.: Rapidly evolving CRISPRs implicated in acquired resistance of microorganisms to viruses. Environ. Microbiol. **10**(1), 200–207 (2008)

29. Wang, S., Meade, A., Lam, H.-M., Luo, H.: Evolutionary timeline and genomic plasticity underlying the lifestyle diversity in rhizobiales. mSystems **5**(4), e00438 (2020)
30. Weiner, S., Bansal, M.S.: Improved duplication-transfer-loss reconciliation with extinct and unsampled lineages. Algorithms **14**(8), 231 (2021)
31. Zhang, L.: On a Mirkin-Muchnik-Smith conjecture for comparing molecular phylogenies. J. Comput. Biol. **4**(2), 177–187 (1997)
32. Zmasek, C.M., Eddy, S.R.: A simple algorithm to infer gene duplication and speciation events on a gene tree. Bioinformatics **17**, 821–828 (2001)

Quantifying Hierarchical Conflicts
in Homology Statements

Krister M. Swenson[1,2](✉) [ID], Afif Elghraoui[3] [ID], Faramarz Valafar[4] [ID],
Siavash Mirarab[5] [ID], and Mathias Weller[1,6] [ID]

[1] Centre National de la Recherche Scientifique (CNRS), Paris, France
mathias.weller@cnrs.fr
[2] LIRMM, University of Montpellier, Montpellier, France
swenson@lirmm.fr
[3] Department of Electrical and Computer Engineering, San Diego State University,
San Diego, CA, USA
aelghraoui@sdsu.edu
[4] Department of Electrical and Computer Engineering, University of California,
San Diego, La Jolla, CA, USA
faramarz@sdsu.edu
[5] School of Public Health, San Diego State University, San Diego, CA, USA
smirarab@ucsd.edu
[6] LIGM, Université Gustave Eiffel, Paris, France

Abstract. A fundamental step in any comparative whole genome analysis is the annotation of homology relationships between segments of the genomes. Traditionally, this annotation has been based on coding segments, where orthologous genes are inferred and then syntenic blocks are computed by agglomerating sets of homologous genes into homologous regions. More recently, whole genomes, including intergenic regions, are being aligned *de novo* as whole genome alignments (WGA). In this article we develop a test to measure to what extent sets of homology relationships given by two different software are hierarchically related to one another, where matched segments from one software may contain matched segments from the other and *vice versa*. Such a test should be used as a sanity check for an agglomerative syntenic block software, and provides a mapping between the blocks that can be used for further downstream analyses. We show that, in practice, it is rare that two collections of homology relationships are perfectly hierarchically related. Therefore we present an optimization problem to measure how far they are from being so. We show that this problem, which is a generalization of the assignment problem, is NP-Hard and give a heuristic solution and implementation. We apply our distance measure to data from the Alignathon competition, as well as to *Mycobacterium tuberculosis*, showing that many factors affect how hierarchically related two collections are, including sensitivities to guide trees and the use or omission of an outgroup. These findings inform practitioners on the pitfalls of homology relationship inference, and can inform development of more robust inference tools.

L. Jin and D. Durand (Eds.): RECOMB-CG 2022, LNBI 13234, pp. 146–167, 2022.
https://doi.org/10.1007/978-3-031-06220-9_9

Keywords: Homology · Syntenic block · T-Star Packing · Assignment problem

1   Introduction

The increasing ease of whole genome sequencing and assembly has opened a new era of comparative genomics. With the data available today, not only can the phylogenetic histories of all the genes between a set of genomes be analyzed, but also the interaction between these genes, linking gene regulation and function to the positions of groups of genes. These analyses require a reliable grouping of homologous genomic segments from the multiple genomes in question.

Thus, the inference of sets of homologous genomic segments is of fundamental importance. Such a *homology statement* comes in the form of a set of genomic segments that contains at least one, but potentially multiple, segments from several genomes. Each pair of segments from the set shares common ancestry over some proportion of their intervals, which varies depending on the scale and level of precision required by the application.

The most basic segment on which statements are made has traditionally been the gene, detected through either manual or automatic means. The number of tools designed to infer homology relationships between annotated genes has grown, provoking the formation of the Quest for Orthologs (QfO) consortium dedicated to the evaluation and comparison of these tools [12].

General genomic intervals, that can contain both coding and noncoding positions, have also been used as homology statements. In this case, researchers have considered bidirectional best hits as evidence for orthology [22,31]. More recently, "whole genome alignment" methods partition entire genomes into blocks that can be aligned into multiple sequence alignments (MSAs), *de novo*, with no special input from the user. The Alignathon collaborative competition was developed to evaluate and compare these methods [9].

Study of the large scale changes between genomes has inspired a more vague notion of homology between genome segments. Even before the discovery of the double helix, groups were studying homology of large segments of genomes from the salivary glands of drosophila [29]. More precise lengths of roughly *conserved chromosomal segments* began to be studied using linkage maps [23]. In the postgenomic era, basic homology segments are agglomerated into *syntenic blocks*, possibly separated by micro-rearrangements. GRIMM-synteny was developed for the study of large scale chromosomal changes, in response to the whole genome sequencing efforts in human and mouse [27]. Since then, many syntenic block inference tools have been introduced but, despite twenty years of development, a unified definition of syntenic block has yet to be found. Indeed, most tools rely on operational definitions rather than biological or mathematical ones [11,30].

There are several tests used for comparing and evaluating homology statements. For orthology statements between coding sequences, the QfO project has established tests that: 1) compare trees inferred from orthologous families to agreed-upon species trees, 2) compare the subsets of orthologs from curated gene families, and 3) use consistency in gene ontology annotation [1]. For statements between syntenic blocks, comparisons of inclusion and exclusion of segments between methods have been done, and blocks from a single method have been compared to known gene clusters [19]. Other sources of ground truth, such as RNA-seq data, have been used to confirm co-regulation between genes occurring in a proposed block [33].

For general statements (on both coding and noncoding DNA), the Alignathon competition used three different measures [9]. If homology statements are given as a set of (potentially gapped) equal-length segments from several genomes, then each homologous pair of positions between two genomes as given by one method, can be queried in another method. The number of such shared positions is a measure of similarity and, when one of the methods is taken as ground truth, the number of shared positions can be used to measure precision and recall. The `mafComparator` tool estimates these values by sampling positions [9]. For sets of aligned homology statements (*i.e.* MSAs), *probabilistic sampling-based alignment reliability* (PSAR) was used to assess each aligned column [16]. PSAR fixes all rows of the alignment but one, and samples from the many ways to align that row within the fixed alignment. After this is repeated for each row, an alignment reliability score for each pair of positions in a column can be assigned. When aligned homology statements are augmented with a phylogeny, another statistical test called `StatSigMA` can be used [28]. For each edge of the phylogeny, the rows of the alignment are split into two alignments. The two alignments are then tested for exhibiting "unrelated behaviour" using Karlin-Altschul log likelihood scores. If the test for all branches passes, then the homology statement is validated.

For homology statements that come in the form of syntenic blocks, Ghiur-cuta and Moret outline some necessary conditions for a valid agglomeration of homologous units into such blocks [11].

There exists very few methods that compare homology statements in the form of sets of genomic segments, unmarried to connotations of orthologous genes, and independent of multiple sequence alignments. To our knowledge, the Jaccard distance (e.g. as computed by `mafComparator`) applied to pairwise homology statements, is the only known comparison that falls into this category.

In this article we introduce a simple definition of *homology block* (Sect. 2.1) and formally characterize the conditions under which a set of homology blocks are valid (Sect. 2.2). We show what it means for collections of blocks to be hierarchically related and use this to develop a method for measuring disagreement between two different collections of blocks. In Sect. 2.3 we show a necessary condition for two collections of blocks to be in a hierarchical relationship (in the form of Lemma 1), based on a graph representing the overlap between the sets. For different parts of the genomes in question, our test allows for the collections of blocks to be hierarchically related in both ways; in some parts of the genomes the first set could be more general than the second, while in other parts of the genomes the opposite can be true. We introduce an optimization problem, called MINIMUM DELETION INTO DISJOINT STARS (MDDS), which gives a lower bound on the number of positions that must be ignored so that the two collections of blocks could be related through a hierarchical relationship. Not only does a solution to MDDS give a measure as to the degree of hierarchical dissonance between two collections, but it serves as an unambiguous mapping between the blocks of the two. This mapping could be used for further downstream comparisons, an illustration of which is shown in Appendix C.

We show that the MDDS problem is NP-Hard, before presenting a polynomial time heuristic based on an exact algorithm for solving MDDS on a tree. In Sect. 4 we define the *homology discordance ratio* and use this measure as a distance between block collections built on Alignathon data and on a set of 94 *Mycobacterium tuberculosis* isolates. On the tuberculosis strains we study the relationship between blocks built using an outgroup or no outgroup, using annotations or no annotations, using maf2synteny to agglomerate or not, as well as study the effect of the guide tree on Cactus MSA blocks. On the simulated data from the Alignathon project, we highlight differences between our method and the Jaccard distance (as computed with mafComparator).

For the entirety of the article we focus on the general case of homology statements, although most of the discussion also applies to the restricted case of orthology statements.

2   Methodological Foundations

2.1   Overlapping Homology Statements and the Block Graph

We use $g_i[k..\ell]$ to denote a *segment* between positions k and ℓ of genome g_i and we let T denote the universe of all segments over all possible genomes and position-pairs. Define the overlap $op(s_1, s_2)$ between two segments s_1 and s_2 as the number of positions where they overlap in the same genome. For example $op(g_1[1..5], g_1[3..9]) = 3$ but $op(g_1[1..5], g_2[3..9]) = 0$.

Definition 1 (homology statement block). *A homology statement block (called a block for short) B is a set of segments $B \subset T$ such that all pairs of segments in B have zero overlap.*

The right panel of Fig. 1 depicts two collections of homology statement blocks $\mathcal{A} = \{A1, A2, A3\}$ and $\mathcal{B} = \{B1, B2, B3\}$. The blocks of $\mathcal{A}$ are $A1 = \{g_1[1..12], g_2[13..24]\}$, $A2 = \{g_1[14..28], g_2[26..41]\}$, $A3 = \{g_1[53..66], g_2[101..113]\}$, while the blocks of $\mathcal{B}$ are $B1 = \{g_1[1..28], g_2[13..41]\}$, $B2 = \{g_1[46..68], g_2[94..115]\}$, and $B3 = \{g_1[34..39], g_2[42..46], g_2[116..121]\}$.

Before discussing the semantic interpretation of a homology statement block, we first introduce a graph that represents the overlap between blocks. The overlap $op(B1, B2) = \sum_{s_1, s_2 \in B1 \times B2} op(s_1, s_2)$ between blocks $B1$ and $B2$ is the total overlap between all pairs of segments in the two. A collection of blocks $\mathcal{B} = \{B1, B2, \ldots\}$ is considered to be *clean* if the overlap between any pairs of blocks in $\mathcal{B}$ is zero. Both collections depicted in Fig. 1 are clean.

For two collections of blocks $\mathcal{A}$ and $\mathcal{B}$, we build a bipartite *block graph* $BG(\mathcal{A}, \mathcal{B})$ where there is an edge between A and B for any $A \in \mathcal{A}$ and $B \in \mathcal{B}$ if

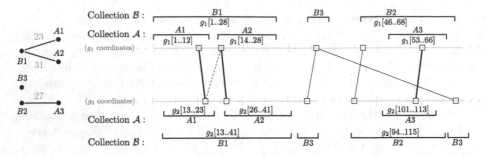

Fig. 1. To the right, the collections of blocks $\mathcal{A} = \{A1, A2, A3\}$ and $\mathcal{B} = \{B1, B2, B3\}$ appearing in genomes g_1 and g_2, along with the graph $BG(\mathcal{A}, \mathcal{B})$. Segments are depicted with brackets and lined up according to their positions on the chromosomes. They are labeled by their tuple (when space permits) and the block to which they belong. The configuration of positive and negative witness pairs shows that $\mathcal{B}$ generalizes $\mathcal{A}$. Some of the genome positions are highlighted with boxes, and two such positions are connected by a solid line if they appear as a positive witness in $\mathcal{B}$, and that line is bold if they are also in $\mathcal{A}$. The dashed line represents one of the (many) negative homology witnesses between $A1$ and $A2$ that are negative witness pairs for $\mathcal{A}$ but not for $\mathcal{B}$. **To the left,** the graph $BG(\mathcal{A}, \mathcal{B})$ appears with edges labeled by overlap length in gray. All of the connected components are stars.

and only if blocks A and B overlap (*i.e.* $op(A, B) > 0$). Thus, the block graph $BG(\mathcal{A}, \mathcal{A})$ for a clean collection $\mathcal{A}$ is a perfect matching. $E(G)$ is the set of edges of the graph G. We associate to each edge $AB \in E(G)$ a weight function $\omega : E(G) \to \mathbb{N}$ such that $\omega(AB) = op(A, B)$. The left side of Fig. 1 shows the block graph for the collections to the right.

2.2   Homology Witnesses and Block Hierarchies

A homology block can be interpreted as a *positive* and *negative* statement of homology (*i.e.* statements about common ancestry). On the positive side, the block $\{g_1[1..5], g_2[11..16]\}$ says that positions 1 through 5 in genome g_1 are somehow homologous to positions 11 through 16 in genome g_2 (in this case the segments are not the same length, so we assume that each position from the segment $g_1[1..5]$ is homologous to either a position in $g_2[11..16]$ or to none other in g_2). On the negative side, the block could be interpreted as saying that *no* position in $g_1[1..5]$ is homologous to any other position in g_1 or any other position in g_2, outside of $g_2[11..16]$.

In this section we suppose that we know the truth about the ancestral relationships between the base-pair positions of the genomes in question. With this supposed knowledge, we can categorize pairs of *homology witness* positions as positive or negative, depending on their evolutionary relationship. Using these relationships, we define properties that a valid collection of homology blocks must respect. These definitions are extended to encompass hierarchical relationships between collections of blocks.

Consider any pair of positions $g_i[x]$ and $g_j[y]$ from genomes g_i and g_j. This pair is called a *positive homology witness* if the two positions descend from the same ancestral position, otherwise the pair is called a *negative homology witness* (positive homology witnesses represent pairs of positions that are typically called "homologous" positions). Note that the true relationship between positions is unknown, yet it imposes constraints on what we consider a valid collection of blocks according to the following definition.

Consider any position-pair $(g_i[x], g_j[y])$ such that $g_i[x]$ is contained in a segment from a block B in a collection $\mathcal{B}$. If $(g_i[x], g_j[y])$ is a positive homology witness, then either

1. $g_j[y]$ appears in B, and we say that the pair is a *positive witness in B*, or
2. $g_j[y]$ appears in no block of $\mathcal{B}$.

By this definition no position-pair $(g_i[x], g_j[y])$ with $g_i[x]$ and $g_j[y]$ in different blocks of $\mathcal{B}$, can be a positive homology witness and, since all position-pairs are either positive or negative homology witnesses, $(g_i[x], g_j[y])$ must be a negative homology witness. Any position-pair $(g_i[x], g_j[y])$, where $g_i[x]$ and $g_j[y]$ are in different blocks of $\mathcal{B}$ or in no block of $\mathcal{B}$, is called a *negative witness for $\mathcal{B}$*.

Note that, for a clean collection $\mathcal{B}$, no position-pair can be both a positive and negative witness in/for $\mathcal{B}$. There may also be position-pairs that are neither positive nor negative witnesses in/for $\mathcal{B}$, such as those pairs that have one position contained in a block of $\mathcal{B}$ and the other outside all blocks of $\mathcal{B}$. Finally, note that not all position-pairs appearing between segments in a homology block need necessarily be positive homology witnesses.

Positive witness pairs limit what can exist in two different blocks; a block containing one position of a positive homology witness imposes the constraint that the other position must either be in the same block, or in no block. On the other hand, we will see in the following that negative homology witnesses existing between two different blocks in a collection enforce constraints on the hierarchical relationships that this collection can have with another block collection.

Consider the collections of blocks in Fig. 1 and note that whenever a positive homology witness is a positive witness in $\mathcal{A}$, it must also be a positive witness in $\mathcal{B}$, whereas not all positive witnesses in $\mathcal{B}$ exist in $\mathcal{A}$. Conversely, every negative witness for $\mathcal{B}$ is also a negative witness for $\mathcal{A}$. In this sense, the blocks of $\mathcal{B}$ are "more general" than the blocks of $\mathcal{A}$. This motivates the following definition, for which we focus on subcollections of blocks $\mathcal{A}' \subseteq \mathcal{A}$ and $\mathcal{B}' \subseteq \mathcal{B}$.

Definition 2 (generalization). *A clean (sub)collection of blocks $\mathcal{B}'$ generalizes a clean (sub)collection $\mathcal{A}'$ if and only if every positive witness in $\mathcal{A}'$ is also a positive witness in $\mathcal{B}'$ and every negative witness for $\mathcal{B}'$ is also a negative witness for $\mathcal{A}'$.*

Note that any clean collection generalizes itself.

While some subcollections of $\mathcal{B}$ may generalize subcollections of $\mathcal{A}$, other subcollections of $\mathcal{A}$ may generalize subcollections of $\mathcal{B}$. Partition them $\mathcal{A} = \mathcal{A}_1 \cup \mathcal{A}_2 \cup \cdots \cup \mathcal{A}_k$ and $\mathcal{B} = \mathcal{B}_1 \cup \mathcal{B}_2 \cup \cdots \cup \mathcal{B}_k$ according to the connected components of $BG(\mathcal{A}, \mathcal{B})$ (e.g. $\mathcal{A}_1 \cup \mathcal{B}_1$ is the set of vertices in the first connected component).

Definition 3 (hierarchical). *We say that $\mathcal{A}$ and $\mathcal{B}$ have a* hierarchical *relationship if and only if $\mathcal{A}_i$ generalizes $\mathcal{B}_i$, or $\mathcal{B}_i$ generalizes $\mathcal{A}_i$, for $1 \le i \le k$.*

The existence of hierarchical relationships between collections of blocks are interesting to us for at least two reasons. Consider two block inference methods, MethodA and MethodB, producing different clean collections of blocks $\mathcal{A}$ and $\mathcal{B}$ respectively. If MethodB is meant to agglomerate blocks from MethodA, then we would expect $\mathcal{B}$ to generalize $\mathcal{A}$. This is useful for the verification of agglomeration methods, and as a sanity check for the practitioner. In this case, if MethodB also trims spurious blocks or segments from MethodA, $\mathcal{B}$ may not generalize $\mathcal{A}$, but $\mathcal{A}$ and $\mathcal{B}$ would still be hierarchically related. Another reason for interest in the hierarchical relationship may be that, if $\mathcal{B}$ generalizes $\mathcal{A}$, then we can define a mapping from each block $A \in \mathcal{A}$ to a block $B \in \mathcal{B}$. This mapping can be used for further comparisons between the collections. The refinement of orthology assignments, as illustrated in Appendix C is an example of one such comparison.

2.3   Relating Block Hierarchy to Stars in the Block Graph

While simple hierarchical relationships are easy to detect, real-world data are not so well behaved, and require a formalism to measure the extent to which a relationship is hierarchical. The types of connected components in the block graph give us insight into collections that cannot have a hierarchical relationship.

Lemma 1. *Let $\mathcal{A}$ and $\mathcal{B}$ be clean collections of blocks such that $\mathcal{B}$ generalizes $\mathcal{A}$ and $BG(\mathcal{A}, \mathcal{B})$ is connected. Then, all vertices of $\mathcal{A}$ have degree one in $BG(\mathcal{A}, \mathcal{B})$, that is, $BG(\mathcal{A}, \mathcal{B})$ is a star with center in $\mathcal{B}$.*

Proof. Let A be a block in $\mathcal{A}$, and assume that it has at least two distinct neighbors $B1, B2 \in \mathcal{B}$ in $BG(\mathcal{A}, \mathcal{B})$, that is, both $B1$ and $B2$ overlap A. Thus, there are positions $g_i[x]$ and $g_j[y]$ appearing in A such that $g_i[x]$ appears in $B1$ and $g_j[y]$ appears in $B2$. Since $\mathcal{B}$ is clean, we also know that these positions are distinct. Since $(g_i[x], g_j[y])$ appears in different blocks in $\mathcal{B}$, we know that it is a negative homology witness for $\mathcal{B}$. However, since $g_i[x]$ and $g_j[y]$ appear in the same block in $\mathcal{A}$, the pair is not a negative witness for $\mathcal{A}$. Thus, $(g_i[x], g_j[y])$ is a negative witness for $\mathcal{B}$ but not for $\mathcal{A}$, contradicting the fact that $\mathcal{B}$ generalizes $\mathcal{A}$.   □

We say that a graph is *hierarchical* if it is a collection of vertex-disjoint stars, that is, if no component has two vertices of degree greater than one. It is easy to check if a graph meets this criterion. Note that the condition of a graph $BG(\mathcal{A}, \mathcal{B})$ being hierarchical is necessary for $\mathcal{A}$ and $\mathcal{B}$ to have a hierarchical relationship, but it is not sufficient. Also note that, Lemma 1 outlines a property on each individual connected component, allowing some parts of $\mathcal{A}$ to generalize parts

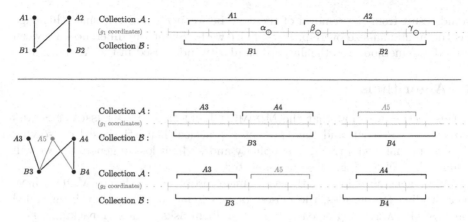

Fig. 2. Two subsets of blocks from collections $\mathcal{A}$ and $\mathcal{B}$ that are not hierarchically related. The **top panel** depicts a subset of blocks for part of the genome g_1, but not the other genomes. Positions α and β form a negative homology witness for $\mathcal{A}$, but not for $\mathcal{B}$, while β and γ form a negative homology witness for $\mathcal{B}$, but not for $\mathcal{A}$. This contradicts properties of a hierarchy and, therefore, yields the non-star topology to the left. The **bottom panel** depicts a subset of blocks for part of the genomes g_1 and g_2. These segments contradict properties of a hierarchy in a different way, and yield the non-star topology to the left. This kind of scenario would arise when orthologs are matched in one way from MethodA, and in another way for MethodB. Note that even if the block $A5$ was not in collection $\mathcal{A}$, the contradiction still holds and the connected component is not a star.

of $\mathcal{B}$, while allowing other parts of $\mathcal{B}$ to generalize parts of $\mathcal{A}$. Thus, a natural corollary to Lemma 1 is that if $\mathcal{A}$ and $\mathcal{B}$ are hierarchically related, then $BG(\mathcal{A}, \mathcal{B})$ is hierarchical.

For a graph that is *not* a collection of stars, one may want to measure to what degree it deviates from being so. Lemma 1 inspires the search for *star packings* on $G = BG(\mathcal{A}, \mathcal{B})$.

— MINIMUM DELETION INTO DISJOINT STARS (MDDS) —

Input: a bipartite graph G with weight function $\omega : E(G) \to \mathbb{N}$
Output: $E' \subseteq E(G)$ such that the subgraph of G formed by the edge set $(E(B) \setminus E')$ is a collection of vertex-disjoint stars
Measure: $\sum_{e \in E'} \omega(e)$

A solution to MDDS gives a lower bound on the number of overlapping positions that must be ignored so that $\mathcal{A}$ and $\mathcal{B}$ can be hierarchically related. For example, Fig. 2 shows two connected components that are not stars. Consider the graph from the upper panel and assume that in the non-depicted genomes (*i.e.* g_i for $i > 1$) there is no overlap of the segments of $A1$ with those of $B2$, or of segments of $A2$ with those of $B1$. Then, the solution to MDDS on this component would result from the removal of the edge between $B1$ and $A2$, since this edge has the minimum-size overlap. In Sect. 4.1 we highlight the differences between our MDDS method, and the Jaccard similarity index used in the Alignathon.

3   Algorithms

In this section we show that the MINIMUM DELETION INTO DISJOINT STARS problem is NP-Hard, and then present a practical heuristic based on solving MDDS optimally on a tree. For simplicity and without loss of generality, we will assume that all block graphs are connected.

Other generalizations of the assignment problem, similar to MDDS, have been studied for decades, the closest of which has most recently been called the T-STAR PACKING problem [4]. This problem asks for a star packing where the size of the star is limited by an input parameter T. When the measure is the number (or weight) of edges, Hell and Kirkpatrick show that the T-STAR PACKING is NP-Hard by reduction to the version of the problem that asks for a decomposition of a given graph into subgraphs isomorphic to the star with T edges [13]. Since the only difference between MDDS and the edge-weighted T-STAR PACKING problem is the parameter T, it is tempting to adapt the same series of reductions to MDDS by setting T to the maximum degree over all vertices in the graph. This approach is not clearly feasible, however, since the reduction from 3-DIMENSIONAL MATCHING to the decomposition version of the problem creates vertices of degree higher than T [17].

Babenko and Gusakov give a $\frac{9}{4}\frac{T}{T+1}$ approximation algorithm for the T-STAR PACKING problem based on a reduction to the max-network flow problem [4]. We could use this elaborate approximation algorithm by fixing T to the maximum degree of the input graph, but we choose instead to implement the much simpler heuristic presented in Sect. 3.2.

3.1   NP-Hardness of MDDS

We will show that the decision version of MDDS is NP-hard by reducing the well-known 3-SAT problem to it. Our construction uses similar techniques as the NP-hardness proof of the TRANSITIVITY EDGE DELETION problem [32].

Construction 1 (see Fig. 3). *Let φ be an instance of 3-SAT with variables $X := \{x_1, x_2, \ldots, x_n\}$ and clauses $\mathcal{C} := \{C_1, C_2, \ldots, C_m\}$ such that each clause contains exactly three literals. For each variable x_i, let n_i denote the number of clauses that contain x_i or $\neg x_i$ and let $\gamma_i^0, \gamma_i^1, \ldots, \gamma_i^{n_i-1}$ be any sequence of these clauses. We construct an edge-weighted graph (G, ω) as follows:*

1. *For each variable x_i create a cycle Q_i containing $6n_i$ vertices $v_i^0, v_i^1, \ldots, v_i^{6n_i-1}$ and give all edges weight m.*
2. *For each clause $C_k \in \mathcal{C}$, create a single vertex u_k.*
3. *For each i, j let ℓ be such that $\gamma_i^j = C_\ell$ and, if x_i occurs non-negated in C_ℓ, then add the edge $\{v_i^{6j}, u_\ell\}$ with weight 1, otherwise add the edge $\{v_i^{6j+2}, u_\ell\}$ with weight 1.*

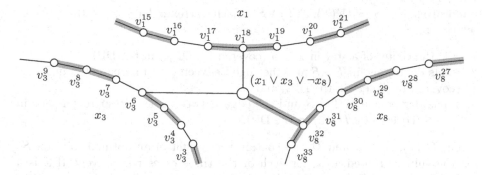

Fig. 3. Example of Construction 1. The clause $C := (x_1 \vee x_3 \vee \neg x_8)$ corresponding to the center vertex is equal to $\gamma_1^3 = \gamma_3^1 = \gamma_8^5$, that is, it is the 4th clause containing x_1, the 2nd clause containing x_3 and the 6th clause containing x_8. A truth-assignment setting x_1 to TRUE and x_3 and x_8 to FALSE corresponds to the star cover indicated by gray highlights. Note that taking the edge between v_1^{18} and C instead of the edge between v_8^{32} and C corresponds to satisfying the clause C by x_1 instead of $\neg x_8$.

Note that the image of ω is $\{1, m\}$, the total weight of all edges is $18m^2 + 3m$, and G is bipartite, since any edge from part 3. of the construction, connecting a (variable) cycle to a (clause) vertex u_ℓ, connects to an even numbered vertex in the cycle.

Besides NP-hardness, our reduction implies exponential lower bounds assuming widely believed complexity-theoretic hypotheses. The "Exponential-Time Hypothesis" (combined with results by Impagliazzo et al. [15]) roughly states that 3SAT on formulas with m clauses cannot be decided in $2^{o(m)}$ time. This lower bound transfers since the constructed graph G has only $21m$ edges.

Theorem 1. MINIMUM DELETION INTO DISJOINT STARS *is NP-hard and cannot be solved in* $2^{o(|E(G)|)}$ *time on graphs* G, *even if* G *is restricted to maximum degree three, assuming the Exponential-Time Hypothesis.*

3.2 A Heuristic for MDDS

In light of the hardness result presented in Sect. 3.1, we devised a heuristic that first computes a maximum-weight spanning tree T on each connected component of BG. It then transforms each T into a star packing by computing MDDS on T.

We present a dynamic programming algorithm solving MDDS on a tree T. To this end, we root T at an arbitrary vertex, and compute a dynamic programming table for vertices in a post-order traversal. Consider a set S of edges that, after removal from T, yields a collection of disjoint stars. We denote the result of this removal as $T - S = (V(T), E(T) \setminus S)$. Each vertex x has one of three states relative to S:

1. x is the center of a star in $T - S$ (covered by $D_*(x)$ in the DP),
2. x has degree one in $T - S$ and the edge between x and its parent is in $T - S$ (covered by $D_+(x)$ in the DP), and
3. x has degree one in $T - S$ and the edge between x and its parent is not in $T - S$ (covered by $D_-(x)$ in the DP).

Then, $D_*(x)$, $D_+(x)$, and $D_-(x)$ contain the weight of an optimal solution S_x for the subtree rooted at x, for each of the three cases respectively. If x is a leaf of T, then set $D_*(x) := D_-(x) := D_+(x) := 0$. Otherwise, let $v_1, v_2, \ldots, v_m$ denote the children of x in T. We visit the children in this order and accumulate each partial subsolution, starting with $D_*^0(x) := D_+^0(x) := D_-^0(x) := 0$ and proceeding as follows for each $1 \leq i \leq m$:

$$D_*^i(x) := D_*^{i-1}(x) + \min \left(D_+(v_i), \ \omega(xv_i) + \min(D_*(v_i), D_-(v_i)) \right)$$

That is, if x is the center of a star, then the edge xv_i must be in S if either v_i is the center of a star or the edge between v_i and its parent x is not in $T - S$.

$$D_+^i(x) := D_+^{i-1}(x) + \omega(xv_i) + \min(D_*(v_i), D_-(v_i))$$

That is, if x is a leaf of a star centered at the parent of x, then the edge xv_i must be in S.

$$D_-^i(x) := \min \left(\begin{array}{l} D_-^{i-1}(x) + \omega(xv_i) + \min(D_*(v_i), D_-(v_i)), \\ D_+^{i-1}(x) + D_*(v_i) \end{array} \right)$$

The case of $D_-^i(x)$ is a bit more subtle. Since x is not the center of a star, all but at most one edge between x and its children are in S, so if xv_i is not in S then $D_+^{i-1}(x)$ forces all xv_j to be in S, for $1 \leq j < i$. Finally, the subsolutions rooted at x are, then, given by:

$$D_+(x) := D_+^m(x) \qquad D_-(x) := D_-^m(x) \qquad D_*(x) := D_*^m(x)$$

4   Quantifying Hierarchical Conficts

We applied our MDDS heuristic to homology statements on a set of prokaryotes, and on a set of eukaryotes. The solution to MDDS provides an estimate of the minimum number of positions that must be ignored so that the necessary conditions for a hierarchy, highlighted by Lemma 1, are achieved. Before applying the heuristic of Sect. 3.2 we cleaned the syntenic blocks according to Appendix B, and preprocessed the graphs for segmental duplications according to Appendix C.

4.1   Discordance Ratio and Distinction from Jaccard Index

Define *coverage(B)* of a collection of blocks B as the total number of positions covered by all segments in blocks of B. We report the *hierarchical discordance ratio* between collections A and B as $d(A, B) = w/(coverage(A) + coverage(B))$, where w is the weight of the MDDS on $BG(A, B)$. A discordance ratio of 0.1 means that we have to ignore at least 10% of the total coverage of the blocks (in both methods) in order to have a hierarchical relationship between them.

Alignathon used `mafComparator` to compute the straightforward Jaccard similarity index between collections of blocks. In this case, the elements of the sets in question are the pairwise alignments of positions implied by the blocks. So if a pair of positions are aligned in one class of blocks but not the other, this will contribute one to the denominator.

Consider collections A and B such that A only contains blocks with segments from $\{g_i[200x+1..200x+100] \mid 0 \le x < \lfloor \frac{\ell(g_i)}{200} \rfloor\}$, and B only contains blocks with segments from $\{g_i[200x + 101..200x + 200] \mid 0 \le x < \lfloor \frac{\ell(g_i)}{200} \rfloor\}$, for all genomes g_i with length $\ell(g_i)$. In other words, the collections can only have blocks of length 100 that do not overlap with each other. In this case the Jaccard similarity measure will be zero no matter the length of the genomes, indicating the most severe dissimilarity, whereas the two collections are hierarchically related, showing no conflicts, and the block graph is composed only of degree zero vertices. In that sense, our comparison method is tolerant to collections that conservatively make no assertion about a region.

We consider the two measures complementary in that they capture different qualities of the overlap properties of block collections. We see in Sect. 4.3 instances from the Alignathon data where the Jaccard similarity is low, yet the two collections are hierarchically related, and *vice versa*.

4.2   *Mycobacterium Tuberculosis* Clinical Isolates

For the prokaryotes, we used a set of 94 *Mycobacterium tuberculosis* strains [5,21] with homology statements given by the methods listed in Table 1. These sets of blocks are those produced in [10], where the methods were compared to assess their impact on inferring rearrangement phylogenies. Note that all `Cactus` blocks used in this subsection had segments with fewer than 50 positions filtered out.

The collections of blocks for four of the methods, along with their `maf2synteny` counterparts are compared in Fig. 4. As expected, each collection of blocks had a very low discordance ratio with its counterpart agglomerated by `maf2synteny`. Further, the agglomerated blocks always have lower discordance to all the other methods, when compared to their unagglomerated counterparts. Of the unagglomerated methods, `SibeliaZ` is the least discordant.

There were a couple of surprises. The first is that the most discordant pairs are between the gene-based annotation method and the *de novo* inference methods `Cactus` and `SibeliaZ`. Contrary to the other methods, the agglomerated annotation blocks show a small improvement against `Cactus`, and a surprising degradation (going up from 4% to 5%) in the discordance ratio for `SibeliaZ`.

Table 1. Homology statement determination methods applied to the *M. tuberculosis* genomes

Method	Description
Cactus(SNP)	Cactus [3] alignment guided by a ML tree based on Concatenated substitutions with respect to reference strain H37Rv (NCBI accession NC_000962.3)
Cactus(SibeliaZ)	Cactus alignment guided by a MLWD [14] adjacency tree Computed from SibeliaZ+M2S synteny blocks
Cactus(Mash)	Cactus alignment guided by a B(I)ONJ tree based on the genomes' Mash [25] distance matrix
SibeliaZ	Locally collinear blocks produced in the first step of the SibeliaZ pipeline [20]
Annotation	Simultaneous annotation and orthology assignment by 95% amino acid sequence identity and 95% Alignment coverage
Modifiers	**Description**
+out	Synteny blocks computed while including the outgroup Strain *M. canettii* (NCBI accession NC_019951.1)
+M2S	Agglomerated with maf2synteny [18]

This implies that either 1) many blocks from the Cactus method bridge between coding regions, or 2) many duplicate regions are assigned in discordant ways. The second surprise is that the unagglomerated Cactus methods, with different guide trees, are more discordant from each other than they are with SibeliaZ. It has been reported that Cactus's sensitivity to guide trees also has implications on the downstream phylogenetic analyses [10].

In Fig. 5, the checkered pattern shows that the inclusion of an outgroup affects Cactus blocks more than the choice of a guide tree. The inclusion of the outgroup strain also decreases the discordance between the Cactus blocks on different guide trees. For example, Cactus(Mash) has discordance ratios of 0.044 and 0.061 against Cactus(SNP) and Cactus(SibeliaZ), but for Cactus(Mash)+out these values are 0.022 and 0.026. Table 2 shows the discordance between a method and its version with the outgroup. Cactus is most highly affected by the inclusion of the outgroup. While SibeliaZ is somewhat affected, Annotation is barely affected. Agglomerating the blocks with maf2synteny diminishes the discordance in all cases but Annotation.

4.3   Alignathon

The Alignathon competition was created to compare "whole genome alignment" methods [9]. Authors of WGA software were invited to submit the collections of blocks computed by their program, which were compared using the measures

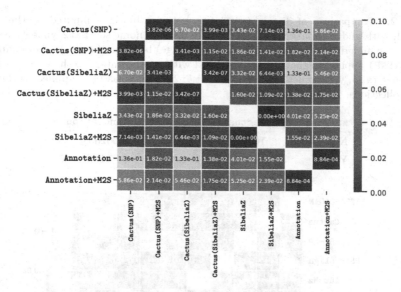

Fig. 4. The discordance ratio between each pair of block collections. Each of methods Cactus(SNP), Cactus(SibeliaZ), SibeliaZ, and Annotation along with their maf2synteny counterpart.

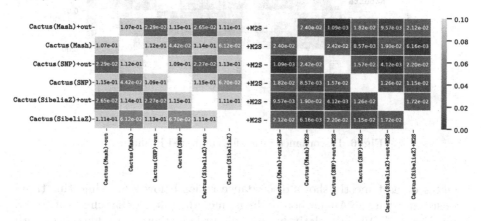

Fig. 5. The inclusion of an outgroup affects Cactus blocks more than the choice of a guide tree.

described in the introduction. The project fabricated two synthetic datasets that were used to evaluate the block collections, one that mimicked the properties of set of Primates, and another that mimicked the properties of a set of Mammals.

We applied our MDDS heuristic to each pair of block collections. Note that we were limited to the collections available on the Alignathon downloads page, so were unable to compare to some methods, such as Mercator/Pecan [26]. The results for the Primate dataset are depicted in Fig. 6 while results for the mammal dataset are depicted in Fig. 7. Being evolutionarily closely related, the

Table 2. Comparison of discordance ratios between blocks computed by the same method, with and without an outgroup in the input genome set. maf2synteny usually reduces the high discordance (shown as a percentage) between the blocks. For example, Cactus(SNP) applied to the TB sets with and without the outgroup shows a very high divergence ratio, yet a much lower one after maf2synteny has been applied to the two block collections.

maf2synteny?	Cactus(SNP)	SibeliaZ	Annotation
No	10.09%	3.57%	0.00263%
Yes	1.57%	2.37%	0.0547%

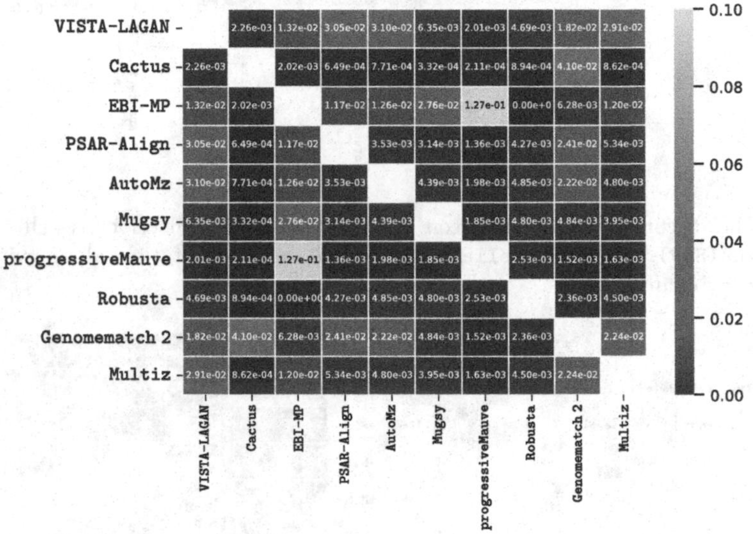

Fig. 6. Discordance ratios for simulated Primates.

Primates dataset mostly shows discordance ratios below 2%. While this trend is consistent with the Alignathon findings, including GenomeMatch2 (SoftBerry, Mount Kisco, NY) being relatively more discordant, there were differences with the Jaccard index reported by Alignathon. VISTA-LAGAN [8], for instance, stands out as generally more discordant than the others, being rather dissimilar to PSAR-Align [16], AutoMz, and Multiz [6]. EBI-MP stands out as having both the best, and the worse discordance ratios of the dataset; despite having a ratio of over 12% against progressiveMauve [7], it also is the only method in the set to be hierarchically related to another one (Robusta [24]). Cactus has very low discordance with all methods except GenomeMatch2.

The simulated mammal dataset contained genomes that were separated by a larger evolutionary distance, and this was reflected in surprisingly large discordance ratios. We observe several discrepancies with the Jaccard distances reported by Alignathon ([9] – Fig. 8B). GenomeMatch3 was extremely dissimilar to all

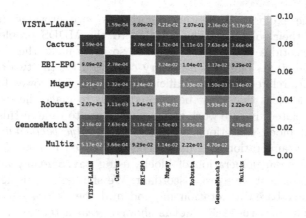

Fig. 7. Discordance ratios for simulated Mammals.

methods but Mugsy [2], yet we observe high hierarchical discordance ratios only against Robusta and Multiz. Cactus has low hierarchical discordance against all other methods, whereas it had high Jaccard distances against Mugsy, GenomeMatch3, and EBI-EPO. On the other hand, Robusta seemed to have poor comparisons for both the Jaccard and hierarchical measures.

5 Discussion and Conclusions

In this article we addressed the question of how to relate two collections of homology statement blocks to each other. We established a relationship between collections where we allowed overlapping parts of those collections to be hierarchically related. In the absence of these conditions, we developed a method that gives a lower bound on the number of positions that must be ignored in order for the two to be hierarchically related.

The notion of being hierarchically related depends on semantics that we imposed on the blocks, which speak to the pairwise homology relationships between the constituent genomic positions appearing in the blocks. As Ghiurcuta and Moret [11] used "homology statements" to define their "syntenic blocks", we used "positive homology witness" pairs to limit which segments can be contained within a homology statement block; while they required every homology statement within a segment to occur in all other segments of the block, we allowed positions that do not appear in a positive witness pair in the block, as long as they do not occur in another block. We went further by associating semantic meaning to the fact that two positions appear in different blocks. This allowed us to define what it means for some (sub)collection of blocks to generalize another (sub)collection.

On the algorithmic side, we showed the MINIMUM DELETION INTO DISJOINT STARS problem to be NP-Complete. Our heuristic for MDDS is based on a dynamic program that solves MDDS exactly on a tree. Future improvements will include the exploration of other algorithms with provable guarantees to the quality of their solutions. The solution to the MDDS problem gives the number of nucleotides that must be ignored so as to make the components of the block graph stars. This is a necessary condition for the two collections to be hierarchically related, but not sufficient, and thus is a lower bound on the number of nucleotides that must be ignored so as to make the two collections hierarchically related. Future work will explore ways to tighten this bound.

We studied block collections on a set of 94 *Mycobacterium tuberculosis* strains, built by annotation and non-annotation based means. We showed on this data that the agglomeration of blocks using `maf2synteny` almost always yielded collections that were less discordant. We showed surprising discordance between the gene-based annotation method and the *de novo* block inference methods `Cactus` and `SibeliaZ`. `Cactus` showed great heterogeneity, dependent on the guide tree that was used to construct the blocks.

When performing a phylogenetic analysis on the blocks, one is tempted to incorporate an outgroup for the sake of rooting the tree. We showed the inclusion of that outgroup had drastic effects on blocks, producing blocks that were less sensitive to the `Cactus` guide tree. This was concordant with our results from a phylogenetic study [10].

We studied block collections from the Alignathon project. The simulated Primates dataset showed that `EBI-MP` had both the best discordance ratio, and the worst, among all pairwise comparisons, being hierarchically related to the `Robusta` blocks while having ratio over 0.13 with `progressiveMauve`. For the less closely related simulated mammalian genomes, we showed several discrepancies between the Jaccard index reported by Alignathon and our discordance ratio, the most notable one being that while `Cactus` had a poor Jaccard index against a few methods, it had very low hierarchical discordance with all other methods (except `GenomeMatch2`).

While WGA tools and syntenic block agglomeration methods have continued to be developed, the methods to compare and analyze them has lagged behind, and the definitions of syntenic blocks are usually procedural or based on co-linearity. In this article we outlined constraints on homology blocks based on the homology relationships between pairs of positions in the genome. These constraints put as much importance on the ends of the blocks as it does their contents; if two genomic segments are put into different blocks, we interpret this as a statement that should only be contradicted in a generalization of the blocks. Our new measure should inform future block inference tool development, and serve as a sanity check for the practitioner studying large scale structure of sets of genomes.

Acknowledgement. The authors would like to thank the helpful suggestions by the reviewers. AE and FV are supported by the NIAID grant R01AI105185. KS is partially supported by the grant ANR-20-CE48-0001.

Availability of Code. All of the code associated with this paper is publicly availble at the following URL: https://bitbucket.org/thekswenson/homology-evaluation.

A   NP-Hardness of MDDS

Note that the notion of star and *induced* star coincide on bipartite graphs since, for any bipartite G, the vertices of any star-subgraph of G also form an induced star in G. Further, no collection of node-disjoint stars can contain the triangle C_3 or the path on 4 vertices P_4 as a subgraph and it can be seen that this condition is also sufficient.

Observation 1. *A bipartite graph G is a collection of stars if and only if G does not contain a P_4 subgraph.*

For the correctness proof, we will make two assumptions on the structure of the input formula φ, without loss of generality. First, we assume that no variable occurs in all clauses. If a variable x does occur in all clauses, then we simply add a new variable y and the singleton clause on y. Second, we assume that each clause in φ has exactly three literals. If a clause C has at most two literals, we can simply double the occurrence of any literal in C.

Lemma 2. *Let φ be an instance of 3SAT and let $(G = (V, E), \omega)$ be the result of applying Construction 1 to φ. Then, φ is satisfiable if and only if (G, ω) has a star packing of weight at least $12m^2 + m$.*

Proof. For each variable x_i of φ, let us define the edge sets

$$T_i := \bigcup_{0 \le j < 2n_i} \{v_i^{3j} v_i^{3j+1}, v_i^{3j+2} v_i^{3j \oplus 3}\} \quad \text{and} \quad F_i := \bigcup_{0 \le j < 2n_i} \{v_i^{3j+1} v_i^{3j+2}, v_i^{3j+2} v_i^{3j \oplus 3}\}$$

where $3j \oplus 3 := (3j + 3) \bmod 6n_i$. Note that any v_i^j has degree two in subgraph (V, T_i) if and only if $j \equiv 0 \mod 3$ and any v_i^j has degree two in (V, F_i) if and only if $j \equiv 2 \mod 3$. Further, $\omega(T_i) = \omega(F_i) = 4mn_i$. We prove the two directions of the lemma separately.

$\Rightarrow$: Let φ be satisfiable, that is, there is a set L of literals over variables in φ such that each clause C_k intersects L in at least one literal ℓ_k and L contains exactly one of x_i and $\neg x_i$ for all i. If ℓ_k is the literal x_i in clause $C_k = \gamma_i^j$, then let $e_k := u_k v_i^{6j}$ and, if ℓ_k is the literal $\neg x_i$, then let $e_k := u_k v_i^{6j+2}$. Note that all e_k are distinct, $e_k \in E(G)$ for all k. Let S^{clause} contain e_k for all clauses C_k of φ and note that $\omega(S^{\text{clause}}) = m$. Further, for all variables x_i of φ, let $S_i^{\text{var}} := T_i$ if $x_i \in L$ and $S_i^{\text{var}} := F_i$, otherwise (that is, $\neg x_i \in L$). Finally, let the selected edges be $S := S^{\text{clause}} \cup \bigcup_i S_i^{\text{var}}$ (see gray edges in Fig. 3), noting that $\omega(S) = 12m^2 + m$. It remains to show that (V, S) does not contain a P_4 as a subgraph. Towards a contradiction, assume that (V, S) contains a P_4 $p := (a, b, c, d)$. By construction, neither $\bigcup_i S_i^{\text{var}}$ nor S^{clause} contains a P_4, and p must contain edges from both of these sets. Thus, p contains u_k for some clause C_k. Since any u_k has degree one in (V, S), we can assume without loss of generality that $ab = e_k$. By definition of e_k, there are i and j such that either $b = v_i^{6j}$ and $x_i \in L \cap C_k$ or $b = v_i^{6j+2}$ and $\neg x_i \in L \cap C_k$. Since $6j \equiv 0 \bmod 3$ and $6j + 2 \equiv 2 \bmod 3$ we know that in both cases b has degree two in (V, S_i^{var}), and both of its neighbors have degree one in (V, S_i^{var}) and, thus, in (V, S). This contradicts (a, b, c, d) being a path in (V, S).

$\Leftarrow$: Let S be a maximum-weight subset of E such that $\omega(S) \geq 12m^2 + m$ and (V, S) does not contain a P_4 as a subgraph. First, let S^{clause} denote the set of edges of S incident with a clause node u_k. Second, for each x_i, let S_i^{var} denote the set of edges of S on the variable cycle corresponding to x_i and note that, for each P_3 (a, b, c) in (V, S_i^{var}), both a and c have degree one in (V, S_i^{var}) since, otherwise, (V, S) contains a P_4. For each i, the connected components of (V, S_i^{var}) are paths of lengths 1, 2, or 3 and we denote the number of P_1s, P_2s, and P_3s in (V, S_i^{var}) by r_i, s_i and t_i, respectively. By construction, each P_2 is adjacent to at most one clause vertex u_k in (V, S), and since (V, S) does not contain P_4, each P_3 is also adjacent to at most one clause vertex u_k in (V, S).

Claim. $\sum_i t_i = 6m$.

Proof. By decomposing the $18m$ vertices of the variable cycles into P_3 subgraphs separated by single edges, the upper bound of $6m$ is attained. It suffices to show $\sum_i t_i \geq 6m$ so, towards a contradiction, assume that $\sum_i t_i < 6m$. Then, there is a variable x_i such that $t_i < 2n_i$ implying $|S_i^{\mathrm{var}}| \leq 4n_i - 1$ by construction. Let S' result from S by removing all edges incident with vertices of the variable cycle corresponding to x_i and adding the edges in T_i. Since x_i does not occur in all clauses, we removed edges of total weight strictly less than $m + (4n_i - 1)m = 4mn_i$ and we added edges of total weight $m|T_i| = 4mn_i$. Since neither (V, S) nor (V, T_i) contains a P_4, neither does (V, S'), thus contradicting optimality of S. ∎

Corollary 1. *Each subgraph (V, S_i^{var}) decomposes into disjoint copies of P_3.*

Corollary 2. *Let v_i^j and $v_i^{j'}$ be nodes of degree two in some subgraph (V, S_i^{var}). Then, $|j - j'| \equiv 0 \bmod 3$.*

Corollary 3. *Let $u_k v_i^j$ be an edge in S^{clause}. Then, v_i^j has degree two in (V, S_i^{var}).*

Note that each P_3 in $(V, \bigcup_i S_i^{\mathrm{var}})$ has weight exactly $2m$, so S contains exactly m edges of S^{clause}. Further, by Corollary 3, all clause vertices u_k have degree at most one since they are not adjacent to degree-one vertices. Together, this means that all clause vertices u_k are incident to *exactly* one edge in S.

We now construct an assignment β and show that it satisfies φ. To this end, let $\beta(x_i) = \mathrm{TRUE}$ if and only if S contains the edge $u_k v_i^{6j}$ for some $j, k \in \mathbb{N}$. Note that, if S contains the edge $u_k v_i^{6j}$ for any $j, k \in \mathbb{N}$ then, by Corollary 1, v_i^{6j} has degree two in (V, S_i^{var}). Then, by Corollary 2, S cannot contain the edge $u_{k'} v_i^{6j+2}$ for any $j', k' \in \mathbb{N}$. Thus, β is well-defined. It remains to show that β satisfies φ. To this end, let C_k be any clause in φ, let $u_k z$ be the unique edge incident with u_k in S and let x_i be the variable whose variable cycle contains z.

If x_i occurs non-negated in C_k, then $z = v_i^{6j}$ for some $j \in \mathbb{N}$ by construction. But then, $\beta(x_i) = \text{TRUE}$ and x_i satisfies C_k. If x_i occurs negated in C_k, then $z = v_i^{6j+2}$ for some $j \in \mathbb{N}$ by construction. But then, $\beta(x_i) = \text{FALSE}$ and x_i satisfies C_k. In both cases, C_k is satisfied. □

B   Collections of Block that are not Clean

Many of the software we studied produced blocks that were not clean, containing blocks with overlapping segments. We removed overlapping segments by visiting pairs of blocks in an arbitrary order, removing the overlap between their overlapping segments. Although the order in which overlaps are removed can effect the final set of blocks, we made the process deterministic by visiting the pairs in a fixed order.

C   Segmental Duplications

If some method makes orthology predictions that may contain multiple segments from the same genome (*e.g.* clusters of orthologous groups that contain paralogs from a single genome), the block graph may provide insight into how to refine the orthology groups using blocks from another method. This section outlines such a case.

When a block $A \in \mathcal{A}$ contains multiple segments from multiple genomes, blocks from another set $B1, B2 \in \mathcal{B}$ could overlap in ways that create non-star graph topologies. Figure 8 shows one such example.

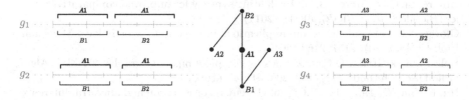

Fig. 8. The block $A1 \in \mathcal{A}$ has two (duplicated) segments in genomes g_1 and g_2. The blocks $B1, B2 \in \mathcal{B}$ each overlap with one of the two copies. This configuration creates the non-star topology depicted in the middle. The block $A1$ can easily be split into two so that the graph becomes only stars. This results in a refinement of the blocks of $\mathcal{A}$, based on the blocks of $\mathcal{B}$.

Blocks $B1$ and $B2$ each overlap one of the two duplicate copies of $A1$ in genomes g_1 and g_2. The block $A1$ can be split into two blocks $A1'$ and $A1''$ such that the collections $\{A1', A1'', A2, A3\}$ and $\mathcal{B}$ are hierarchically related. The two connected components of $BG(\{A1', A1'', A2, A3\}, \mathcal{B})$ are both stars with vertices $\{A2, B2, A1'\}$ and $\{A1'', B1, A3\}$. This transformation can be generalized to vertices of higher degree, as long as the overlapping segments can be split in this way.

References

1. Altenhoff, A.M.: Standardized benchmarking in the quest for orthologs. Nat. Methods **13**(5), 425–430 (2016)
2. Angiuoli, S.V., Salzberg, S.L.: Mugsy: fast multiple alignment of closely related whole genomes. Bioinformatics **27**(3), 334–342 (2011)
3. Armstrong, J., et al.: Progressive Cactus is a multiple-genome aligner for the thousand-genome era. Nature **587**(7833), 246–251 (2020). (Nature Publishing Group)
4. Babenko, M., Gusakov, A.: New exact and approximation algorithms for the star packing problem in undirected graphs. In: Symposium on Theoretical Aspects of Computer Science (STACS2011), vol. 9, pp. 519–530 (2011)
5. Berney, M., et al.: Essential roles of methionine and S-adenosylmethionine in the autarkic lifestyle of Mycobacterium tuberculosis. Proc. Natl. Acad. Sci. **112**(32), 10008–10013 (2015). (National Academy of Sciences Section: Biological Sciences)
6. Blanchette, M., Kent, W.J., Riemer, C., Elnitski, L., Smit, A.F., Roskin, K.M., Baertsch, R., Rosenbloom, K., Clawson, H., Green, E.D., et al.: Aligning multiple genomic sequences with the threaded blockset aligner. Genome Res. **14**(4), 708–715 (2004)
7. Darling, A.E., Mau, B., Perna, N.T.: progressiveMauve: multiple genome alignment with gene gain, loss and rearrangement. PloS One **5**(6), e11147 (2010)
8. Dubchak, I., Poliakov, A., Kislyuk, A., Brudno, M.: Multiple whole-genome alignments without a reference organism. Genome Res. **19**(4), 682–689 (2009)
9. Earl, D., et al.: Alignathon: a competitive assessment of whole-genome alignment methods. Genome Res. **24**(12), 2077–2089, e11147 (2014)
10. Elghraoui, A., Mirarab, S., Swenson, K.M., Valafar, F.: Evaluating impacts of syntenic block detection strategies on rearrangement phylogeny using Mycobacterium tuberculosis isolates. bioRxiv (2022)
11. Ghiurcuta, C.G., Moret, B.M.: Evaluating synteny for improved comparative studies. Bioinformatics **30**(12), i9–i18 (2014)
12. Glover, N., et al.: Advances and applications in the quest for orthologs. Molecular Biol. Evol. **36**(10), 2157–2164 (2019)
13. Hell, P., Kirkpatrick, D.G.: Packings by complete bipartite graphs. SIAM J. Algebraic Discr. Methods **7**(2), 199–209, e11147 (1986)
14. Hu, F., Lin, Y., Tang, J.: MLGO: phylogeny reconstruction and ancestral inference from gene-order data. BMC Bioinform. **15**(1), 354 (2014)
15. Impagliazzo, R., Paturi, R., Zane, F.: Which problems have strongly exponential complexity? J. Comput. Syst. Sci. **63**(4), 512–530 (2001)
16. Kim, J., Ma, J.: PSAR: measuring multiple sequence alignment reliability by probabilistic sampling. Nucleic Acids Res. **39**(15), 6359–6368 (2011)
17. Kirkpatrick, D.G., Hell, P.: On the complexity of general graph factor problems. SIAM J. Comput. **12**(3), 601–609 (1983)
18. Kolmogorov, M., et al.: Chromosome assembly of large and complex genomes using multiple references. Genome Res. **28**(11), 1720–1732 (2018)
19. Marcet-Houben, M., Gabaldón, T.: EvolClust: automated inference of evolutionary conserved gene clusters in eukaryotes. Bioinformatics **36**(4), 1265–1266 (2020)
20. Minkin, I., Medvedev, P.: Scalable multiple whole-genome alignment and locally collinear block construction with SibeliaZ. Nat. Commun. **11**(1), 6327 (2020). (Nature Publishing Group)

21. Modlin, S.J., et al.: Drivers and sites of diversity in the DNA adenine methylomes of 93 Mycobacterium tuberculosis complex clinical isolates. eLife **9**, e58542 (2020). (eLife Sciences Publications Ltd.)
22. Mural, R.J., et al.: A comparison of whole-genome shotgun-derived mouse chromosome 16 and the human genome. Science **296**(5573), 1661–1671 (2002)
23. Nadeau, J.H., Taylor, B.A.: Lengths of chromosomal segments conserved since divergence of man and mouse. Proc. Natl. Acad. Sci. **81**(3), 814–818 (1984)
24. Notredame, C.: Robusta: a meta-multiple genome alignment tool (2012). http://www.tcoffee.org/Projects/robusta
25. Ondov, B.D., et al.: Mash: fast genome and metagenome distance estimation using MinHash. Genome Biol. **17**(1), 132 (2016)
26. Paten, B., Herrero, J., Beal, K., Fitzgerald, S., Birney, E.: Enredo and Pecan: genome-wide mammalian consistency-based multiple alignment with paralogs. Genome Res. **18**(11), 1814–1828 (2008)
27. Pevzner, P., Tesler, G.: Genome rearrangements in mammalian evolution: lessons from human and mouse genomes. Genome Res. **13**(1), 37–45 (2003)
28. Prakash, A., Tompa, M.: Measuring the accuracy of genome-size multiple alignments. Genome Biol. **8**(6), 1–11 (2007)
29. Sturtevant, A.H., Novitski, E.: The homologies of the chromosome elements in the genus drosophila. Genetics **26**(5), 517 (1941)
30. Svetlitsky, D., Dagan, T., Ziv-Ukelson, M.: Discovery of multi-operon colinear syntenic blocks in microbial genomes. Bioinformatics **36**(Supplement_1), i21–i29 (2020)
31. Waterston, R.H., et al.: Initial sequencing and comparative analysis of the mouse genome. Nature **420**(6915), 520–562 (2002)
32. Weller, M., Komusiewicz, C., Niedermeier, R., Uhlmann, J.: On making directed graphs transitive. J. Comput. Syst. Sci. **78**(2), 559–574 (2012)
33. Winter, S., et al.: Finding approximate gene clusters with GECKO 3. Nucleic Acids Res. **44**(20), 9600–9610 (2016)

On Partial Gene Transfer and Its Impact on Gene Tree Reconstruction

Sumaira Zaman[1] and Mukul S. Bansal[1,2(✉)]

[1] Department of Computer Science and Engineering, University of Connecticut, Storrs, CT 06269, USA
sumaira.zaman@uconn.edu
[2] Institute for Systems Genomics, University of Connecticut, Storrs, CT 06269, USA
mukul.bansal@uconn.edu

Abstract. Horizontal transfer of genetic material between different organisms is one of the most important evolutionary processes in microbial evolution. Such horizontal transfer events can result in the transfer of genomic fragments containing multiple complete genes, complete single genes, or partial genes. However, partial gene transfer (PGT) remains poorly understood and generally underappreciated. Indeed, existing phylogenetic approaches for studying microbial evolution and horizontal gene transfer largely ignore PGT, leading to potential biases and errors in downstream inferences.

In this work, we (i) perform a systematic study of the impact of PGT on the ability to correctly reconstruct the evolutionary histories of gene families (i.e., gene trees) and (ii) propose a simple, yet effective approach, called *trippd*, to detect if a given gene family has been affected by PGT. Our analysis, using simulated and real biological datasets, reveals many interesting insights related to when and how PGT affects gene tree reconstruction, demonstrates the utility of trippd, and sheds light on the importance of detecting and accounting for PGT when studying microbial evolution.

1 Introduction

Horizontal gene transfer (HGT) is known to play an important role in microbial evolution and many different computational techniques have been developed to infer HGTs; see, e.g., [28] for a review. While most methods for inferring and studying HGT view single genes as the "unit" of HGT, it is well known that multiple genes can be transferred in a single transfer event [5,9,13,23] and that many transfers result in the transfer of only partial genes (i.e., fraction of a gene) [3,6,7,33,35]. Partial gene transfer (PGT), in particular, remains poorly understood and existing phylogenetic approaches for studying microbial evolution and horizontal gene transfer largely ignore PGT. Such PGTs can occur not only when the transferred genomic fragments themselves are small but also when boundaries of larger genomic fragments containing one or more complete genes overlap flanking genes. Moreover, integration of new genetic material into a genome often occurs through homologous recombination in flanking regions [24].

L. Jin and D. Durand (Eds.): RECOMB-CG 2022, LNBI 13234, pp. 168–186, 2022.
https://doi.org/10.1007/978-3-031-06220-9_10

While many approaches have been developed for studying recombination in genomes, e.g., [10,18–20,34], such methods have been observed to have high false-positive rates for breakpoint detection [2], decreasing their utility for PGT detection. To our knowledge, the two approaches most directly applicable to the problem of detecting PGTs within gene families are T-REX [3] and PhyML-Multi [4]. T-REX [3] uses a sliding window technique and infers PGT by constructing window trees and comparing them to a known species tree to infer possible transfer events. However, T-REX assumes all discordance is due to replacing transfer (or homologous recombination) and cannot be directly applied to gene families with a history of gene duplication or additive transfer. PhyML-Multi [4] uses a more sophisticated HMM based approach and can partition the given gene family alignment into a user-specified number of partitions with distinct evolutionary histories. Notably, PhyML-Multi does not rely on a known species tree or on inference of actual transfer events, both of which are known to be error-prone [1,11,17], and can therefore be directly applied to any gene family alignment to detect possible PGT. However, the utility of PhyML-Multi for PGT detection has not been sufficiently explored and its effectiveness for this problem has not been previously studied. Furthermore, the impact of PGT on gene tree reconstruction itself remains poorly understood and generally underappreciated. Previously, Posada and Crandall [25] systematically evaluated the impact of recombination on phylogeny inference. However, that work did not focus directly on PGTs and used small, 8-taxon trees with only a single recombination event per tree.

In this work, we advance the study of PGT and gene family evolution by (i) performing a systematic assessment of the impact of PGT on gene tree reconstruction, (ii) evaluating the ability of PhyML-Multi to accurately detect PGTs, and (iii) proposing a conceptually simple and easy-to-use approach, called *trippd*, based on alignment tri-partitioning, to identify gene families affected by nonnegligible PGT. Among many interesting findings, we demonstrate that PGTs can significantly impact gene tree reconstruction and identify the scenarios under which PGTs may or may not significantly affect gene tree reconstruction accuracy; despite considerable conceptual and methodological differences, some of these findings are also consistent with previous results from [25]. Our evaluation of PhyML-Multi as the basis for PGT detection reveals that such an approach has a very high false-positive rate of PGT detection. At the same time, our experimental analysis shows how our new approach, trippd, can help address this limitation of PhyML-Multi, achieving a false-negative rate comparable to that of the PhyML-Multi based approach while having a negligible false-positive rate. An application of trippd to two biological datasets demonstrates the prevalence of PGT in real gene families.

Overall, this work sheds fresh light on the importance of detecting PGTs and accounting for them in microbial evolutionary analyses, reveals new insights into when and how gene tree reconstruction is impacted by PGT, and proposes a simple approach that can help end-users easily identify gene families affected by sufficient PGT to impact gene tree reconstruction. Scripts implementing trippd, along with all simulated datasets, are freely available from https://github.com/suz11001/Tripartition.

2   Materials and Methods

We use an extensive simulation study to assess the impact of PGT on gene tree reconstruction accuracy and to evaluate the effectiveness of the PhyML-Multi based approach and of our proposed PGT detection approach trippd.

2.1   Simulated Datasets

We used the phylogenetic simulation framework SaGePhy [16] to generate a large collection of simulated datasets consisting of gene families affected by PGT. Each gene family is represented by a gene family alignment, where the alignment is composed of a *genic-region*, consisting of sequences evolved down a gene tree, and a *PGT-region* consisting of sequences evolved down the same gene tree but with a certain rate of replacing transfer (homologous recombination). In other words, each gene family alignment represents two or more distinct evolutionary histories, appended together, with one representing the evolution of the gene tree and the other(s) representing the evolutionary history of a gene sequence region (or locus) affected by PGT. The resulting gene family datasets represent a wide range of evolutionary conditions, with varying gene lengths, PGT-region to genic-region ratios (i.e., fraction of gene sequence affected by PGTs), rates of PGT, sequence evolution rate, etc. We divide these datasets into three broad categories: *baseline* datasets, *multi-PGT* datasets, and *PGT-location* datasets. We describe the construction of these datasets below:

Baseline Datasets. Our baseline collection consists of 14 distinct datasets, each representing a distinct combination of evolutionary parameter settings and consisting of 100 gene families generated under the corresponding parameter settings. To generate each dataset, we first simulated 100 species trees with 100 leaves each using a birth-death process and then simulated a gene tree inside each species tree using specific rates of gene duplication, replacing HGT, additive HGT, and gene loss. (The exact parameter values used for simulating species trees and gene trees, along with all simulated data, are freely available from the GitHub page linked above.) This yields 100 gene trees per dataset, and this same set of 100 gene trees was used for simulating the 100 gene families in each dataset. These gene trees had between between 20 and 236 leaves, with an average of 98.2, and each gene tree was of height 1. To simulate PGTs within each gene tree, we used SagePhy to simulate 3 different "subgene" trees, each with a different rate of PGT, within each of the 100 gene trees. Each subgene tree represents a history of PGT via homologous recombination within the corresponding gene tree. Specifically, each subgene tree was evolved down the gene tree under a certain rate of replacing subgene transfer and no other events. We used replacing transfer rates of 0.2, 0.4, and 0.6 (per unit branch length) to simulate low, medium, and high rates of partial gene transfer, resulting in 3 subgene trees per gene tree, each with the same height and number of leaves as the corresponding gene tree. These three resulting sets of subgene trees correspond to, on average, 2.03, 3.87, and 5.55 PGTs per gene family, respectively.

The resulting set of 100 gene trees and 300 subgene trees was then used to simulate sequences under different evolutionary scenarios, resulting in the 14 baseline simulated gene family datasets. For these datasets, only one PGT-region is included in each gene family alignment and this PGT-region is always appended at the end of the genic-region. To generate the 14 baseline datasets, we considered the three PGT evolution rates (0.2, 0.4, and 0.6) as discussed above and, in addition, varied the following sequence-related parameters: (i) total sequence length (500, 1000, and 2000nt; for reference, the average prokaryotic gene length is roughly 1000nt [14]), (ii) substitution rates (0.1, 0.5, 1, 2, and 5 substitutions per site per unit branch length, capturing a wide range of evolutionary distances from closely related to distantly related), and (iii) fraction of sequence length represented by PGT-region (10%, 20%, 30%, 40%, 50% and 60%). We created one dataset with default parameter values of 0.4 for PGT evolution rate, 1000nt for sequence length, 0.5 for substitution rate, and 40% for fraction of sequence length represented by PGT-region. To study the impact of different parameters on gene tree reconstruction and PGT detection, we generated additional datasets by varying one parameter value at a time and keeping other parameters at their default values. This resulted in $2 + 2 + 4 + 5 = 13$ additional datasets, yielding a total of 14 baseline datasets. All sequences were generated using Seq-Gen [26] under the GTR model with gamma distributed rates and default settings for other Seq-Gen parameters.

Multi-PGT Datasets. To assess how gene tree reconstruction is impacted by the presence of *multiple* PGT-regions within the same gene family, we created 4 additional datasets, each containing 2 PGT-regions. Specifically, we used default values for PGT evolution rate, total sequence length, and substitution rates, but varied the fraction of sequence length represented by PGT-regions as well as the specific fractions corresponding to each of the two PGT-regions. The 5 Multi-PGT datasets correspond to the following splits of PGT-region length between the two PGT-regions: {20%, 20%}, {30%, 10%}, {40%, 20%}, {60%, 10%}.

PGT-Location Datasets. To further assess the impact of PGT-region location within gene family alignments, we created 3 additional datasets corresponding to offsets of 34 base pairs (bps), 84 base pairs, and 134 base pairs from the end of the sequence alignment. These datasets otherwise use default parameter settings for all parameters. This small number of PGT-location datasets is sufficient to assess the impact of PGT-region location (Sect. 4.2).

2.2 Biological Datasets

To assess the prevalence of PGTs in real microbial gene families, we used samples from two large published biological datasets: a dataset consisting of over 4700 gene families from 100 broadly sampled species (11 eukaryotic, 12 archaeal, and 67 bacterial) [8], and a dataset of 8,277 gene families from 103 *Aeromonas* strains [13,27]. The first dataset represents a scenario where, due to the great evolutionary divergence between included taxa, we do not expect to see much PGT. In contrast, the second dataset represents a scenario where the taxa under consideration are closely related and so a high prevalence of PGT is expected due to the ease of homologous recombination.

For each dataset, we first filtered the collection of gene families present in each original dataset by removing all gene families that had fewer than 40 genes or alignments shorter than 150 amino acids or 450nt. After applying this filtering we were left with 823 gene families for the 100-taxon dataset and 3,357 gene families for the 103-taxon *Aeromonas* dataset. Since the *Aeromonas* dataset is quite large, we randomly sampled 500 gene families from the remaining 3,357. During subsequent analysis of the resulting datasets, we found that some gene families had very large gaps (greater than one-third of the total alignment length) in the alignment of one or more sequences. We therefore removed all gene families with such large gaps, leaving us with 784 and 466 gene families for the 100-taxon dataset and 103-taxon *Aeromonas* dataset, respectively.

2.3   Gene Tree Construction and Comparison

To study the impact of PGTs on gene tree reconstruction accuracy, we compared the topologies of the three main tree types for each gene family in each dataset: The *true gene tree* for that gene family (as simulated using SaGePhy), the *pre-PGT gene tree* reconstructed using the genic-region of the corresponding sequence alignment, and the *post-PGT gene tree* reconstructed using the full sequence alignment (appended genic- and PGT-regions). A pre-PGT gene tree represents the best tree we could reasonably reconstruct given only the sequence alignment and knowledge of the presence of PGTs in that gene family. A post-PGT gene tree represents the tree we would reconstruct if we were unaware of the presence of PGTs in that gene family.

All pre-PGT and post-PGT gene trees were reconstructed using RAxML v8.2.11 [31] (with 100 rapid bootstrap samples (-f a -N 100) and under the GTRCAT model). Divergence between any pair of (unrooted) gene tree topologies was measured using Robinson-Fould's distance [29]. Specifically, we count the number of splits present in only one of the two trees being compared. We refer to the resulting number as the RF-score and use $RF(T_1, T_2)$ to denote the RF-score between trees T_1 and T_2. Note that the RF-score counts unique splits of *both* trees (i.e., we do not divide the computed score by 2).

2.4   Using PhyML-Multi to Detect PGTs

PhyML-Multi [4] is an existing state-of-the-art approach designed to identify plausible recombination breakpoints within a given sequence alignment and to reconstruct phylogenetic trees for each identified recombination-free region. Next, we briefly describe how PhyML-Multi can be used to detect PGTs. Our new approach, trippd, is introduced later in Sect. 3.

For our analysis, we used the more rigorous HMM-based implementation of PhyML-Multi and used suitable parameter settings expected to maximize inference accuracy. Specifically, we specified the number of expected partitions/trees to be 2 (which is the correct expected number for all baseline datasets), used the TN93 model of evolution (the closest one to GTR, since GTR is not available within PhyML-Multi), used 4 rate categories, allowed PhyML-Multi to estimate the transition/transversion ratio, proportion of invariable sites, and gamma shape parameter, and used BIONJ to build starting trees (instead of providing user-specified starting trees).

The output from PhyML-Multi includes breakpoints for the number of specified partitions along with the PhyML maximum likelihood (ML) tree corresponding to each partition. For a fair comparison with trippd, we ignored the output PhyML trees and instead used the breakpoints/partitions identified by PhyML-Multi to generate the corresponding RAxML tree for each partition using the same RAxML parameter settings as described above.

Note that PhyML-Multi will always find the specified number of partitions (and trees) for the given sequence alignment, even if no homologous recombination has occurred. Thus, PhyML-Multi cannot be directly used to determine if PGT has occurred. We therefore use a simple histogram intersection test to determine if any phylogenetic differences for the sequence partitions identified by PhyML-Multi may, in fact, be due to PGT. We describe this test below. A similar test is also used as part of trippd.

Histogram Intersection Test for PGT Presence and Absence. Given the two partitions of a gene family alignment output by PhyML-Multi, we employ a simple classification procedure to determine if any inferred phylogenetic differences between the two partitions are likely to have been caused by PGT. As part of this test, we compute 100 bootstrap replicates for each of the two partitions using RAxML (under GTRCAT, as above). Let A and B denote the two partitions and $\{A_1, ..., A_{100}\}$ and $\{B_1, ..., B_{100}\}$ denote the corresponding bootstrap replicate trees, respectively. We also compute a maximum likelihood tree (using RAxML) for the full, unpartitioned sequence alignment for that gene family. Let $\mathcal{R}$ denote this maximum likelihood tree.

We then compute the RF-scores between each bootstrap replicate A_i and $\mathcal{R}$, and between each bootstrap replicate B_i and $\mathcal{R}$, for each $i \in \{1, ..., 100\}$. This generates two discrete distributions of 100 RF-scores for the two partitions. The classification is based on the histogram intersection of these two distributions: If the intersection is less than a certain threshold, fixed at 50% in our experiments, then the phylogenetic difference between partitions A and B is assumed to be due to PGT, and otherwise assumed to be due to inference uncertainty or other random effects. The key idea is that if both partitions are a result of the same evolutionary process, i.e., no PGT, then the differences between the bootstrap trees for each partition and the overall ML tree should be similar for the two partitions. An illustration appears in Fig. 1A.

3   Trippd: Tri-Partition Based PGT Detection

In our experiments, we found that PhyML-Multi showed a high false positive rate for identifying gene families affected by PGT (Sect. 4.2). We therefore devised a simple, proof-of-concept approach that, in our experiments, nearly matches the accuracy of PhyML-Multi in correctly detecting PGT (i.e., has low false negative rate) while also achieving a very low false positive rate. Our new approach, called trippd (short for tri-partition based PGT detection, and pronounced "tripped") has three key features: (i) unlike PhyML-Multi, it does not rely on breakpoint detection and is therefore robust to errors in detecting the breakpoints/partitions correctly, (ii) it does not require any advance knowledge of the number of partitions or PGT-regions, and (iii) it leverages insights from our experimental evaluation of the impact of PGTs on gene tree reconstruction and is especially designed to classify gene family alignments as those having *sufficient* or *insufficient* PGT to impact gene tree reconstruction. trippd is illustrated in Fig. 1B, and a step-by-step description of trippd follows:

Alignment tri-partitioning. The given gene family alignment is partitioned into three equal (or roughly equal) parts, each consisting of one-third of the sites in the alignment. We refer to these partitions as window-1, window-2, and window-3.

ML window tree reconstruction. Use RAxML to compute a maximum likelihood tree for each of the three windows.

Identifying most similar and most dissimilar pairs of windows. Compute the RF-score between each pair of ML window trees. Identify the pairs with smallest RF-score, denoted ww_{min}, and largest RF-score, denoted ww_{max}. Note that if $ww_{min} = ww_{max}$ then subsequent steps need not be executed and PGT is assumed to be absent.

Bootstrap replicates for each window. Compute several (100 in our experiments) bootstrap replicates for each of the three windows using RAxML. Denote these as $\{w_1^i, ..., w_b^i\}$ for window-i, where $1 \leq i \leq 3$ and b denotes the number of bootstrap replicates per window.

Computing distributions of RF-scores. Given the bootstrap replicates for any two windows i and j, define $D(i, j)$, to be the distribution of RF-scores $RF(w_k^i, w_k^j)$, where $k \in \{1, \ldots, b\}$. Compute $D(ww_{min})$ and $D(ww_{max})$.

Histogram intersection test. Apply a simple test (similar to the one for PhyML-Multi described in Sect. 2.4) to determine if differences between $D(ww_{min})$ and $D(ww_{max})$ are likely due to PGT or not. Specifically, compute the histogram intersection of $D(ww_{min})$ and $D(ww_{max})$ and check if the intersection is no more than a certain threshold, fixed at 50% in our experiments. An intersection smaller than or equal to the threshold percentage indicates presence of PGT. Intersection greater than the threshold indicates lack of significant PGT.

Our choice of using only three static windows in trippd is based on several observations and considerations: For example, we find in our experiments (see

Results) that a PGT-region that spans less than a third of the total sequence does not have measureable impact on gene tree reconstruction. At the same time, any PGT-region longer than a third of the total sequence length would overlap significantly with at least one of the three static windows, impacting at least one of the window trees. Having three windows, rather than just two, also allows for multiple pairwise window comparisons. The three-window approach is also relatively robust to the size of the PGT-region, allowing for the PGT-region to dominate and the genic-region to be relatively short, as long as the genic-region still makes up a majority of at least one of the three windows. Finally, this approach is also relatively robust to the presence of multiple PGT-regions, as long as there are at least two windows in which either the genic-region or one of the PGT-regions constitutes the majority of the sequence.

Selecting Histogram Intersection Test Threshold. We used a simple simulation study to determine a reasonable (not optimized) threshold for the histogram intersection test. Specifically, we used the baseline dataset with default parameter values to measure false-positive and false-negative inference at thresholds of 0% (i.e., complete separation between the two distributions), 25% and 50%. At the very strict threshold of 0%, we observed a very high false negative rate of about 0.5 and no false positives. At 25% the false negative rate improved only slightly. At 50% we observed a large reduction in the false negative rate, while still observing no false positives. We therefore fixed the threshold at 50%. We did not further optimize this threshold to maintain robustness to varying evolutionary conditions. An evaluation of its robustness appears in Sect. 4.

4 Results

4.1 Impact of PGT on Gene Tree Reconstruction Accuracy

We first assessed the impact of PGT on gene tree reconstruction using the baseline and multi-PGT simulated datasets described earlier. Recall that the 14 baseline datasets encompass a wide range of evolutionary scenarios, allowing for an assessment of the impact of PGT rate, total sequence length, sequence evolution (substitution) rate, and ratio of PGT-to genic-region length. In addition, the 4 multi-PGT datasets make it possible to assess the impact of multiple distinct PGT-regions within the same gene family alignment.

For each gene family within each dataset, we reconstruct two gene trees by applying RAxML to the simulated sequence data: A *pre-PGT* gene tree reconstructed using only the geneic-region of the sequence, and a *post-PGT* gene tree reconstructed using the entire sequence alignment (consisting of both the genic- and PGT-regions). Thus, a *pre-PGT* gene tree represents the best tree we can reasonably reconstruct if all PGTs were correct detected and accounted for, while a *post-PGT* gene tree represents the gene tree we would normally reconstruct if we do not account for possible PGT. The results of our analysis are shown in Fig. 2, where we plot the average RF-scores between each pre-PGT

A. PGT Detection With PhyML-Multi

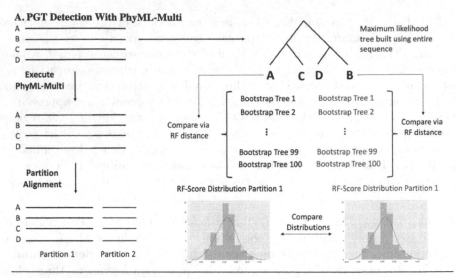

B. PGT Detection With Trippd

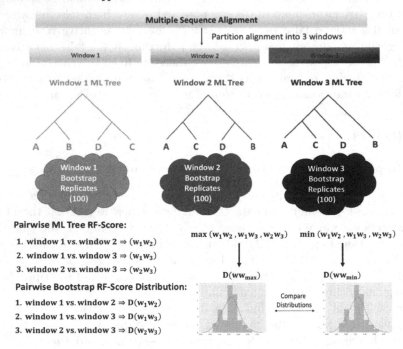

Fig. 1. Overview of PhyML-Multi and trippd for PGT detection. Both approaches start with a given multiple sequence alignment for the gene family. (A) The PhyML-Multi based approach works by using PhyML-Multi to partition the alignment into two regions, using RAxML to compute multiple bootstrap replicates for the two regions, comparing the resulting trees to the maximum likelihood (ML) tree for the entire sequence alignment, and using a simple histogram intersection test to determine if the resulting distributions of RF-scores suggest different evolutionary histories for the two regions. (B) tripped executes the step-by-step approach described in Sect. 3.

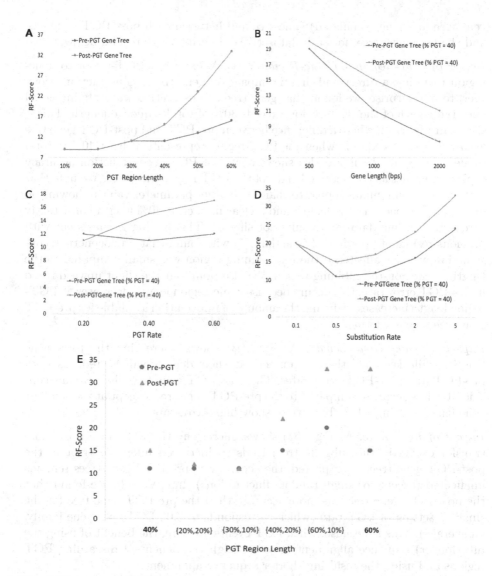

Fig. 2. Impact of PGT on gene tree reconstruction accuracy. The plots show the impact of various evolutionary parameters on the reconstruction error of pre-PGT (blue) and post-PGT (orange) gene trees. (A) shows the impact of PGT region length (as percentage of total gene length), (B) of total gene length, (C) of PGT rate, (D) of the rate of sequence evolution, and (E) of having multiple PGT regions of different lengths within a single gene family. Plots (A)–(D) are based on baseline datasets and plot (E) is based on multi-PGT datasets. For Plot (E), the first and last columns (40% and 60%) show results for the corresponding baseline (single-PGT) datasets for reference. Reconstruction error is measured in terms of RF-score by comparing reconstructed pre- and post-PGT gene trees against corresponding true gene trees. All results are averaged over the 100 gene families in the corresponding dataset. (Color figure online)

gene tree and true (simulated) gene tree and between each post-PGT gene tree and the true gene tree, for each dataset. We describe these results below.

Impact of PGT-Region to Genic-Region Ratio. As expected, PGT-region to genic-region ratio has a direct and drastic impact on gene tree reconstruction. However, to our surprise, we found that gene tree reconstruction was only impacted once the length of the PGT-region exceeds 30% of total sequence length. This is shown in Fig. 2(A), where a difference between pre-PGT and post-PGT gene tree accuracy is observed only when the PGT-region represents at least 40% of total sequence length. The figure also shows how post-PGT reconstruction accuracy rapidly degrades as the relative length of the PGT region increases. We note that this observation remains robust to changes in other parameter values, showing no or minimal impact at 30% length and a clear impact at 40% length consistently across all baseline datasets (results not shown). This finding is consistent with previous results of Posada and Crandall [25] who found that phylogenetic reconstruction was not affected if the recombining region was small compared to the length of the non-recombining region. We also point out the slight upward trend in pre-PGT error rate; this occurs because genic-region length decreases as PGT-region length increases, reducing the amount of information available for pre-PGT gene tree reconstruction.

Impact of Total Gene Length. As Fig. 2(B) shows, increasing the total gene length, while keeping other parameters at their default values, reduces both pre-PGT and post-PGT error rates. The post-PGT gene tree also remains considerable less accurate compared to the pre-PGT gene tree, except at the smallest gene length setting where both trees show high error-rate.

Impact of PGT Rate. As Fig. 2(C) shows, increasing the PGT rate (i.e., more transfer events in the subgene tree) leads to increased inference error in the post-PGT gene tree. As expected, the accuracy of pre-PGT gene trees remains unaffected (except for small random fluctuations). Interestingly, we found that the post-PGT gene tree was more accurate than the pre-PGT gene tree for the smallest setting of PGT rate (which corresponds to 2.03 PGTs per gene family, on average). This is because, when PGT events are rare, the benefit of using the full (longer) sequence alignment may outweigh the benefit of discarding PGT regions and using the resulting shorter sequence alignment.

Impact of Sequence Evolution Rate. The impact of sequence evolution rate is similar to that of total gene length, affecting both pre-PGT and post-PGT gene trees similarly. This is shown in Fig. 2(D) where both pre- and post-PGT gene trees are either simultaneously worsened or simultaneously improved as substitution rate changes. Somewhat surprisingly, we found that the error rates of both pre-PGT and post-PGT gene trees were nearly identical for the smallest setting of substitution rate. This is likely because at low evolutionary rates there may not be sufficient information in the sequence alignment to confidently reconstruct either type of gene tree. As the figure shows, and as expected, error rates also start to increase at higher substitution rates.

Impact of Multiple PGT-Regions. Figure 2(E) shows pre-PGT and post-PGT gene tree reconstruction accuracies for the four multi-PGT datasets. Unsurprisingly, the accuracy of pre-PGT gene trees increases with increasing length of the genic-region. However, careful analysis of post-PGT gene tree error rates reveals an important, unexpected insight: We find that the error-rate of the post-PGT gene trees is impacted not by the total length of PGT-regions, but rather by the length of the single longest PGT-region. For instance, as the figure shows, post-PGT error rates for the {20%, 20%} and {30%, 10%} multi-PGT datasets are the same as their pre-PGT error rates, and much lower than the corresponding baseline dataset post-PGT error rate for PGT-region length 40%, despite the total PGT-region length being 40% in both these multi-PGT datasets. Likewise, the post-PGT error-rate for the {40%, 20%} multi-PGT dataset is much lower than for the corresponding baseline dataset with PGT-region length 60%.

A key insight from the above results is that when PGT regions are small (say less than a third of the total sequence length) or when PGTs occur very rarely, and even if multiple such PGT-regions appear within the same gene family, it may be beneficial to use the full gene family sequence alignment for gene tree reconstruction. At the same time, these results clearly demonstrate the significant adverse impact of longer and frequent PGTs on gene tree reconstruction.

We note that the results above show results averaged across all 100 gene families in the corresponding baseline dataset, even though not all 100 gene families in each dataset may have PGTs. Given the randomness inherent in any simulation framework, we found that, in datasets with the default PGT-rate of 0.4, 75 out of the 100 gene families had at least one PGT. These numbers were 71 and 88, out of 100, for datasets with PGT rates of 0.2 and 0.6, respectively. The results shown in Fig. 2 are only minimally impacted even when limited only to gene families with at least one PGT (detailed results not shown). We also point out that post-PGT gene tree reconstruction accuracy does not depend on the "location" of the PGT-region within the sequence alignment since the gene tree reconstruction methods assume each site evolves independently. We therefore did not separately evaluate reconstruction accuracy on PGT-location datasets.

4.2   PGT Detection Accuracy

We used the baseline dataset with default parameter values (i.e., with 0.4 PGT evolution rate, 1000nt total sequence length, 0.5 substitution rate, and 40% PGT-region to genic-region ratio) to assess the ability of the PhyML-Mutli based approach and trippd to correctly detect the presence or absence PGTs. Since baseline datasets have the PGT-region appended at the end of the alignment, we also used the PGT-location datasets to further assesss the impact (if any) of PGT location within the sequence alignment. We also simulated additional datasets without any PGTs to further assess the false-positive rate of PGT detection for these methods.

Detecting PGTs Using the PhyML-Multi Approach. Recall that the default baseline dataset consists of 75 gene families with at least one PGT and 25 gene famlies without any PGT. We found that the PhyML-Multi based approach, using a histogram intersection test threshold of 50%, was correctly able to classify 63 of the 75 gene families as having PGT. However, PhyML-Multi also incorrectly classified 11 of the 25 gene families without any PGTs as having PGT. For additional false positive testing, we ran PhyML-Multi on an additional dataset of 100 gene families with no PGTs and found that the method incorrectly detected PGTs in 65 out of the 100 gene families. Thus, the PhyML-Multi based approach shows a false negative rate of 0.16 (12/75) and a false positive rate of about 0.5 (more precisely, 0.44 (11/25) for the baseline dataset and 0.65 for the additional simulated dataset). Importantly, we found that these results are robust to the specific histogram intersection test threshold used and Table 1 shows the clear tradeoff between false-positive and false-negative rates of this approach as the threshold is decreased or increased.

Observe that the accuracy of his PhyML-Multi-based approach depends on PhyML-Multi's ability to correctly identify PGT boundaries/breakpoint(s). We found that, out of the 75 baseline dataset gene families with PGTs, PhyML-Multi was able to correctly detect the breakpoint to within 5 basepairs for 54 gene families. Thus, the breakpoint could not be accurately detected for 28% of the gene families.

Table 1. PGT detection accuracy using the PhyML-Multi based approach and trippd. False-positive and false-negative rates for both methods are shown when applied to the default baseline dataset and to the additional simulated dataset of 100 gene families with no PGTs. Results are shown for three different histogram intersection test thresholds, where the default threshold is 50%.

PhyML-Multi Based Approach		
Baseline dataset	Threshold = 40%	Threshold = 50%	Threshold = 60%
False Positive Rate	0.36 (9/25)	0.44 (11/25)	0.52 (13/25)
False Negative Rate	0.24 (18/75)	0.16 (12/75)	0.13 (10/75)
No PGT dataset	Threshold = 40%	Threshold = 50%	Threshold = 60%
False Positive Rate	0.57	0.65	0.70
False Negative Rate	N/A	N/A	N/A

trippd		
Baseline dataset	Threshold = 40%	Threshold = 50%	Threshold = 60%
False Positive Rate	0 (0/25)	0 (0/25)	0.16 (4/25)
False Negative Rate	0.27 (20/75)	0.2 (15/75)	0.11 (8/75)
No PGT dataset	Threshold = 40%	Threshold = 50%	Threshold = 60%
False Positive Rate	0	0.02	0.09
False Negative Rate	N/A	N/A	N/A

Detecting PGTs Using Trippd. As the lower half of Table 1 shows, an application of trippd to the same datasets shows much better PGT detection accuracy. In particular, we find that tripped has a drastically lower false positive rate and a comparable false negative rate as compared to PhyML-Multi. For instance, at the 50% histogram intersection test threshold, we found that tripped had a false positive rate of 0 on the baseline dataset and just 0.02 on the additional simulated dataset with no PGTs, compared to 0.44 and 0.65, respectively, for the PhyML-Multi approach. The false negative rate was also a relatively low 0.2, which is roughly comparable to the 0.16 false negative rate for the PhyML-Multi approach. In fact, at a threshold of 60% both false positive and false negative rates of tripped are lower than those for PhyML-Multi.

To assess the impact of PGT-region length (default 40%) on the detection accuracy of trippd, we applied it to the baseline datasets with PGT lengths of 20%, 30%, 50% and 60% of the total gene length. We found that tripped was able to correctly classify 56, 58, 48, and 59 gene families, respectively, out of 75, as having PGTs. This corresponds to false-negative rates of 0.25, 0.22, 0.36, and 0.21, respectively; only slightly higher than for the default baseline dataset. Importantly, false-positive rates remained extremely low at 0.04, 0.07, 0, and 0, respectively.

We also assessed the impact of substitution (sequence evolution) rate (default 0.5) on trippd. Since sequences that are more similar are expected to undergo homologous recombination more easily, we applied trippd to the baseline dataset with a much smaller substitution rate of 0.1 and observed false-negative and false-positive rates of 0.29 and 0.16, respectively. Crucially, the increased false-positive rate is still much lower than the false-positive rate for PhyML-Multi. We also applied trippd to the baseline dataset with a very high substitution of 5. As expected, performance degrades substantially and the false-negative rate increases to 0.63. This is not surprising since the error-rates of the trees constructed for each of the three window are likely to be very high under this setting. Notably, the false-positive rate still remains very low, at 0, for this setting.

Interestingly, we observed that there were 11 gene families with PGT (out of the 75) that were consistently incorrectly classified as not having PGT, regardless of PGT-region length. We discovered that these 11 gene families had only a single PGT event. Thus, most of the gene families for which trippd fails to detect the presence of PGT are those in which only a very small amount of PGT has occurred. Furthermore, we found that among these 11 gene families, 8 had a lower post-PGT RF-score than pre-PGT RF-score. This indicates that for many of the gene families where tripped fails to correctly detect the presence of PGTs, it may, in fact, be beneficial to use the entire gene sequence alignment for gene tree reconstruction.

Impact of Multiple PGT Regions. We also assessed the impact of the presence of multiple PGT regions on the detection accuracy of trippd. Since trippd relies on phylogenetic discordance between pairs of windows, we chose the most challenging of all multi-PGT datasets, {40%, 20%}, for our evaluation. This specific dataset is particularly challenging for trippd since it has the property that each of the three windows largely represent three different evolutionary histories; window 1 consists entirely of the genic sequence, window 2 consists almost entirely of the first PGT region, and window 3 consists mostly of the second PGT region. On this dataset, 14 of the 100 gene families did not have any PGTs. Using our default histogram intersection test threshold of 50%, we found that 49 of the 86 gene families with

PGT were correctly classified as having PGTs and only one of the 14 gene families without PGT was classified as having PGT. This corresponds to a false positive rate of 0.07 and a false negative rate of 0.43. Thus, as expected, the PGT detection accuracy of trippd suffers when multiple PGTs occur in such a way that all three windows largely represent different evolutionary histories. However, such instances are expected to be relatively rare in practice.

Impact of PGT Location. Finally, we used the three PGT-location datasets to assess the impact of PGT location on trippd. These results are shown in Fig. 3. We find that as the evolutionary histories of window 2 and window 3 become more similar, the false negative rate of trippd increases. Specifically, from the baseline false negative rate of 0.2 (on the default baseline dataset using a threshold of 50%), the rate increases to 0.24 for the dataset with 34 bps offset, 0.35 for the dataset with 84 bps offset, and 0.35 for the 134 bps offset. The false-positive rate is also affected but remains relative low for all settings, with a high of 0.07 for the 84 bps offset dataset. Note that, since the three windows are treated identically by trippd, just these three PGT-location datasets cover all relevant cases. These results show that the PGT detection accuracy of trippd can be affected, though not drastically, if the PGT-region does not appear towards the beginning or end of a gene. However, since horizontal gene transfer often occurs through homologous recombination in flanking regions [24], PGTs may be more likely to occur at the beginnings or ends of genes.

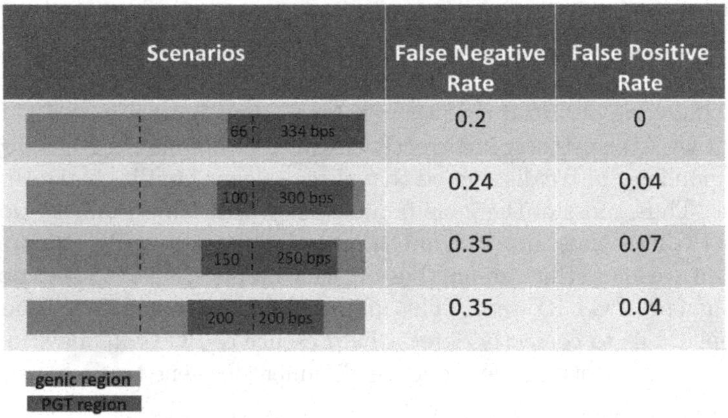

Scenarios	False Negative Rate	False Positive Rate
66 : 334 bps	0.2	0
100 : 300 bps	0.24	0.04
150 : 250 bps	0.35	0.07
200 : 200 bps	0.35	0.04

genic region
PGT region

Fig. 3. Impact of PGT location on trippd. The PGT detection accuracy of tripped, in terms of false positve and false negative rates, is reported for various locations of the PGT region within the gene sequence alignment. The first scenario describes our default baseline case where the PGT region is 40% of the gene length and occurs at the end of the gene. For this baseline case, the PGT region (orange) falls into two windows, 334 bps of the PGT-region is in the last window and the remaining 66 bps is in the middle window. The remaining three scenarios correspond to the three PGT-location datasets with offsets of 34, 84, and 134 bps, respectively.

4.3   Application to Biological Datasets

To assess the impact of trippd in practice, we applied it to the two biological datasets previously described. For the 784 gene families of the 100-taxon broadly sampled dataset, we observed that 62 (7.5%) were identified as having PGT. It is not surprising to see only a small number of gene families with detectable PGTs for this dataset since its species are broadly sampled from the entire tree of life and are therefore very distantly related to each other.

On the 466 gene family *Aeromonas* dataset, trippd identified 151 (32.4%) of the gene families as having PGT. This much higher percentage, compared with the 100-taxon broadly sampled dataset, is expected since the taxa in the *Aeromonas* dataset are much more closely related; thus, homologous recombination is expected to be both abundant and more easily detectable (due to relative recency) in this dataset.

Recall that trippd shows a very low false-positive rate of PGT detection. Thus, our results on these biological datasets indicate that PGTs, particularly those that are capable of affecting gene tree reconstruction, occur frequently in real biological datasets. trippd can easily help identify such cases for further analysis or filtering. Note, however, that these results about PGT prevalence are preliminary and should therefore be interpreted with caution.

5   Discussion and Conclusion

In this work, we used a simulation study to assess the impact of partial gene transfer on gene tree reconstruction and presented a simple computational approach, trippd, based on alignment tri-partitioning to detect the presence of PGTs in gene family alignments. Our study of the impact of PGT reveals several important insights: We find that there can be significant adverse impacts of PGT on gene tree reconstruction accuracy. In such cases, it can be helpful to identify and remove the PGT region(s) from the alignment and reconstruct the gene tree on the reduced alignment. However, our results also show that if PGT regions are small (no more than a third of the total sequence length), or if only a very small number of PGTs have occurred, then gene tree reconstruction is unlikely to be impacted and it is likely beneficial to use the full gene family sequence alignment for gene tree reconstruction. We also find that multiple small PGTs do not significantly impact gene tree reconstruction accuracy and that adverse impacts depend on the length of the longest PGT-region. Our experiments with using PhyML-Multi to detect PGTs show that such an approach is effective at detecting PGTs, showing low false negative rate, but that it also has a very high false positive rate. Furthermore, we find a clear tradeoff between false positive rate and false negative rate for such an approach. The new approach, tripped, attempts to address this limitation and we demonstrate how tripped matches the false-negative rate of the PhyML-Multi based approach while having a negligible false-positive rate. Having a low false-positive rate is important for any effective PGT detection method since incorrect detection of PGTs can inflate or overestimate the impact of PGT in a dataset and lead to corrective measures (such as using only an identified "non-PGT" region of the alignment) that ultimately lower the accuracy of reconstructed gene trees.

We view trippd as a preliminary, proof-of-concept approach, and it has several important limitations worth addressing. Most importantly, trippd can only *detect* the presence of PGT and not *identify* actual PGT regions. It may be possible to combine the strengths of recombination/breakpoint detection approaches such as PhyML-Multi and of tripped to both detect and identify PGT regions with high accuracy. Furthermore, it would be helpful to not only identify the different regions of an alignment but also to identify which region represents the underlying genic region and which represent PGT regions. The accuracy of trippd is also somewhat sensitive to PGT length, PGT location, and substitution rates, and methodological refinements could help address this limitation.

Several aspects of our simulation study can also be improved. In particular, our current study assumes that the same region of the underlying gene sequence undergoes repeated homologous recombination. A more reasonable model would be to allow each homologous recombination event to independently affect any region of the recipient gene. Likewise, it may help to appropriately model when homologous recombination between two gene sequences can occur (e.g., based on sequence similarity).

It is also possible that species-tree-aware approaches for gene tree reconstruction [1,8,12,15,21,22,30,32] are more robust to the presence of PGTs and the impact of PGT on such approaches is worth investigating further. Finally, while our preliminary experimental analysis indicates that methods used to study genomic recombination, such as those implemented in RDP4 [20], have high false-positive rates of PGT detection (results not shown), it may be useful to evaluate the utility of such methods for PGT detection and identification more systematically.

Funding Information. This work was supported in part by NSF award IIS 1553421 to MSB.

References

1. Bansal, M.S., Wu, Y.C., Alm, E.J., Kellis, M.: Improved gene tree error correction in the presence of horizontal gene transfer. Bioinformatics **31**(8), 1211–1218 (2015). https://doi.org/10.1093/bioinformatics/btu806
2. Bay, R.A., Bielawski, J.P.: Recombination detection under evolutionary scenarios relevant to functional divergence. J. Mol. Evol. **73**, 273–286 (2011)
3. Boc, A., Makarenkov, V.: Towards an accurate identification of mosaic genes and partial horizontal gene transfers. Nucleic Acids Res. **39**(21), e144 (2011)
4. Boussau, B., Guéguen, L., Gouy, M.: A mixture model and a hidden markov model to simultaneously detect recombination breakpoints and reconstruct phylogenies. Evol. Bioinform. Online **5**, 67–79 (2009)
5. Brinkmann, H., Göker, M., Koblízek, M., Wagner-Döbler, I., Petersen, J.: Horizontal operon transfer, plasmids, and the evolution of photosynthesis in Rhodobacteraceae. ISME J. **12**, 1994–2010 (2018)

6. Chan, C.X., Beiko, R.G., Darling, A.E., Ragan, M.A.: Lateral transfer of genes and gene fragments in prokaryotes. Genome Biol. Evol. **1**, 429–438 (2009). https://doi.org/10.1093/gbe/evp044

7. Chan, C.X., Darling, A.E., Beiko, R.G., Ragan, M.A.: Are protein domains modules of lateral genetic transfer? PLoS ONE **4**(2), e4524 (2009). https://doi.org/10.1371/journal.pone.0004524

8. David, L.A., Alm, E.J.: Rapid evolutionary innovation during an archaean genetic expansion. Nature **469**, 93–96 (2011)

9. Dunning, L.T., et al.: Lateral transfers of large DNA fragments spread functional genes among grasses. Proc. Natl. Acad. Sci. **116**(10), 4416–4425 (2019). https://doi.org/10.1073/pnas.1810031116

10. Etherington, G.J., Dicks, J., Roberts, I.N.: Recombination analysis tool (RAT): a program for the high-throughput detection of recombination. Bioinformatics **21**(3), 278–281 (2004)

11. Gogarten, J.P., Doolittle, W.F., Lawrence, J.G.: Prokaryotic evolution in light of gene transfer. Mol. Biol. Evol. **19**(12), 2226–2238 (2002)

12. Jacox, E., Chauve, C., Szollosi, G.J., Ponty, Y., Scornavacca, C.: ecceTERA: comprehensive gene tree-species tree reconciliation using parsimony. Bioinformatics **32**(13), 2056 (2016). https://doi.org/10.1093/bioinformatics/btw105

13. Kloub, L., Gosselin, S., Fullmer, M., Graf, J., Gogarten, J.P., Bansal, M.S.: Systematic detection of large-scale multigene horizontal transfer in prokaryotes. Mol. Biol. Evol. **38**(6), 2639–2659 (2021). https://doi.org/10.1093/molbev/msab043

14. Koonin, E.V., Wolf, Y.I.: Genomics of bacteria and archaea: the emerging dynamic view of the prokaryotic world. Nucleic Acids Res. **36**(21), 6688–6719 (2008). https://doi.org/10.1093/nar/gkn668

15. Kordi, M., Bansal, M.S.: TreeSolve: Rapid Error-Correction of Microbial Gene Trees. In: Martín-Vide, C., Vega-Rodríguez, M.A., Wheeler, T. (eds.) AlCoB 2020. LNCS, vol. 12099, pp. 125–139. Springer, Cham (2020). https://doi.org/10.1007/978-3-030-42266-0_10

16. Kundu, S., Bansal, M.S.: SaGePhy: an improved phylogenetic simulation framework for gene and subgene evolution. Bioinformatics **35**(18), 3496–3498 (2019)

17. Lewis, P.O., Chen, M.H., Kuo, L., Lewis, L.A., Fucikova, K., Neupane, S., Wang, Y.B., Shi, D.: Estimating bayesian phylogenetic information content. Syst. Biol. **65**(6), 1009–1023 (2016). https://doi.org/10.1093/sysbio/syw042

18. Lole, K.S., et al.: Full-length human immunodeficiency virus type 1 genomes from subtype c-infected seroconverters in India, with evidence of intersubtype recombination. J. Virol. **73**(1), 152–160 (1999)

19. Martin, D.P., Lemey, P., Posada, D.: Analysing recombination in nucleotide sequences. Mol. Ecol. Resour. **11**(6), 943–955 (2011). https://doi.org/10.1111/j.1755-0998.2011.03026.x

20. Martin, D.P., Murrell, B., Golden, M., Khoosal, A., Muhire, B.: RDP4: detection and analysis of recombination patterns in virus genomes. Virus Evol **1**(1) (2015)

21. Morel, B., Kozlov, A.M., Stamatakis, A., Szollosi, G.J.: GeneRax: a tool for species-tree-aware maximum likelihood-based gene family tree inference under gene duplication, transfer, and loss. Mol. Biol. Evol. **37**(9), 2763–2774 (2020). https://doi.org/10.1093/molbev/msaa141

22. Nguyen, T.H., Doyon, J.-P., Pointet, S., Arigon Chifolleau, A.-M., Ranwez, V., Berry, V.: Accounting for gene tree uncertainties improves gene trees and reconciliation inference. In: Raphael, B., Tang, J. (eds.) WABI 2012. LNCS, vol. 7534, pp. 123–134. Springer, Heidelberg (2012). https://doi.org/10.1007/978-3-642-33122-0_10

23. Petersen, J., Wagner-Dobler, I.: Plasmid transfer in the ocean - a case study from the roseobacter group. Front. Microbiol. **8**, 1350 (2017). https://doi.org/10.3389/fmicb.2017.01350

24. Polz, M.F., Alm, E.J., Hanage, W.P.: Horizontal gene transfer and the evolution of bacterial and archaeal population structure. Trends Genet. **29**(3), 170–175 (2013)

25. Posada, D., Crandall, K.: The effect of recombination on the accuracy of phylogeny estimation. J. Mol. Evol. **54**, 396–402 (2002)

26. Rambaut, A., Grass, N.C.: Seq-Gen: an application for the Monte Carlo simulation of DNA sequence evolution along phylogenetic trees. Bioinformatics **13**(3), 235–238 (1997). https://doi.org/10.1093/bioinformatics/13.3.235

27. Rangel, L.T., Marden, J., Colston, S., Setubal, J.C., Graf, J., Gogarten, J.P.: Identification and characterization of putative Aeromonas spp. T3SS effectors. PLOS ONE **14**(6), 1–20 (2019). https://doi.org/10.1371/journal.pone.0214035

28. Ravenhall, M., Škunca, N., Lassalle, F., Dessimoz, C.: Inferring horizontal gene transfer. PLOS Comput. Biol. **11**(5), 1–16 (2015). https://doi.org/10.1371/journal.pcbi.1004095

29. Robinson, D.F., Foulds, L.R.: Comparison of phylogenetic trees. Math. Biosci. **53**(1), 131–147 (1981)

30. Sjostrand, J., Tofigh, A., Daubin, V., Arvestad, L., Sennblad, B., Lagergren, J.: A bayesian method for analyzing lateral gene transfer. Syst. Biol. **63**(3), 409–420 (2014). https://doi.org/10.1093/sysbio/syu007

31. Stamatakis, A.: RAxML version 8: a tool for phylogenetic analysis and post-analysis of large phylogenies. Bioinformatics **30**(9), 1312–1313 (2014)

32. Szollosi, G.J., Rosikiewicz, W., Boussau, B., Tannier, E., Daubin, V.: Efficient exploration of the space of reconciled gene trees. Syst. Biol. **62**(6), 901–912 (2013)

33. Tuomanen, E.I., Hollingshead, S.K., Becker, R., Briles, D.E.: Diversity of PspA: mosaic genes and evidence for past recombination in streptococcus pneumoniae. Infect. Immun. **68**(10), 5889–5900 (2000). https://doi.org/10.1128/IAI.68.10.5889-5900.2000

34. Weiller, G.F.: Phylogenetic profiles: a graphical method for detecting genetic recombinations in homologous sequences. Mol. Biol. Evol. **15**(3), 326–335 (1998)

35. Zhaxybayeva, O., Lapierre, P., Gogarten, J.P.: Genome mosaicism and organismal lineages. Trends Genet. **20**(5), 254–260 (2004)

Genome Rearrangements

Sorting by k-Cuts on Signed Permutations

Andre Rodrigues Oliveira[1]([✉])(iD), Alexsandro Oliveira Alexandrino[1](iD),
Géraldine Jean[2](iD), Guillaume Fertin[2](iD), Ulisses Dias[3](iD), and Zanoni Dias[1](iD)

[1] Institute of Computing, University of Campinas, Campinas, Brazil
{andrero,alexsandro,zanoni}@ic.unicamp.br
[2] LS2N, UMR CNRS 6004, Nantes University, Nantes, France
{geraldine.jean,guillaume.fertin}@univ-nantes.fr
[3] School of Technology, University of Campinas, Limeira, Brazil
ulisses@ft.unicamp.br

Abstract. Sorting by Genome Rearrangements is a classic problem in Computational Biology. Several models have been considered so far, each of them defines how a genome is modeled (for example, permutations when assuming no duplicated genes, strings if duplicated genes are allowed, and/or use of signs on each element when gene orientation is known), and which rearrangements are allowed. Recently, a new problem, called Sorting by Multi-Cut Rearrangements, was proposed. It uses the k-Cut rearrangement which cuts a permutation (or a string) at $k \geq 2$ places and rearranges the generated blocks to obtain a new permutation (or string) of same size. This new rearrangement may model *chromoanagenesis*, a phenomenon consisting of massive simultaneous rearrangements. Similarly as the Double-Cut-and-Join, this new rearrangement also generalizes several genome rearrangements such as reversals, transpositions, revrevs, transreversals, and block-interchanges. In this paper, we extend a previous work based on unsigned permutations and strings to *signed* permutations. We show the complexity of this problem for different values of k, that the approximation algorithm proposed for unsigned permutations with any value of k can be adapted to signed permutations, and a 1.5-approximation algorithm for the specific case $k = 4$.

Keywords: Genome rearrangements · Sorting permutations · Approximation algorithms · Algorithmic complexity

1 Introduction

Genome rearrangements are widely studied in the field of Comparative Genomics. They refer to large scale mutations that affect the position and

This work was supported by the National Council of Technological and Scientific Development, CNPq (grants 425340/2016-3 and 202292/2020-7), the Coordenação de Aperfeiçoamento de Pessoal de Nível Superior - Brasil (CAPES) - Finance Code 001, and the São Paulo Research Foundation, FAPESP (grants 2013/08293-7, 2015/11937-9, and 2019/27331-3).

L. Jin and D. Durand (Eds.): RECOMB-CG 2022, LNBI 13234, pp. 189–204, 2022.
https://doi.org/10.1007/978-3-031-06220-9_11

orientation of genes in a genome. When comparing two genomes $\mathcal{G}_1$ and $\mathcal{G}_2$, one goal is to compute the minimum number of rearrangements that transform $\mathcal{G}_1$ into $\mathcal{G}_2$, which is called the rearrangement distance between genomes $\mathcal{G}_1$ and $\mathcal{G}_2$.

When a genome has no duplicated genes, we model it using a permutation, where each element represents a gene. When gene orientation information is available, we use a '+' or '−' sign on each element to represent its orientation, and the permutation is said to be signed. Otherwise, we use unsigned permutations. Since the target genome can be represented as the identity permutation $\iota^n = (1\ 2\ \ldots\ n)$, the problem of transforming one genome into another can be considered as the one of sorting a permutation using rearrangements.

The most common rearrangement events in the literature are the reversal and the transposition [4,10,11]. The reversal acts on a segment of the genome, inverting this segment and changing the orientation of the genes in it. The transposition acts on two adjacent segments, exchanging the position of these two segments. Among more particular rearrangement events are revrevs, which act on two adjacent segments inverting both segments but not exchanging them, and transrevs, which act like transpositions with an inversion in of one of the two segments. More general rearrangements were also proposed, such as the block-interchange and the Double-Cut-and-Join (DCJ) [8,15]. The block-interchange generalizes a transposition: it exchanges two segments that are not necessarily adjacent; the DCJ generalizes operations on multichromosomal genomes, including reversals.

For the reversals and transpositions, when computing the rearrangement distance, we assume that events happen sequentially. However, in a (chaotic) process called *chromoanagenesis*, a large number of rearrangements can occur in a single event [12,14], that is, in a single event the genome is cut in many blocks that are rearranged. These events were mainly observed in tumors and congenital diseases [12].

Alekseyev and Pevzner [2] introduced the multi-break rearrangement in circular genomes. A k-break is a rearrangement that breaks k adjacencies of the genome, creating $2k$ open ends, and then combines these $2k$ open ends, creating k new adjacencies. For instance, a DCJ is a 2-break. Alekseyev and Pevzner [2] presented an algorithm to solve the k-break rearrangement distance that takes polynomial time considering the size of the genome but takes exponential time considering the value of k. Alekseyev [1] extended the investigation of multi-break rearrangements with linear genomes and showed lower bounds for computing the distance between genomes and an analysis of the breakpoint reuse rate for linear genomes.

Recently, Bulteau *et al.* [6] introduced the multi-cut (k-cut) rearrangement, which models the most common genome rearrangements [10] and *chromoana-genesis*. In the k-cut rearrangement, the genome is cut in k positions, forming $k + 1$ blocks, which are then rearranged in any way except for the first and last blocks (which can be empty) that must remain in the same positions. However, the relative position and orientation of the genes inside blocks remain the same. Note that using this definition, we are restricted to genomes without gene orientation information, since there are no block inversions. In general, the k-cut distance problem is NP-Hard [6]. Considering specific values for k, we have that the 3-cut distance is NP-Hard [7], which is equivalent to the transposition distance, while the 4-cut distance (equivalent to the block-interchange distance [8]) can be solvable in polynomial time.

In this work, we extend the k-cut rearrangement definition to permit block reversals, which allows us to compare genomes considering both genome order and gene orientation. We show that this problem is NP-Hard for $k \geq 5$, and we adapt the results from Bulteau $et\ al.$ [6] to show an ℓ-approximation algorithm for $k \geq 5$, where $\ell = \frac{2k}{k-1}$ if k is odd, and $\ell = \frac{2k}{k-2}$ if k is even. Besides, we present a 1.5-approximation algorithm for signed permutations when $k = 4$ using the cycle graph structure.

2   Basic Definitions

In this work we use permutations to represent the genomes, where each element represents a particular gene or a contiguous sequence of genes. Each element has a sign, '+' or '−', indicating its orientation. Given two genomes with no repeated genes and sharing the same set of genes, we represent one of them as the $identity\ permutation$ $\iota^n = (+1\ +2\ \dots\ +n)$, and we map the other according to ι^n as the permutation $\pi = (\pi_1\ \dots\ \pi_n)$. In the following, we assume two positional and fixed elements on any signed permutation π: $\pi_0 = +0$ (right before π_1) and $\pi_{n+1} = +(n+1)$ (right after π_n). We also denote the set of all signed permutations of size n by $S_n^{\pm}$.

A k-cut, denoted by $\rho_\sigma^{(i_1, i_2, \dots, i_k)}$, $1 \leq i_1 \leq i_2 \leq \dots \leq i_k \leq n+1$ and $\sigma = (\sigma_1\ \dots\ \sigma_{k-1}) \in S_{k-1}^{\pm}$, cuts a signed permutation π into $(k+1)$ blocks: $q_0 = [\pi_0 \dots \pi_{i_1-1}]$, $q_1 = [\pi_{i_1} \dots \pi_{i_2-1}]$, $q_2 = [\pi_{i_2} \dots \pi_{i_3-1}], \dots, q_k = [\pi_{i_k} \dots \pi_{n+1}]$, such that π can be reconstructed by concatenating these $(k+1)$ blocks in the same relative order, and rearranges these blocks to obtain a new permutation $\pi' \in S_n^{\pm}$. Given a block q_i, we denote by q_{-i} the $reversal$ of q_i, which consists in reversing the order of the elements in q_i and flipping the sign of every element inside it.

Formally, a k-cut $\rho_\sigma^{(i_1, \dots, i_k)}$ applied to a permutation π, denoted by $\pi \cdot \rho$, results in a permutation π' starting with q_0, followed by the $(k-1)$ middle blocks $q_{\sigma_1}, \dots, q_{\sigma_{k-1}}$, followed by block q_k (blocks q_0 and q_k are fixed and cannot be reversed). We say that a k-cut ρ is $valid$ if $\pi \cdot \rho \neq \pi$.

The Sorting by k-Cuts with Block Reversals problem (SKCBR) seeks for the minimum number of k-cuts needed to transform a signed permutation π into ι^n; we denote this number as $d_k(\pi)$.

3   Breakpoints and Strips

Given a signed permutation $\pi \in S_n^{\pm}$ with n elements, a pair of adjacent elements (π_i, π_{i+1}), $0 \leq i \leq n$, is called a $breakpoint$ if $\pi_{i+1} - \pi_i \neq 1$. The number of breakpoints in a permutation π is denoted by $b(\pi)$.

A *strip* is a maximal sequence of elements from π (including the two extra elements $+0$ and $+(n+1)$) without breakpoints. The number of strips in a permutation is denoted by $s(\pi)$. Every permutation π with $b(\pi)$ breakpoints satisfies $s(\pi) = b(\pi) + 1$. A strip is called *positive* if each of its elements has positive sign, and *negative* otherwise. Note that the leftmost strip in π, starting with element $+0$, and the rightmost strip in π, ending with element $+(n+1)$, always are positive strips.

Note that, for fixed n, ι^n is the permutation satisfying $b(\iota^n) = 0$ and $s(\iota^n) = 1$, so to transform a permutation $\pi \in S_n^{\pm}$ into ι^n, we need to decrease the number of breakpoints from $b(\pi)$ to 0 and the number of strips from $s(\pi)$ to 1.

Given a permutation π with $s(\pi) > 1$, we say that a pair of strips (u, v), where $u = \langle u_1, \ldots, u_i \rangle$ and $v = \langle v_1, \ldots, v_j \rangle$, is *contiguous* if one of the following is true: u and v are both positive or both negative strips, and either $v_1 = u_i + 1$ or $u_1 = v_j + 1$; or u is positive, v is negative, and either $u_i = |v_j| - 1$ or $u_1 = |v_1| + 1$.

Given a permutation π and a k-cut ρ, let $\Delta_b(\pi, \rho) = b(\pi) - b(\pi \cdot \rho)$ denote the difference in the number of breakpoints after ρ is applied to π.

Lemma 1. *For any signed permutation π and any k-cut ρ, $\Delta_b(\pi, \rho) = b(\pi) - b(\pi \cdot \rho) \leq k$.*

Proof. Note that a k-cut ρ cuts exactly k adjacent elements from π. In the best case, each of these k adjacent elements is distinct and is a breakpoint in π, while in $\pi' = \pi \cdot \rho$ the k new pairs of adjacent elements are not breakpoints, so $b(\pi \cdot \rho) = b(\pi) - k$. If any of the adjacent elements affected by ρ is not unique, or is not a breakpoint, or remains a breakpoint in π', then $b(\pi \cdot \rho) < b(\pi) - k$. $\square$

Lemma 1 results in the following lower bound for SKCBR.

Lemma 2. *For any signed permutation π, $d_k(\pi) \geq \frac{b(\pi)}{k}$.*

3.1   SKCBR is NP-Hard for $k \geq 5$

The Sorting by k-Cuts problem (SbkC), originally called Sorting by Multi-Cut Rearrangements in Permutations [6], is NP-Hard for any value of $k \geq 5$. Since there are no block reversals allowed in this model, the permutations must be unsigned. We also extend an unsigned permutation π by fixing the elements $\pi_0 = 0$ and $\pi_{n+1} = n + 1$. The definition of a breakpoint is analogous for this problem, that is, a pair of elements (π_i, π_{i+1}) is a breakpoint if $\pi_{i+1} - \pi_i \neq 1$.

More than showing that SbkC is NP-Hard, Bulteau et al. [6, Proposition 1 and Lemma 7] showed that deciding if a permutation π can be sorted using exactly $\frac{b(\pi)}{k}$ k-cuts is NP-Hard for $k \geq 5$.

To prove that the Sorting by k-Cuts with Block Reversals problem is NP-Hard for $k \geq 5$, we use a reduction from the Sorting by k-Cuts problem (SbkC). Next, we define the decision version of both problems.

Problem 1 (SKCBR). Given a value k, a signed permutation π and an integer z, decide whether is possible to sort π using up to z k-cuts allowing block reversals.

Problem 2 (SbkC). Given a value k and an unsigned permutation π, decide whether is possible to sort π using exactly $\frac{b(\pi)}{k}$ k-cuts without allowing block reversals.

Theorem 1. *The SKCBR problem is NP-Hard for $k \geq 5$.*

Proof. Given an instance (k, π) for the SbkC problem, we create an instance (k, π', z) for the SKCBR problem, where $z = \frac{b(\pi)}{k}$ and π' is a signed permutation such that $\pi'_i = +\pi_i$, for $0 \leq i \leq n+1$. Note that since the definition of breakpoint for both problems is analogous, we have that $b(\pi) = b(\pi')$.

Now, we show that the answer to the instance (k, π) for the SbkC problem is yes if, and only if, the answer to the instance (k, π', z) for the SKCBR problem is also yes.

($\Rightarrow$) Suppose that the instance (k, π) for the SbkC problem can be sorted using $\frac{b(\pi)}{k}$ k-cuts without allowing block reversals, then we can also sort (k, π', z) using the same operations, since all elements have $+$ sign and no reversals are applied.

($\Leftarrow$) Now, suppose that the instance (k, π', z) for the SKCBR problem can be sorted using $z = \frac{b(\pi)}{k} = \frac{b(\pi')}{k}$ k-cuts.

We show that all these k-cuts do not have block reversals and, consequently, the same operations can be used to sort the instance (k, π) for SbkC.

Let $S = (\rho_1, \rho_2, \ldots, \rho_m)$, where $m = \frac{b(\pi')}{k}$, be a sequence of k-cuts that sorts π'. Since the number of operations in S is exactly the lower bound for π' (by Lemma 2), every k-cut in S must remove exactly k breakpoints. Suppose by contraction that there is a k-cut that has block reversals. Let ρ_i be such k-cut with minimum i. Note that until ρ_i is applied all elements of the permutation have a positive sign, because no block reversal has previously been applied. Let π^i be the permutation after applying $(\rho_1, \rho_2, \ldots, \rho_i)$. Let $(\pi^i_x, \ldots, \pi^i_y)$ be a block that was reversed by ρ_i, such that x is minimum. All elements in $(\pi^i_x, \ldots, \pi^i_y)$ have negative sign. Since the operations cannot alter the fixed element π^i_0, we know that $x > 0$ and the element π^i_{x-1} has positive sign. This means that there is a breakpoint (π^i_{x-1}, π^i_x) between an adjacency affected by ρ^i, which contradicts the fact ρ^i removed the maximum number of breakpoints possible by a k-cut. Therefore, $S = (\rho_1, \rho_2, \ldots, \rho_m)$ has only k-cuts without block reversals, and the same sequence can be used to sort (k, π) for the SbkC problem. $\square$

4 An Approximation Algorithm for SKCBR

In this section, we adapt the results from Bulteau *et al.* [6] on unsigned permutations to show an approximation algorithm for SKCBR. In the following, we explain the algorithm and we show its approximation factor for $k \geq 5$ on signed permutations. Furthermore, we present results for $2 \leq k \leq 4$.

Our approximation algorithm follows the same idea of the 2-approximation algorithm from Bulteau *et al.* [6, Theorem 10], but is designed for signed permutations. Given a signed permutation π and a parameter k, we use a parameter

$z = \lfloor \frac{k-1}{2} \rfloor$, that denotes how many strips will be affected on each step. The following process is repeated until π becomes ι: let $m = \min(z, s(\pi) - 1)$, and ρ be a k-cut that cuts $(k - 2m)$ times after the first strip $u = \langle 0, \ldots \rangle$ and at the extremities of m other strips $v_1, \ldots, v_m$ such that any pair of strips in $\{(u, v_1)\} \cup \{(v_{i-1}, v_i) : 2 \leq i \leq m\}$ is contiguous. This k-cut will rearrange the $k - 1$ blocks in such a way that blocks with strips $v_1, \ldots, v_m$ are placed in the left (after the block q_0 that is the strip u), in the same order as their indices, and blocks with a negative strip must be reversed, which guarantees that at least m breakpoints are removed at each step (if $m \neq z$ then all breakpoints are removed and $\pi = \iota$).

For example, given the permutation $\pi = (+1\ -4\ -5\ +3\ -2\ +6)$ and $k = 5$, we have that π has 5 breakpoints and 6 strips: $\langle 0, +1 \rangle$, $\langle -4 \rangle$, $\langle -5 \rangle$, $\langle +3 \rangle$, $\langle -2 \rangle$, and $\langle +6, +7 \rangle$, so $z = \lfloor \frac{5-1}{2} \rfloor = 2$ and $m = \min(2, 5) = 2$. The first 5-cut will join the first strip with strips $\langle +3 \rangle$ and $\langle -2 \rangle$, so it cuts once ($k - 2m = 5 - 4 = 1$) right after the first strip and before and after these two other strips, which results in $\rho_{(-4\ +2\ +1\ +3)}^{(2,4,5,5,6)}$ (note that one of the blocks is empty). Now let $\pi' = \pi \cdot \rho = (+1\ +2\ +3\ -4\ -5\ +6)$, that has 3 breakpoints 4 strips: $\langle 0, +1, +2, +3 \rangle$, $\langle -4 \rangle$, $\langle -5 \rangle$, and $\langle +6, +7 \rangle$. In this step we have $m = \min(2, 3) = 2$, and the 5-cut will join the first strip with strips $\langle -4 \rangle$ and $\langle -5 \rangle$, which results in $\rho_{(-2\ -4\ +1\ +3)}^{(4,4,5,5,6)}$. Now $\pi' \cdot \rho = \iota$, with no breakpoints and one strip, and the algorithm stops. Note that the first 5-cut removed two breakpoints from π and the second 5-cut removed three breakpoints from π'.

The following lemma gives the approximation factor of our algorithm when $k \geq 5$.

Lemma 3. *When $k \geq 5$, the algorithm has an approximation factor of $\frac{2k}{k-1} \leq 2.5$ if k is odd, and $\frac{2k}{k-2} \leq 3$ otherwise.*

Proof. The approximation factor follows from the lower bound from Lemma 2.

If $k \geq 5$ is odd, then each k-cut applied removes $\frac{k-1}{2}$ breakpoints, which results in $\frac{2b(\pi)}{k-1}$ k-cuts needed to transform π into ι^n, so the approximation factor is $\frac{2b(\pi)}{k-1} \times \frac{k}{b(\pi)} = \frac{2k}{k-1} \leq 2.5$.

When $k \geq 6$ is even, then each k-cut applied removes $\frac{k-2}{2}$ breakpoints, which results in $\frac{2b(\pi)}{k-2}$ k-cuts needed to transform π into ι^n, so the approximation factor is $\frac{2b(\pi)}{k-2} \times \frac{k}{b(\pi)} = \frac{2k}{k-2} \leq 3$.   □

We note that when $k = 2$, any valid k-cut is a *reversal*. Sorting by Reversals has a polynomial algorithm on signed permutations [11].

When $k = 3$, any valid k-cut $\rho_\sigma^{(i_1, i_2, i_3)}$, $i_1 \neq i_2 \neq i_3$, is either a *revrev*, if $\sigma = (-1\ -2)$, a *transreversal* if either $\sigma = (-2\ +1)$ or $\sigma = (+2\ -1)$, a *transposition* if $\sigma = (+2\ +1)$, and a *reversal* otherwise; any valid k-cut $\rho_\sigma^{(i_1, i_2, i_3)}$, where either $i_1 = i_2$ or $i_2 = i_3$ is also a *reversal*. Sorting by Reversals, Transpositions, Transreversals and RevRevs is NP-hard [3], and with an argument similar to the one used in Lemma 3, we prove that the algorithm described in this section has an approximation factor of 3 (when $k = 3$). However, we note that there is a 1.75-approximation algorithm based on the cycle graph structure for this problem [13].

We present a 1.5-approximation algorithm for $k = 4$ in Sect. 7, also based on the cycle graph structure (described in the next section). Note that when $k = 4$, the complexity of SKCBR is unknown.

5   Cycle Graph and Complement Cycle Graph

Our next approximation algorithm will use the *cycle graph*, a well-known graph structure that is used to represent a permutation in sorting problems [4]. Given a signed permutation π, we construct its cycle graph $G(\pi)$ as follows. The vertex set is formed by vertices $+0$ and $-(n+1)$, plus vertices $-\pi_i$ and $+\pi_i$ for $1 \le i \le n$. The edge set is formed by two colored edge subsets: black and gray edges. The *gray edge* set is formed by $\{(+(i-1), -i) : 1 \le i \le n+1\}$ and the *black edge* set is formed by $\{(-\pi_i, +\pi_{i-1}) : 1 \le i \le n+1\}$.

We draw the *cycle graph* of a permutation π by placing its vertices on a horizontal line, starting (from left to right) with $+0$, ending with $-(n+1)$, and the remaining vertices between them in the same order as they appear in π, where vertex $-\pi_i$ is always drawn to the left of $+\pi_i$. Black edges are drawn as straight edges and gray edges are drawn as curved edges. Since π is a signed permutation, each vertex has one gray edge and one black edge, so there is a unique decomposition of $G(\pi)$ into alternating cycles, i.e. into cycles that alternate between black and gray edges. The number of cycles in $G(\pi)$ is denoted by $c(G(\pi))$.

Given a cycle graph $G(\pi)$, we assign the label i to the black edge that links $-\pi_i$ to $+\pi_{i-1}$. A cycle C in $G(\pi)$ is represented in a unique fashion by the ordered list of its black edge labels, starting with the label with highest value, say c_1 (which means that c_1 is the rightmost element in the horizontal displacement of vertices), followed by the labels that are encountered while visiting C, knowing that c_1 is traversed from right to left. Hence, we also use signs together with black edge labels, just to indicate the traversed orientation of black edges: when a black edge is traversed from right to left we add a '$+$' sign to its label; otherwise, we add a '$-$' sign to its label. By the unique representation we adopted, c_1 will thus always have a '$+$' sign.

Given a signed permutation π and its cycle graph $G(\pi)$, we refer to the *size* of a cycle $C \in G(\pi)$ as the number of black edges inside it; a cycle of size ℓ is called *trivial* if $\ell = 1$, *short* if $\ell \in \{2, 3\}$, and *long* otherwise. We also refer to short and long cycles as *non-trivial* cycles. Besides, a cycle of size ℓ is called *odd* if ℓ is odd and it is called *even* otherwise.

Another graph structure we use is the *complement cycle graph* [9], that is similar to the cycle graph. Given a signed permutation π, we construct its complement cycle graph $\bar{G}(\pi)$ as follows. The vertex set is the same as in $G(\pi)$. The edge set is formed by two colored edge sets: white and gray edges. The *gray edge* set is the same as in $G(\pi)$, and the *white edge* set is formed by

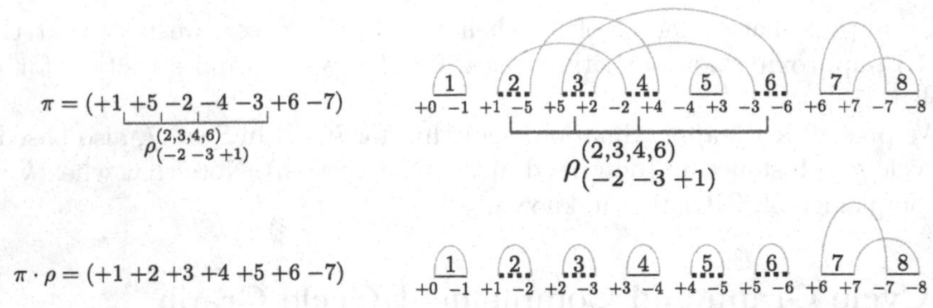

(a) signed permutations π and $\pi \cdot \rho$ (b) cycle graphs $G(\pi)$ and $G(\pi \cdot \rho)$

Fig. 1. An example of a 4-cut affecting a permutation π and its cycle graph. **(a)** the 4-cut $\rho^{(2,3,4,6)}_{(-2\,-3\,+1)}$ applied to $\pi = (+1\ +5\ -2\ -4\ -3\ +6\ -7)$, that results in $\pi = (+1\ +2\ +3\ +4\ +5\ +6\ -7)$. **(b)** how this 4-cut ρ affects $G(\pi)$ (top) resulting in $G(\pi \cdot \rho)$ (bottom): it removes four black edges (those with dashed lines), rearranges the vertices according to $\pi \cdot \rho$, and creates four new black edges (also with dashed lines). Note that $c(G(\pi)) = 5$ and two cycles from $G(\pi)$ are affected by ρ: $C_1 = (+4, -2)$ and $C_2 = (+6, +3)$, so the other three cycles $C_3 = (+1)$, $C_4 = (+5)$ and $C_6 = (+8, -7)$ are not affected and remain unchanged in $G(\pi \cdot \rho)$, except for their black edge labels that may change. After the 4-cut is applied, $c(G(\pi \cdot \rho)) = 7$, so after ρ is applied to π two new cycles are created.

$\{(+0, -(n+1))\} \cup \{(-\pi_i, +\pi_i) : 1 \leq i \leq n\}$. Note that any complement cycle graph $\bar{G}(\pi)$ of a permutation π has a Hamiltonian cycle following its gray and white edges [9, Observation 23]. This can be seen by the construction of these two edges: gray edges link $+i$ to $-(i+1)$, for $0 \leq i \leq 0$, and white edges link $-i$ to $+i$, for $1 \leq i \leq n$, with an extra white edges that links $+0$ to $-(n+1)$. This means that if we start at vertex $+0$, and following its gray edge, we can construct a Hamiltonian path $P = [+0, -1, +1, -2, \ldots, -n, +n, -(n+1)]$, where each two consecutive vertices (x, y) in P is a white edge, if $x < 0$, or it is a gray edge otherwise. Since $(-(n+1), +0)$ is also a white edge, and these two vertices are the endpoints of P, it follows that $\bar{G}(\pi)$ has a Hamiltonian cycle.

Given a permutation π, a k-cut $\rho^{(i_1,\ldots,i_k)}_\sigma$ applied to π affects $G(\pi)$ as follows. Up to k black edges will be removed (namely, those with labels $i_1, \ldots, i_k$); the $(k-1)$ blocks of vertices (between q_0 and q_{k+1}) will be rearranged according to σ; another k black edges will be created according to $\pi \cdot \sigma$. Note that this means that the set of gray edges from $G(\pi)$ is never affected by a k-cut, and any cycle whose black edges are not affected by ρ remains with the same set of black and gray edges. See Fig. 1 for an example.

A gray edge g in a non-trivial cycle $C \in G(\pi)$ is called *convergent* if it links the leftmost vertex of one black edge with the rightmost vertex of another black edge; g is called *divergent* otherwise. When traversing a cycle, a divergent gray edge makes its adjacent black edges be traversed in opposite orientations, as shown in Fig. 2. A cycle C is called *fully divergent* if every gray edge in C is divergent.

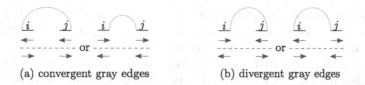

(a) convergent gray edges   (b) divergent gray edges

Fig. 2. (a) the two types of convergent gray edges and (b) the two types of divergent gray edges that can appear in a cycle graph. Arrows above the dashed line indicate the traversed orientation if we go from black edge of label i to j, and arrows under the dashed line from black edge of label j to i. Note that with convergent gray edges all black edges have the same traversed orientation, while the orientations are opposite with divergent gray edges.

Given a vertex v_i in $G(\pi)$, let $p(v_i)$ denote the position (from left to right) of v_i in the cycle graph representation. For instance, $p(0) = 1$ and $p(-(n+1)) = 2n+2$. Given two gray edges $g_1 = (v_a, v_b)$ and $g_2 = (v_c, v_d)$, such that $p(v_a) < p(v_b)$ and $p(v_c) < p(v_d)$, we say that they are *crossing* if either $p(v_a) < p(v_c) < p(v_b) < p(v_d)$ or $p(v_c) < p(v_a) < p(v_d) < p(v_b)$.

Lemma 4. *For any permutation π, and any 4-cut ρ, $\Delta_c(\pi, \rho) = c(G(\pi \cdot \rho)) - c(G(\pi)) \leq 3$.*

Proof. Recall that a 4-cut ρ is applied to up to four black edges, and assume that ρ is applied to a permutation π such that $c(G(\pi)) = x$. After the cuts take place, we observe that each cycle that is not affected by ρ has the same set of black and gray edges, and each cycle C affected by the $t \leq 4$ cuts can be seen as a set of t paths (each path starts and ends with a gray edge), so it follows that there are up to four paths and up to $(x-1)$ cycles left. In the best scenario, after the blocks are rearranged each new black edge links two grays edges of a same path, resulting in four new cycles. Since the 4-cut is applied to at least one cycle, it follows that $\Delta_c(\pi, \rho) = c(G(\pi \cdot \rho)) - c(G(\pi)) \leq 4 + x - 1 - x = 3$. □

Lemma 5. *Given a permutation π and a 4-cut ρ, if $\Delta_c(\pi, \rho) = 3$ then ρ is applied to four distinct black edges of a same (long) cycle; $\Delta_c(\pi, \rho) \leq 2$ otherwise.*

Proof. These observations follow from Lemma 4: Let π be a permutation and let ρ be a 4-cut applied to π. After the four cuts in black edges from $y > 1$ cycles we have up to $c(G(\pi)) - 2$ cycles left, and even if we transform the $z \leq 4$ paths into z cycles it follows that $\Delta_c(\pi, \rho) \leq (4 + c(G(\pi)) - 2) - c(G(\pi)) = 2$. Besides, if we apply a 4-cut to black edges of $y \geq 1$ cycles such that $x > 1$ cuts affect the same black edge we have $x - 1$ empty blocks, and consequently $z \leq 4 - x + 1$ paths, so $\Delta_c(\pi, \rho) \leq (3 + c(G(\pi)) - 1) - c(G(\pi)) = 2$. □

The following lemma defines a lower bound for SKCBR.

Lemma 6. *For any signed permutation $\pi \in S_n^{\pm}$, if $G(\pi)$ has no short cycles and there is a 4-cut ρ such that $\Delta_c(\pi, \rho) = 3$, then $d_4(\pi) \geq \frac{(n+1)-c(G(\pi))}{3}$; otherwise, $d_4(\pi) \geq \frac{(n+2)-c(G(\pi))}{3}$.*

Proof. To transform a permutation π with n elements into ι^n, we need to go from $c(G(\pi))$ to $n + 1$ cycles, and by Lemma 4, we have that $d_4(\pi) \geq \frac{(n+1)-c(G(\pi))}{3}$. However, if there is no 4-cut that increases the number of cycles in $G(\pi)$ by three, then at least the first 4-cut will increase the number of cycles by up to two. Besides, if $G(\pi)$ has a short cycle, we know by Lemma 5 that at least one 4-cut will increase the number of cycles by up to two. In both cases it follows that $d_4(\pi) \geq 1 + \frac{(n+1)-c(G(\pi))-2}{3} = \frac{(n+2)-c(G(\pi))}{3}$. □

6 Increasing the Number of Cycles in $G(\pi)$ with 4-Cuts

In following lemmas, we explain how 4-cuts can be used to increase the number of cycles by two or three each, depending on the number of cycles and types of gray edges.

We first show that any convergent gray edge admits a 4-cut that increases the number of cycles by at least two. Before that, we show a property of convergent and divergent gray edges, that can also be seen in the overlap graph [5].

Lemma 7. *Any convergent or divergent gray edge in $G(\pi)$ crosses with another gray edge.*

Proof. Note that given two black edges sharing a divergent gray edge d, there is a path between the other two end points of those black edges to form a cycle F, so at least one gray edge from F crosses with d.

For convergent gray edges, we first recall that any complement cycle graph $\bar{G}(\pi)$ of a signed permutation π has a Hamiltonian cycle following its gray and white edges. Now assume that a cycle graph $G(\pi)$ has a cycle C with a convergent gray edge $g = (g_1, g_2)$ that does not cross with any other gray edge. Note that only a gray edge from a trivial cycle has no vertices between the extremities of that gray edge. Since g is convergent, it follows that C is a non-trivial cycle, and there are at least two vertices between g_1 and g_2. But since g is not crossing with another gray edge, any vertex v_i between g_1 and g_2 (in $G(\pi)$ or $\bar{G}(\pi)$) must have a gray edge with another vertex v_j also between g_1 and g_2, so $\bar{G}(\pi)$ cannot have a Hamiltonian cycle, which is a contradiction with the construction of $\bar{G}(\pi)$. □

Lemma 8. *For any signed permutation $\pi \in S_n^\pm \backslash \{\iota^n\}$, if $G(\pi)$ has a convergent gray edge, then there is a 4-cut that increases the number of cycles by at least two.*

Proof. Let $G(\pi)$ be a cycle graph and let g_1 be a convergent gray edge of a cycle C that shares vertices with two distinct black edges of labels i and j, with $i < j$. Recall that, by Lemma 7, there is a gray edge, say g_2, that shares vertices with two distinct black edges of labels x and y, $x < y$, crossing with g_1.

We note that if (i, j) and (x, y) are equal, then C is a convergent cycle of size two, which contradicts the fact that g_1 and g_2 intersect, so it follows that there are at least three distinct black edge labels among i, j, x, and y. It can be easily seen that we can always apply a 4-cut to these (three or four) black edges

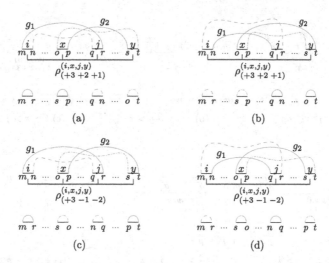

Fig. 3. Examples of a convergent gray edge g_1 ($g_1 = (m, r)$ in (a) and (c); $g_1 = (n, q)$ in (b) and (d)) crossing with another gray edge g_2 that is convergent in (a-b) (namely $g_2 = (p, s)$ in (a) and $g_2 = (o, t)$ in (b)) and divergent in (c-d) (namely $g_2 = (p, t)$ in (c) and $g_2 = (o, s)$ in (d)); dashed and dash-dotted curves indicate either a gray edge or an alternating path of black and gray edges between the two vertices in their extremities. In this case there is a 4-cut that increases the number of cycles by two.

in such a way that the two crossing gray edges g_1 and g_2 are transformed into two trivial cycles (thus increasing the number of cycles by two) if g_1 and g_2 are not in the same cycle, and increasing the number of cycles by two or three. □

There are multiple cases where a cycle graph $G(\pi)$ has a 4-cut according to Lemma 8, depending on the relative order among black edge labels and whether g_2 is convergent or divergent, and we show four examples in Fig. 3.

Lemma 9. *If a cycle C is fully divergent, then (i) C is even and (ii) any two consecutive black edges are traversed in opposite directions.*

Proof. To prove that C is even, one can observe that a divergent gray edge links the same side (left or right) of two distinct black edges, so if C has only divergent gray edges the number of gray edges (and consequently black edges) must be even to cover both sides of every black edge. The second statement follows from the fact that any divergent gray edge makes its adjacent black edges to be traversed in opposite orientations, and since every gray edge is divergent, this follows for any pair of consecutive black edges. □

Note that any cycle of size ℓ that is fully divergent is written $C = (+c_1, -c_2, +c_3, \ldots, -c_\ell)$. Now we show how to increase the number of cycles with a 4-cut if every non-trivial cycle of $G(\pi)$, $\pi \neq \iota^n$, is fully divergent. We split into two cases: when $G(\pi)$ has a long cycle or when $G(\pi)$ has only short cycles.

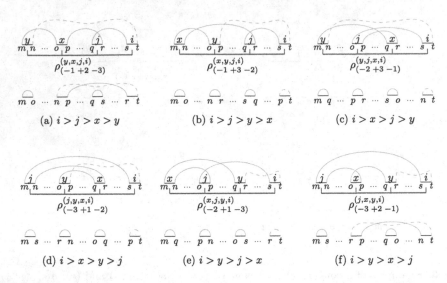

Fig. 4. (a–f) the six possible cases in a fully divergent cycle $C = (+i, -j, +x, -y, \ldots)$, with $z \geq 4$ black edges, depending on the relative order of the first four black edge labels (following its unique representation); the dashed curve indicates either a gray edge, if $x = 4$, or an alternating path of black and gray edges between the two vertices in their extremities. For each of them there is a 4-cut that increases the number of cycles by at least two.

Lemma 10. *Let $\pi \neq \iota^n$ be a signed permutation from $S_n^{\pm}$, and let $G(\pi)$ be its cycle graph. If every non-trivial cycle in $G(\pi)$ is fully divergent, and if there is a long cycle C in $G(\pi)$, then there exists a 4-cut that increases the number of cycles by at least two.*

Proof. Suppose that every gray edge in non-trivial cycles of $G(\pi)$ is divergent. By Lemma 9, it follows that any non-trivial cycle in $G(\pi)$ is an even fully divergent cycle. Now assume that $G(\pi)$ has a long cycle C with $z \geq 4$ black edges. We show in Fig. 4 the six possible cases depending on the relative order of black edge labels among the first four black edges of C following its unique representation. In four of these six cases, there is a 4-cut that transforms C into three trivial cycles and one cycle of size $(z-3)$, thus increasing the number of cycles by three (Fig. 4(b-e)); in the other two cases, there is a 4-cut that transforms C into two trivial cycles and one cycle of size $(z-2)$, thus increasing the number of cycles by two (Fig. 4(a) and (f)).   □

From now on, we assume that $G(\pi)$ has no long cycles, i.e. all non-trivial cycles have size two, and that every non-trivial cycle is fully divergent. Before showing how to increase the number of cycles with a 4-cut in this case, we state the following lemma.

Lemma 11. *Let $\pi \in S_n^{\pm} \setminus \{\iota^n\}$ be a signed permutation such that (i) $G(\pi)$ has no long cycles and (ii) every gray edge in non-trivial cycles is divergent. If there* are at least two non-trivial cycles in $G(\pi)$, then there exists a pair of cycles, say C and D, such that gray edges of C do not cross with gray edges of D.

Proof. The proof follows the same idea as proof of Lemma 7, where $\bar{G}(\pi)$ must have a Hamiltonian cycle. Note that in this case, any non-trivial cycle $C \in G(\pi)$ has two black edges, so in the following we will denote the leftmost and rightmost black edge labels of C as c_L and c_R, so it follows that $c_L < c_R$ and $C = (+c_R, -c_L)$. By contradiction, let us assume that given any pair of non-trivial cycles $D = (+d_R, -d_L)$ and $E = (+e_R, -e_L)$ in $G(\pi)$, either $d_L < e_L < d_R < e_R$ or $e_L < d_L < e_R < d_R$.

Now assume that $G(\pi)$ has $w \geq 2$ non-trivial cycles, and let us denote the cycle with the i-th lowest label for its rightmost black edge as $C^i = (+c_R^i, -c_L^i)$. Using the same argument as above, we have that $c_L^1 < c_L^2 < \ldots < c_L^w < c_R^1 < c_R^2 < \ldots < c_R^w$. Now we recall that $G(\pi)$ cannot have non-trivial cycles (i.e., fully divergent cycles) before c_L^1, after c_R^w, and between c_L^w and c_R^1, otherwise it is possible to have a pair of divergent gray edges not crossing, so in these three regions either there is no cycle, or there is a collection of trivial cycles. But in this case $\bar{G}(\pi)$ is not connected, which is a contradiction with the fact that $\bar{G}(\pi)$ has an Hamiltonian cycle. $\qquad\square$

The following lemma shows how we increase the number of cycles using a 4-cut if every cycle of $G(\pi)$ is either trivial or a fully divergent cycle of size two.

Lemma 12. *Let $\pi \in S_n^{\pm} \setminus \{\iota^n\}$ be a signed permutation, and let $G(\pi)$ be its cycle graph. If there are no long cycles in $G(\pi)$, and if every gray edge of non-trivial cycles is divergent, then a 4-cut increases the number of cycles: (i) by two if there are at least two non-trivial cycles, or (ii) by one otherwise (i.e., if $G(\pi)$ has only one non-trivial cycle).*

Proof. If $G(\pi)$ has only one non-trivial cycle of size two, whose gray edges are both divergent, we can apply a 4-cut to this cycle, which increases the number of cycles by one, as shown in Fig. 5(c).

Now consider that there are at least two short cycles in $G(\pi)$. By Lemma 11, we know that there is at least one pair of short cycles $C_1 = (x, y)$ and $C_2 = (i, j)$ such that gray edges of C_1 do not cross with gray edges of C_2. In this scenario, either these cycles are "side by side" (i.e., $\{x, y\} \not\subset [j..i]$, as shown in Fig. 5(a)), or one cycle is "nested" within the black edges of the other (i.e., $\{x, y\} \subset [j..i]$, as shown in Fig. 5(b)). In both cases, there is a 4-cut that transforms these two cycles into four trivial cycles, which increases the number of cycles by 2 (see Fig. 5(a-b) for an illustration). $\qquad\square$

7   A 1.5-Approximation Algorithm for SKCBR When $k = 4$

In this section, we present our 1.5-approximation algorithm for Sorting Signed Permutations by 4-Cuts with Block Reversals using the lemmas presented at Sect. 6. Algorithm 1 shows the steps of our approximation algorithm.

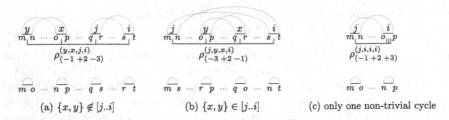

Fig. 5. (a-b) the two possible configurations of two fully divergent cycles $C_1 = (+i, -j)$ and $C_2 = (+x, -y)$ such that gray edges from C_1 do not intersect with gray edges of C_2. In both cases, there is a 4-cut that increases the number of cycles by two. (c) if the only non-trivial cycle of $G(\pi)$ is a fully divergent cycle of size two, then we apply a 4-cut to its two black edges, which increases the number of cycles by one.

Let us argue that Algorithm 1 stops and transforms a signed permutation π into ι^n. The algorithm starts with an empty sorting sequence S and may apply one 4-cut to π at line 4, increasing the number of cycles of $G(\pi)$ by three. After that, the algorithm enters in the while loop at lines 8–20, and iterates while its condition is satisfied. At each iteration at least one of the three if statements will always be satisfied, and a 4-cut will be applied increasing the number of cycles by two or three. After no more than $\frac{n}{2}$ iterations, the algorithm will leave the while loop. At this point either $c(G(\pi)) = n + 1$, which means that $\pi = \iota^n$, or $c(G(\pi)) = n$, and one extra 4-cut will be applied inside the if statement at line 21, resulting in $c(G(\pi)) = n + 1$, so $\pi = \iota^n$. The following lemma shows the approximation factor of Algorithm 1.

The time complexity of Algorithm 1 is $O(n^4)$: finding four distinct black edges to apply the first 4-cut at line 3 using brute force takes $O(n^4)$ time; finding a 4-cut inside the while loop on line 8 takes up to $O(n^2)$ time, and the loop iterates up to $O(n)$ times, resulting in $O(n^3)$; and finding the only non-trivial cycle in $G(\pi)$ to apply a 4-cut in the last if statement in the algorithm takes $O(n)$ time.

Lemma 13. *Algorithm 1 has an approximation factor of* 1.5.

Proof. By Lemma 6 we have that either $d_4(\pi) \geq \frac{(n+1)-c(G(\pi))}{3}$, if $G(\pi)$ has no short cycles and admits a 4-cut that increases the number of cycles by three, or $d_4(\pi) \geq \frac{(n+2)-c(G(\pi))}{3}$ otherwise. Now let us consider two cases, depending on whether line 4 of Algorithm 1 is applied.

First, if a 4-cut is applied at line 4, then this 4-cut increases the number of cycles by three, and we have that $d_4(\pi) \geq \frac{(n+1)-c(G(\pi))}{3}$. If the algorithm enters the if statement at line 21 after the while loop, the 4-cut applied in line 23 increases the number of cycles by one, so together with the first 4-cut from S, they increased the number of cycles by two each on average. Any other 4-cut from S was applied inside the while loop at lines 8–20, so they also increased the number of cycles by at least 2 each. It follows that $|S| \leq \frac{(n+1)-c(G(\pi))}{2}$, so Algorithm 1 has an approximation factor of $\frac{\frac{(n+1)-c(G(\pi))}{2}}{\frac{(n+1)-c(G(\pi))}{3}} = \frac{3}{2} = 1.5$.

Algorithm 1: A 1.5-Approximation Algorithm for Sorting by 4-Cuts with Block Reversals.

Input: A signed permutation $\pi \in S_n^{\pm}$ and its cycle graph $G(\pi)$
Output: A sequence of 4-cuts S that transforms π into ι^n

1 Let $S = \emptyset$
2 **if** $G(\pi)$ *has only long cycles* **then**
3   **if** *there exists a 4-cut ρ, applied to four distinct black edges of a same (long) cycle*
    $C \in G(\pi)$, *such that* $\Delta_c(\pi, \rho) = 3$ **then**
4        $\pi = \pi \cdot \rho$
5   Update S with ρ
6 **while** $c(G(\pi)) < n$ **do**
7   **if** $G(\pi)$ *has a convergent gray edge* **then**
8   Let ρ be a 4-cut such that $\Delta_c(\pi, \rho) \geq 2$, as described in Lemma 8.
9   **else if** $G(\pi)$ *has a long fully divergent cycle C* **then**
10   Let ρ be a 4-cut to black edges of C such that $\Delta_c(\pi, \rho) \geq 2$, as described in
       Lemma 10.
11   **else** // $G(\pi)$ has at least two fully divergent cycles C
12   Let ρ be a 4-cut to black edges of C such that $\Delta_c(\pi, \rho) = 2$, as described in
       Lemma 12.
13     $\pi = \pi \cdot \rho$
14   Update S with ρ
15 **if** $c(G(\pi)) = n$ **then**
    // In this case it follows that there is only one non-trivial cycle $C \in G(\pi)$ that
       is short, even, and fully divergent.
16   Let ρ be a 4-cut applied to both black edges of C such that $\Delta_c(\pi, \rho) = 1$, as described
    in Lemma 12.
17     $\pi = \pi \cdot \rho = \iota^n$
18   Update S with ρ
19 **return** S

Otherwise, no 4-cut is applied before the while loop in lines 8-20, so it follows that $d_4(\pi) \geq \frac{(n+2)-c(G(\pi))}{3}$. In this case, S has z 4-cuts such that the first $(z-1)$ were applied inside the while loop at line 8, so each of them increased the number of cycles by at least two, and the last 4-cut from S was applied either in the while loop, or in the if statement at line 21 (in this last case, it increased the number of cycles by one only). It follows that $|S| \leq 1 + \frac{(n+1)-c(G(\pi))-1}{2} = \frac{(n+2)-c(G(\pi))}{2}$, so Algorithm 1 has an approximation factor of $\frac{\frac{(n+2)-c(G(\pi))}{2}}{\frac{(n+2)-c(G(\pi))}{3}} = \frac{3}{2} = 1.5$. $\qquad\square$

8 Conclusion

In this work we investigated for the first time the use of k-cuts, a genome rearrangement proposed recently, on signed permutations. For $k \geq 5$, we showed that it is possible to adapt the algorithm for unsigned permutations ensuring an approximation factor of $\frac{2k}{k-1}$ for odd values of k, and $\frac{2k}{k-2}$ for even values of k. We also presented an improved 1.5-approximation algorithm for $k = 4$ using the well-known graph structure called cycle graph. Some questions remain open, such as the complexity of the Sorting by k-cuts with Block Reversals when $k = 4$, and the use of k-cuts on signed strings (rather than permutations), i.e. in the case where genomes contain repeated genes.

References

1. Alekseyev, M.A.: Multi-break rearrangements and breakpoint re-uses: from circular to linear genomes. J. Comput. Biol. **15**(8), 1117–1131 (2008). https://doi.org/10.1089/cmb.2008.0080
2. Alekseyev, M.A., Pevzner, P.A.: Multi-break rearrangements and chromosomal evolution. Theor. Compu. Sci. **395**(2–3), 193–202 (2008). https://doi.org/10.1016/j.tcs.2008.01.013
3. Alexandrino, A.O., Oliveira, A.R., Dias, U., Dias, Z.: On the complexity of some variations of sorting by transpositions. J. Univ. Comput. Sci. **26**(9), 1076–1094 (2020). https://doi.org/10.3897/jucs.2020.057
4. Bafna, V., Pevzner, P.A.: Sorting by transpositions. SIAM J. Discrete Math. **11**(2), 224–240 (1998). https://doi.org/10.1137/S089548019528280X
5. Bergeron, A.: A very elementary presentation of the Hannenhalli-Pevzner theory. Discrete Appl. Math. **146**(2), 134–145 (2005). https://doi.org/10.1016/j.dam.2004.04.010
6. Bulteau, L., Fertin, G., Jean, G., Komusiewicz, C.: Sorting by multi-cut rearrangements. Algorithms **14**(6), 169 (2021). https://doi.org/10.3390/a14060169
7. Bulteau, L., Fertin, G., Rusu, I.: Sorting by transpositions is difficult. SIAM J. Discrete Math. **26**(3), 1148–1180 (2012). https://doi.org/10.1137/110851390
8. Christie, D.A.: Sorting permutations by block-interchanges. Inf. Process. Lett. **60**(4), 165–169 (1996). https://doi.org/10.1016/S0020-0190(96)00155-X
9. Elias, I., Hartman, T.: A 1.375-Approximation algorithm for sorting by transpositions. IEEE/ACM Trans. Comput. Biol. Bioinform. **3**(4), 369–379 (2006). https://doi.org/10.1109/TCBB.2006.44
10. Fertin, G., Labarre, A., Rusu, I., Tannier, É., Vialette, S.: Combinatorics Of Genome Rearrangements. Computational Molecular Biology, The MIT Press, London (2009). https://doi.org/10.7551/mitpress/9780262062824.001.0001
11. Hannenhalli, S., Pevzner, P.A.: Transforming cabbage into turnip: polynomial algorithm for sorting signed permutations by reversals. J. ACM **46**(1), 1–27 (1999). https://doi.org/10.1145/300515.300516
12. Holland, A.J., Cleveland, D.W.: Chromoanagenesis and cancer: mechanisms and consequences of localized, complex chromosomal rearrangements. Nat. Med. **18**(11), 1630–1638 (2012). https://doi.org/10.1038/nm.2988
13. Lin, G.H., Xue, G.: Signed genome rearrangement by reversals and transpositions: models and approximations. Theor. Comput. Sci. **259**(1–2), 513–531 (2001). https://doi.org/10.1016/S0304-3975(00)00038-4
14. Pellestor, F., Gatinois, V.: Chromoanagenesis: a piece of the macroevolution scenario. Mol. Cytogenet. **13**(1), 1–9 (2020). https://doi.org/10.1186/s13039-020-0470-0
15. Yancopoulos, S., Attie, O., Friedberg, R.: Efficient sorting of genomic permutations by translocation. Inversion and block interchange. Bioinformatics **21**(16), 3340–3346 (2005). https://doi.org/10.1093/bioinformatics/bti535

A New Approach for the Reversal Distance with Indels and Moves in Intergenic Regions

Klairton Lima Brito[1](✉)[iD], Andre Rodrigues Oliveira[1][iD],
Alexsandro Oliveira Alexandrino[1][iD], Ulisses Dias[2][iD], and Zanoni Dias[1][iD]

[1] Institute of Computing, University of Campinas, Campinas, Brazil
{klairton,andrero,alexsandro,zanoni}@ic.unicamp.br
[2] School of Technology, University of Campinas, Limeira, Brazil
ulisses@ft.unicamp.br

Abstract. Genome rearrangement events are widely used to estimate a minimum-size sequence of mutations capable of transforming a genome into another. The length of this sequence is called distance, and determining it is the main goal in genome rearrangement distance problems. Problems in the genome rearrangement field differ regarding the set of rearrangement events allowed and the genome representation. In this work, we consider the scenario where the genomes share the same set of genes, gene orientation is known, and intergenic regions (structures between a pair of genes and at the extremities of the genome) are taken into account. We use two models, the first model allows only conservative events (reversals and moves), and the second model includes non-conservative events (insertions and deletions) in the intergenic regions. We show that both models result in NP-hard problems and we present algorithms with an approximation factor of 2.

Keywords: Genome rearrangements · Complexity analysis · Approximation algorithms

1 Introduction

The reversal event is one of the most investigated rearrangement events in the literature [2,8,15]. This event inverts the order of the genes in a segment and flips their orientations. When we assume that the compared genomes share the same set of genes and do not have duplicated genes, the genomes are usually mapped into permutations where each element represents a gene. Each element of the permutation receives a plus or minus sign to indicate the gene orientation. Thus, the permutation is called a signed permutation. If the orientation is unknown we omit the sign. Considering the reversal event and the representation using a permutation, we have the Sorting Permutations by Reversals and Sorting Signed Permutations by Reversals problems. The former problem is NP-hard [7], and the best-known algorithm has an approximation factor of 1.375 [2], while the latter has an exact polynomial-time algorithm [15]. When we consider the reversal event and we assume that the genomes may have duplicated genes, the problem is NP-hard whether or not the orientation is known [11,18].

Several advances have been made in the genome rearrangement models over time. For instance, improvement in practical and theoretical results, inclusion of novel events, or consideration of new features from the genomes. In particular, studies have pointed out the importance of structures called intergenic regions and the potential improvement in incorporating them into the models [3,4]. The intergenic regions occur between a pair of genes and at the extremities of a linear genome. Besides, they contain some nucleotides inside them, and the number of nucleotides in an intergenic region defines its size. Several works have been presented considering the case where genomes are mapped into permutations and intergenic region sizes are later incorporated. The problem of finding the distance between two genomes taking into account gene order and the size of the intergenic regions in a model that exclusively allows the reversal operation is NP-hard whether the orientation of the genes is known or not [5,16]. The signed case has a 2-approximation algorithm [16], while the unsigned case has a 4-approximation algorithm [5]. When including insertion and deletion affecting the intergenic regions, the problems have algorithms with the same approximation factor as the model with the exclusive use of the reversal event. The unsigned case is NP-hard, while the complexity of the signed case remains unknown [5,16].

When we allow only the transposition operation, which swaps the position of two consecutive segments of the genome, finding the distance is also a NP-hard problem and has a 3.5-approximation algorithm [17]. Another important event is the Double Cut and Join (DCJ) [9], which cuts the genome into two points and reassembles the stretches following a specific criterion. For the DCJ operation, the distance problem after incorporating the intergenic region sizes is NP-hard [12]. However, it is solvable in polynomial-time if insertion and deletion are allowed to affect the intergenic regions [6]. Oliveira *et al.* showed a 3-approximation algorithm considering a scenario where gene orientation is known and the model allows reversal and transposition operations [17]. Besides, the authors introduced the intergenic move event, which is similar to the transposition event, but differs from it because one of the affected segments is an intergenic region. This event becomes relevant in models that take into account intergenic regions. Considering a scenario where gene orientation is known and using a model that allows reversal, intergenic move, and transposition operations, the distance problem is NP-hard [17] and it has a 2.5-approximation algorithm. Allowing intergenic move and transposition operations in a scenario where the orientation of the genes is unknown, the problem is also NP-hard and has a 2.5-approximation algorithm [17].

Nowadays, we have a significant amount of data on real genomes. With the enhancement of sequencing techniques, it is possible to obtain information regarding the orientation of each gene. Thus, we investigate the genome rearrangement distance problem of two models where gene orientation is known. The first model allows the reversal and intergenic move events, while the second includes the insertion and deletion events affecting only intergenic regions, which we simply call indel. For both models, we show that they belong to the NP-hard class and present a 2-approximation algorithm.

The manuscript is organized as follows. Section 2 shows concepts and definitions used to derive the results. Section 3 presents the theoretical results for both problems. Lastly, Sect. 4 concludes the manuscript.

2   Background

In this section, we introduce concepts and definitions. Furthermore, we formally describe the problems investigated in this work.

Given a genome $\mathcal{G} = (\mathcal{R}_1, \mathcal{G}_1, \mathcal{R}_2, \mathcal{G}_2 \ldots, \mathcal{R}_n, \mathcal{G}_n, \mathcal{R}_{n+1})$ with n genes $\{\mathcal{G}_1, \mathcal{G}_2, \ldots, \mathcal{G}_n\}$ and $n+1$ intergenic regions $\{\mathcal{R}_1, \mathcal{R}_2, \ldots, \mathcal{R}_{n+1}\}$, we use these features to represent the genome. Intergenic regions are placed between each pair of genes and at the extremities of the genome containing a specific number of nucleotides. Thus, we denote the size of an intergenic region by the number of nucleotides in it. We use a representation with a permutation $\pi = (\pi_1 \, \pi_2 \, \ldots \, \pi_n)$, such that π_i, with $1 \leq i \leq n$, is an integer number that represents the gene $\mathcal{G}_i$ and a list of integers $\breve{\pi} = (\breve{\pi}_1, \breve{\pi}_2, \ldots, \breve{\pi}_{n+1})$, such that $\breve{\pi}_i$, with $1 \leq i \leq n + 1$, represents the size of the intergenic region $\mathcal{R}_i$. Each element π_i of the permutation π receives a "+" or "−" sign indicating gene orientation in the genome G. Now let us define the genome rearrangement events of intergenic reversal, intergenic move, and intergenic indel.

Definition 1. *Given a genome $\mathcal{G} = (\pi, \breve{\pi})$, an intergenic reversal $\rho_{(x,y)}^{(i,j)}$, with $1 \leq i \leq j \leq n$, $0 \leq x \leq \breve{\pi}_i$, and $0 \leq y \leq \breve{\pi}_{j+1}$, is an event that inverts the order and orientation of the genes from a segment of the genome. The intergenic regions $\breve{\pi}_i$ and $\breve{\pi}_{j+1}$ are split, respectively, into (x, x') and (y, y'), such that $x' = \breve{\pi}_i - x$ and $y' = \breve{\pi}_{j+1} - y$. The segment $(x', \pi_i, \breve{\pi}_{i+1}, \ldots, \breve{\pi}_j, \pi_j, y)$ is inverted and the signs of the elements from π_i to π_j are flipped. Lastly, the segments are reassembled and the intergenic reversal $\rho_{(x,y)}^{(i,j)}$ applied on genome $(\pi, \breve{\pi})$, denoted by $(\pi, \breve{\pi}) \cdot \rho_{(x,y)}^{(i,j)}$, results in a new genome $G' = (\pi', \breve{\pi}')$, such that:*

$$\pi' = \quad (\pi_1 \, \pi_2 \ldots \pi_{i-1} \, \underline{-\pi_j \, -\pi_{j-1} \ldots -\pi_{i+1} \, -\pi_i} \, \pi_{j+1} \, \pi_{j+2} \ldots \pi_n),$$

$$\breve{\pi}' = \quad (\breve{\pi}_1, \breve{\pi}_2, \ldots, \breve{\pi}_{i-1}, \underline{\breve{\pi}_i', \breve{\pi}_j, \ldots, \breve{\pi}_{i+1}, \breve{\pi}_{j+1}'}, \breve{\pi}_{j+2}, \ldots, \breve{\pi}_{n+1}),$$

with $\breve{\pi}_i' = x + y$ and $\breve{\pi}_{j+1}' = x' + y'$.

Definition 2. *Given a genome $\mathcal{G} = (\pi, \breve{\pi})$, an intergenic move $\mu_{(x)}^{(i,j)}$, with $1 \leq i, j \leq n + 1$, $i \neq j$, and $x \in [0..\breve{\pi}_i]$, moves x nucleotides from $\breve{\pi}_i$ to $\breve{\pi}_j$.*

Definition 3. *Given a genome $\mathcal{G} = (\pi, \breve{\pi})$, an intergenic indel $\delta_{(x)}^{(i)}$, with $1 \leq i \leq n + 1$ and $x \geq -\breve{\pi}_i$, either deletes x nucleotides from the intergenic region $\breve{\pi}_i$ if x is negative, or inserts x nucleotides into $\breve{\pi}_i$, otherwise.*

From now on, we will refer to intergenic reversal, intergenic move, and intergenic indel simply as a reversal, move, and indel, respectively.

Given a source genome S and a target genome T, sharing the same set of genes and with no duplicate genes, we map the permutation of the target genome as $\iota = (+1 +2 \ldots +n)$, which is called the *identity permutation*. The labels of the permutation π, from the source genome S, are mapped according to the mapping of the target genome. Thus, the source genome S is mapped into $(\pi, \breve{\pi})$ and the target genome T is mapped into $(\iota, \breve{\iota})$. Note that finding a pair of genomes with the same set of genes and no duplicates, especially in complex organisms, is quite rare. However, this does not negate the use of this approach to deal with genomes with replicated genes. One way to deal with replicated genes is by using *ortholog assignment* [10, 19], and, after this process, we can represent the gene order of the genomes as permutations. Next, we formally describe the problems addressed in this work.

	Sorting by Intergenic Reversals and Intergenic Moves (SBIRM)
Input:	A source and a target genome $(S = (\pi, \breve{\pi}), T = (\iota, \breve{\iota}))$.
Task:	Determine a minimum length sequence of genome rearrangement events (among reversals and moves) capable of transforming $(\pi, \breve{\pi})$ into $(\iota, \breve{\iota})$.

	Sorting by Intergenic Reversals, Intergenic Moves, and Intergenic Indels (SBIRMI)
Input:	A source and a target genome $(S = (\pi, \breve{\pi}), T = (\iota, \breve{\iota}))$.
Task:	Determine a minimum length sequence of genome rearrangement events (among reversals, moves, and indels) capable of transforming $(\pi, \breve{\pi})$ into $(\iota, \breve{\iota})$.

Given an instance $\mathcal{I} = ((\pi, \breve{\pi}), (\iota, \breve{\iota}))$ of the SBIRM or SBIRMI problem, we extend both permutations by adding the elements 0 and $n + 1$ at the beginning (π_0 and ι_0) and at the end (π_{n+1} and ι_{n+1}) of the permutations. We say that an instance $\mathcal{I} = ((\pi, \breve{\pi}), (\iota, \breve{\iota}))$ is *balanced* if $\sum_{i=1}^{n+1} (\breve{\pi}_i - \breve{\iota}_i) = 0$, and *unbalanced* otherwise. Note that the reversal and move operations are conservative events, which means that they do not change the amount of genetic material. However, the indel operation is non-conservative. For this reason, we assume that an instance $\mathcal{I}$ for the SBIRM problem is balanced. Nevertheless, when we deal with an instance $\mathcal{I}$ for the SBIRMI problem, $\mathcal{I}$ can be either balanced or unbalanced.

The main goal of the SBIRM and SBIRMI problems is to find a sequence of genome rearrangement events of minimum length capable of transforming a source genome into a target genome. The size of such sequence is called *distance*, denoted by $d_{\rho\mu}((\pi, \breve{\pi}), (\iota, \breve{\iota}))$ and $d_{\rho\mu\delta}((\pi, \breve{\pi}), (\iota, \breve{\iota}))$ for the SBIRM and SBIRMI problems, respectively.

2.1 Weighted Breakpoint Graph

This section introduces the weighted breakpoint graph used for the development of algorithms in the context of genome rearrangement problems. The weighted breakpoint graph is an adaptation of the breakpoint graph [1,14] and handles the information regarding the genes and size of the intergenic regions [16,17].

Given an instance $\mathcal{I} = ((\pi, \breve{\pi}), (\iota, \breve{\iota}))$ of the SBIRM or SBIRMI problem, $G(\mathcal{I}) = (V, E, w)$ is a graph composed by the set of vertices $V = \{+\pi_0, -\pi_1, +\pi_1, \ldots, -\pi_n, +\pi_n, -\pi_{n+1}\}$ and the set of edges $E = E_b \cup E_g$, which is divided in black edges (E_b) and gray edges (E_g). The weighted function $w : E \to \mathbb{N}_0$ relates the size of the intergenic regions from the source and target genome with weights of the edges in E. There is a black edge $e_i = (-\pi_i, +\pi_{i-1})$ in E_b, with $w(e_i) = \breve{\pi}_i$, for $1 \leq i \leq n+1$; for simplicity, each black edge e_i is labeled as i. There is a gray edge $(+(j-1), -j)$ in E_g, with $w((+(j-1), -j)) = \breve{\iota}_j$, for $1 \leq j \leq n+1$. The black and gray edges represent the genes adjacencies from the source and target genomes, respectively. Note that each vertex in $G(\mathcal{I})$ has degree two by the incidence of one black edge and one gray edge. Thus, there is a unique decomposition of $G(\mathcal{I})$ in cycles with alternating edge colors.

There are different ways to place the vertices and edges of the weighted breakpoint graph when drawing it. For simplicity and to keep a unique cycle identification, we use the standard representation where vertices are placed horizontally from $+\pi_0$ up to $-\pi_{n+1}$. The black edges connect the vertices in a horizontal line, while the gray edges are drawn as arcs above them. The cycle representation uses a list of its black edges labels $(c^1, c^2, \ldots, c^k)$, $k \geq 1$, following the traversing order where the first edge is the rightmost black edge and it is always traversed from right to left. Besides, a "$-$" sign before a black edge label i indicates that i is traversed from left to right. In the following we refer to a gray edge between two black edges of labels c^i and $c^{i \pmod{k}+1}$ as e'_{c^i}.

A cycle $C = (c^1, c^2, \ldots, c^k)$ with k black edges is called a k-cycle. A cycle $C = (c^1)$ that has only one black edge is called *trivial*, and it is called *non-trivial* otherwise. A cycle $C = (c^1, c^2, \ldots, c^k)$ that has a pair of black edges traversed in opposite directions (left to right and right to left) is called *divergent*, and it is called *convergent* otherwise. Considering the weights associated with the edges in $G(\mathcal{I})$, the cycles are classified as *balanced* or *unbalanced*. A cycle $C = (c^1, c^2, \ldots, c^k)$ is balanced if $\sum_{i=1}^{k}[w(e'_{c^i}) - w(e_{c^i})] = 0$, and unbalanced otherwise. In other words, balanced cycles have the same amount of total weight in black and gray edges, while unbalanced cycles do not. An unbalanced cycle $C = (c^1, c^2, \ldots, c^k)$ is *positive* if $\sum_{i=1}^{k}[w(e'_{c^i}) - w(e_{c^i})] > 0$, and it is *negative* otherwise. Note that negative cycles have a total weight in their black edges greater than the total weight in their gray edges. On the other hand, a positive cycle needs to increase the weight in its black edges to become balanced. Note that the weight of the gray edges is conserved, while the weight from the black edges may be affected by the rearrangement events.

We denote by $c(G(\mathcal{I}))$ and $c_b(G(\mathcal{I}))$ the number of cycles and balanced cycles in $G(\mathcal{I})$, respectively. Given a sequence S of operations, we denote by $\Delta c(G(\mathcal{I}), S) = c(G(\mathcal{I} \cdot S)) - c(G(\mathcal{I}))$ and $\Delta c_b(G(\mathcal{I}), S) = c_b(G(\mathcal{I} \cdot S)) - c_b(G(\mathcal{I}))$

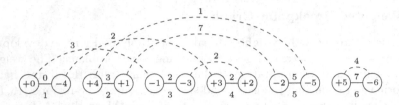

Fig. 1. An example of the weighted breakpoint graph, the black and gray edges are represented by solid and dashed lines, respectively. The number above each edge represents the weight established by the weighted function w. The number below black edge represents their labels.

the variation in the number of cycles and balanced cycles, respectively, generated by applying S to $(\pi, \breve{\pi})$ from $\mathcal{I}$ (denoted by $\mathcal{I} \cdot S$).

Remark 1. The only instance $\mathcal{I} = ((\pi, \breve{\pi}), (\iota, \breve{\iota}))$ of the SBIRM or SBIRMI problems that has $c(G(\mathcal{I})) = n+1$ and $c_b(G(\mathcal{I})) = n+1$ is the instance $((\iota, \breve{\iota}), (\iota, \breve{\iota}))$.

Figure 1 shows an example of the weighted breakpoint graph generated from the instance $\mathcal{I} = ((\pi = (+4 \ -1 \ +3 \ -2 \ +5), \breve{\pi} = (0,3,2,2,5,7)), (\iota = (+1 \ +2 \ +3 \ +4 \ +5), \breve{\iota} = (3,7,2,2,1,4)))$. The graph $G(\mathcal{I})$ has three cycles: $C_1 = (4, 1, -3)$, $C_2 = (5, 2)$, and $C_3 = (6)$, so $c(G(\mathcal{I})) = 3$. The cycle C_3 is trivial while cycles C_1 and C_2 are non-trivial. Note that C_2 is a balanced cycle while C_1 and C_3 are unbalanced cycles, so $c_b(G(\mathcal{I})) = 1$. Besides, C_1 is a positive cycle and C_3 is a negative cycle.

3   Results

In this section, we show that the decision version of SBIRM and SBIRMI problems are NP-hard. After that, we show lower bounds and propose approximation algorithms for the two problems.

3.1   Complexity Analysis

In this section, we use a reduction from the NP-hard 3-Partition problem (3-PART) [13] to prove that the decision version of SBIRM and SBIRMI problems are NP-hard. In the following, we present a formal definition of the decision versions of these two problems.

	SBIRM (DECISION VERSION)
Input:	A natural number k, a source genome $\mathcal{S} = (\pi, \breve{\pi})$, and a target genome $\mathcal{T} = (\iota, \breve{\iota})$.
Question:	Is it possible to transform $(\pi, \breve{\pi})$ into $(\iota, \breve{\iota})$ using at most k operations of reversal or move?

	SBIRMI (DECISION VERSION)
Input:	A natural number k, a source genome $\mathcal{S} = (\pi, \breve{\pi})$, and a target genome $\mathcal{T} = (\iota, \breve{\iota})$.
Question:	Is it possible to transform $(\pi, \breve{\pi})$ into $(\iota, \breve{\iota})$ using at most k operations of reversal, move, or indel?

The decision version of the 3-PART problem consists in determining if it is possible to create a partition of a set of positive integers $A = \{a_1, a_2, \ldots, a_{3n}\}$, such that $\sum_{i=1}^{3n} a_i = Bn$ for some $B \in \mathbb{Z}^+$ and $\frac{B}{4} < a_i < \frac{B}{2}$ for $1 \leq i \leq 3n$, into n triples $A_1, A_2, \ldots, A_n$, such that $\sum_{a_i \in A_j} a_i = B$ for each triple A_j, with $1 \leq j \leq n$.

Theorem 1. *The* SBIRM *and* SBIRMI *problems are NP-hard.*

Proof. Given an instance $A = \{a_1, a_2, \ldots, a_{3n}\}$ of the 3-PART problem, we create an instance $\mathcal{I} = ((\pi, \breve{\pi}), (\iota, \breve{\iota}))$ of the SBIRM or SBIRMI problem as follows:

i) $\pi = \iota = (+1 +2 \ldots +(4n-1))$.
ii) Assign the value $\breve{\pi}_i = a_i$ for $1 \leq i \leq 3n$ and the value $\breve{\pi}_j = 0$ for $3n+1 \leq j \leq 4n$.
iii) Assign the value $\breve{\iota}_i = 0$ for $1 \leq i \leq 3n$ and the value $\breve{\iota}_j = B$ for $3n+1 \leq j \leq 4n$.

Example 1 illustrates the process to create an instance of the SBIRM or SBIRMI problem based on the instance of the 3-PART problem. Now, we show that the instance A of the 3-PART is satisfied $\iff d_{\rho\mu}(\mathcal{I}) \leq 3n$ or $d_{\rho\mu\delta}(\mathcal{I}) \leq 3n$.

Example 1.

$$A = \{a_1, a_2, \ldots, a_{3n}\}$$

$$(\pi, \breve{\pi}) = ((+1 +2 \ldots +(4n-1)), (a_1, a_2, \ldots, a_{3n}, \ 0, 0, \ldots, 0))$$
$$(\iota, \breve{\iota}) = ((+1 +2 \ldots +(4n-1)), (\ 0, \ 0, \ldots, \quad 0, B, B, \ldots, B))$$
$$\mathcal{I} = ((\pi, \breve{\pi}), (\iota, \breve{\iota}))$$

$$G(\mathcal{I})$$

0	0		0	B	B		B

$3n$ negative trivial cycles · · · n positive trivial cycles

($\Rightarrow$) Suppose that it is possible to partition the instance $A = \{a_1, a_2, \ldots, a_{3n}\}$ of the 3-PART problem into n triples, such that $\sum_{a_i \in A_j} a_i = B$ for each triple A_j, with $1 \leq j \leq n$. Then, it is possible to turn the source genome $(\pi, \breve{\pi})$ into the target genome $(\iota, \check{\iota})$ using $3n$ move operations. For each triple A_j, with $1 \leq j \leq n$, apply the following move operations $\mu_{(a_i)}^{(i, 3n+j)}$, for $a_i \in A_j$. Repeating this procedure in all n triples produces a sequence of $3n$ moves operations that results in $c(G(\mathcal{I})) = 4n$ and $c_b(G(\mathcal{I})) = 4n$. Thus, $d_{\rho\mu}(\mathcal{I}) \leq 3n$ and $d_{\rho\mu\delta}(\mathcal{I}) \leq 3n$.

($\Leftarrow$) Now suppose that $d_{\rho\mu}(\mathcal{I}) \leq 3n$ or $d_{\rho\mu\delta}(\mathcal{I}) \leq 3n$. Note that, to reach the desired target genome, the trivial cycles from the black edges 0 to $3n$ form trivial cycles with zero weight in their black edges.

Note that $G(\mathcal{I})$ has $4n$ trivial cycles, where: the first $3n$ trivial cycles (from the left to the right using the standard representation of $G(\mathcal{I})$) have positive weight in their black edges; and the last n trivial cycles have zero weight in their black edges.

We say that a cycle $C = (c^1)$ is *alpha* if C is trivial and it has weight zero in its black and gray edges. We denote by $c_\alpha(G(\mathcal{I}))$ the number of alpha cycles in $G(\mathcal{I})$. Note that any sequence of genome rearrangement operations that turns the source genome into the target genome must necessarily affect the first $3n$ trivial cycles to change the weight in its black edges to zero. Consequently, for the target genome we have that $c_\alpha(G(\mathcal{I})) = 3n$.

Given a sequence S of operations, we denote by

$$\Delta c_\alpha(G(\mathcal{I}), S) = \frac{c_\alpha(G(\mathcal{I} \cdot S)) - c_\alpha(G(\mathcal{I}))}{|S|}$$

the average variation in the number of alpha cycles generated by applying S to $(\pi, \check{\pi})$ from $\mathcal{I}$.

A move $\mu_{(x)}^{(i,j)}$ acts on at most two cycles. If the move μ acts on two black edges of the same cycle, then the weight of that cycle remains unchanged ($\Delta c_\alpha(G(\mathcal{I}), \{\mu\}) = 0$). When it acts on two cycles, we have the following cases:

- If none of the cycles are trivial, then $\Delta c_\alpha(G(\mathcal{I}), \{\mu\}) = 0$. Note that the move operation does not change the size of the cycles.
- If at least one of the cycles is trivial, then the move can transfer the weight in its black edge to another cycle and, in the best case, $\Delta c_\alpha(G(\mathcal{I}), \{\mu\}) \leq 1$.
- If both cycles are alpha, then $\Delta c_\alpha(G(\mathcal{I}), \{\mu\}) = 0$. Note that in this case, the black edges of both cycles have a weight of zero. Consequently, the black edge weight of the cycles remains unchanged.

A reversal $\rho_{(x,y)}^{(i,j)}$ can create at most two alpha cycles ($\Delta c_\alpha(G(\mathcal{I}), \{\rho\}) \leq 2$). However, this occurs only in the particular case where a reversal splits a 2-cycle where both black edges have weight zero. Given that a move operation does not change the size of the cycles, to obtain a 2-cycle with such characteristic, we have that a reversal ρ that merges cycles must be previously applied. Note that any reversal that merges cycles does not create alpha cycles ($\Delta c_\alpha(G(\mathcal{I}), \{\rho\}) \leq 0$), since this operation results in a non-trivial cycle. To obtain a 2-cycle where both black edges have weight zero, we have three possibilities:

- The first case consists of a reversal ρ_1 that merges two alpha cycles (decreasing the number of alpha cycles by two units) followed by a reversal ρ_2 that splits that 2-cycle generating two alpha cycles again. Note that in this case the reversal ρ_2 undoes what ρ_1 created, so $\Delta c_\alpha(G(\mathcal{I}), \{\rho_1, \rho_2\}) = 0$ since $\Delta c_\alpha(G(\mathcal{I}), \{\rho_1\}) = -2$ and $\Delta c_\alpha(G(\mathcal{I}), \{\rho_2\}) = 2$.

- In the second case, we have a reversal ρ_1 that merges two trivial cycles where both black edges have weight greater than zero. By the construction of the 2-cycle, we have that at least one of its black edges has a weight greater than zero. To remove the weight in the black edges of the 2-cycle, at least one move μ_1 must be applied (a reversal to perform this task must split or merge the cycle first). Note that the move μ_1 does not create alpha cycles ($\Delta c_\alpha(G(\mathcal{I}), \{\mu_1\}) \leq 0$) since it transfers the weight from a black edge of a 2-cycle to another cycle. Finally, a reversal ρ_2 that splits the 2-cycles generating two alpha cycles is applied. Note that $\Delta c_\alpha(G(\mathcal{I}), \{\rho_1, \mu_1, \rho_2\}) \leq \frac{2}{3}$ since $\Delta c_\alpha(G(\mathcal{I}), \{\rho_1\}) = 0$, $\Delta c_\alpha(G(\mathcal{I}), \{\mu_1\}) \leq 0$, and $\Delta c_\alpha(G(\mathcal{I}), \{\rho_2\}) = 2$.

- The last possibility is when a reversal ρ_1 merges a trivial cycle whose black edge has weight greater than zero with an alpha cycle, decreasing the number of alpha cycles by one unit. Similar to the previous case, at least one of the black edges in the 2-cycle has a weight greater than zero, and at least one move operation μ_1 must be applied. In the end, a reversal ρ_2 is applied to that 2-cycle generating two alpha cycles. This case results in $\Delta c_\alpha(G(\mathcal{I}), \{\rho_1, \mu_1, \rho_2\}) \leq \frac{1}{3}$ since $\Delta c_\alpha(G(\mathcal{I}), \{\rho_1\}) = -1$, $\Delta c_\alpha(G(\mathcal{I}), \{\mu_1\}) \leq 0$, and $\Delta c_\alpha(G(\mathcal{I}), \{\rho_2\}) = 2$.

Considering the scenario where a reversal creates only one alpha cycle, we also have the fact that a reversal that merges cycles must be previously applied. Thus, we have that $\Delta c_\alpha(G(\mathcal{I}), \{\rho_1, \rho_2\}) \leq \frac{1}{2}$. This implies that if any reversal is used in a sorting sequence S then $|S| > 3n$.

For indels, we note that after a deletion takes place to a black edge, it can increase the number of alpha cycles by one unit, however, after any deletion at least one insertion is required to balance the instance again, and since an insertion cannot increase the number of alpha cycles it results in more than $3n$ operations in the sorting sequence. It follows that a sorting sequence of size $3n$ has only move operations in both SBIRM and SBIRMI problems.

The $3n$ move operations will transfer all the positive weight of the $3n$ black edges of labels $[1..3n]$ to the black edges of labels $[(3n + 1)..4n]$, creating one alpha cycle each. The black edges from labels $[(3n+1)..4n]$ must receive a total weight that corresponds to the sum of weights in exactly three black edges of labels $[1..3n]$, since $\frac{b}{2} > w(b_i) > \frac{b}{4}$. At the end of the sorting process, we just need to trace the weight that was transferred to create each alpha cycle.   □

3.2   Lower Bounds

In this section, we present lower bounds for the SBIRM and SBIRMI problems based on the number of cycles and balanced cycles on the weighted breakpoint graph.

Given an instance $\mathcal{I} = ((\pi, \breve{\pi}), (\iota, \breve{\iota}))$ for the SBIRM or SBIRMI problem, we have that $\Delta c(G(\mathcal{I}), \rho) \in \{1, 0, -1\}$ and $\Delta c_b(G(\mathcal{I}), \rho) \in \{1, 0, -1\}$ for any reversal

ρ. Similarly, we have that $\Delta c(G(\mathcal{I}), \mu) = 0$ and $\Delta c_b(G(\mathcal{I}), \mu) \in \{2, 1, 0, -1, -2\}$ for any move μ and $\Delta c(G(\mathcal{I}), \delta) = 0$ and $\Delta c_b(G(\mathcal{I}), \delta) \in \{1, 0, -1\}$ for any indel δ. That said, we derive the following lower bounds.

Lemma 1. *Given an instance $\mathcal{I} = ((\pi, \breve{\pi}), (\iota, \breve{\iota}))$ for the* SBIRM *problem, then* $d_{\rho\mu}(\mathcal{I}) \geq n + 1 - \frac{c(G(\mathcal{I})) + c_b(G(\mathcal{I}))}{2}$.

Proof. Note that to reach the target genome it is necessary to increase both the number of cycles and balanced cycles to $n + 1$. We already argued that both reversals and moves can increase $c(G(\mathcal{I})) + c_b(G(\mathcal{I}))$ by at most two units, so the lemma follows.   □

Lemma 2. *Given an instance $\mathcal{I} = ((\pi, \breve{\pi}), (\iota, \breve{\iota}))$ for the* SBIRMI *problem, then* $d_{\rho\mu\delta}(\mathcal{I}) \geq n + 1 - \frac{c(G(\mathcal{I})) + c_b(G(\mathcal{I}))}{2}$.

Proof. The proof is similar to the described in Lemma 1.   □

3.3   Reversal and Move Operations

In the following, we present lemmas that will be used to construct an algorithm that transforms the source genome into the target genome using reversal and move operations. Note that reversal and move operations are conservative events. For this reason, we assume that we are dealing with a balanced instance.

Lemma 3. *Given an instance $\mathcal{I} = ((\pi, \breve{\pi}), (\iota, \breve{\iota}))$, such that there is at least one negative trivial cycle $C = (c^1)$, then it is possible to increase the number of balanced cycles by at least one unit after applying one move operation.*

Proof. Since $G(\mathcal{I})$ is a balanced graph, at least one positive cycle D in $G(\mathcal{I})$ must exist. In this case, we apply one move operation that transfers the extra weight from the black edge c^1 to a black edge of D. Thus, C becomes balanced and the lemma follows.   □

Lemma 4. *Given an instance $\mathcal{I} = ((\pi, \breve{\pi}), (\iota, \breve{\iota}))$, such that there is at least one negative or balanced divergent cycle C, then it is possible to increase the number of cycles by one unit and the number of balanced cycles by one unit after applying one reversal operation.*

Proof. Straightforward from Lemma 5 by Oliveira et al. [16].   □

Lemma 5. *Given an instance $\mathcal{I} = ((\pi, \breve{\pi}), (\iota, \breve{\iota}))$ such that all the non-trivial cycles that are either negative or balanced are convergent, there is a sequence of two reversals that increases the number of cycles by one unit and the number of balanced cycles by one unit.*

Proof. Straightforward from Lemma 6 by Oliveira et al. [16].   □

Algorithm 1. The input is an instance $\mathcal{I} = ((\pi, \breve{\pi}), (\iota, \breve{\iota}))$ and the output is a sequence S of reversal and move operations such that $(\pi, \breve{\pi}) \cdot S = (\iota, \breve{\iota})$.

$S \leftarrow [\,]$
while $(\pi, \breve{\pi}) \neq (\iota, \breve{\iota})$ **do**
   if there is a trivial negative cycle in $G(\mathcal{I})$ **then**
      $S \leftarrow S + [\mu_1]$ $\triangleright$ Lemma 3.
   else if there is a negative or balanced divergent cycle in $G(\mathcal{I})$ **then**
      $S \leftarrow S + [\rho_1]$ $\triangleright$ Lemma 4.
   else
      $S \leftarrow S + [\rho_1, \rho_2]$ $\triangleright$ Lemma 5.
   end if
end while
return S

Now consider the Algorithm 1 and note that it has three steps: (i) move operations applied to trivial negative cycles (Lemma 3); (ii) reversals applied to negative or balanced divergent cycles (Lemma 4); (iii) two consecutive reversals applied to negative or balanced convergent cycles (Lemma 5). Thus, we have the following lemma.

Lemma 6. *Given an instance* $\mathcal{I} = ((\pi, \breve{\pi}), (\iota, \breve{\iota}))$, *Algorithm 1 turns* $(\pi, \breve{\pi})$ *into* $(\iota, \breve{\iota})$ *using reversal and move operations.*

Proof. Note that the algorithm ensures that every negative trivial cycle are turned into balanced by Lemma 3. If no negative trivial cycle exists, any negative or balanced divergent cycles is split by Lemma 4. If none of the these two situations apply, $G(\mathcal{I})$ may have positive or balanced trivial cycles, positive divergent cycles, and convergent cycles (positive, negative, or balanced). Thus, all the negative and balanced cycles in $G(\mathcal{I})$ are convergent and Lemma 5 can be applied. Observe that Algorithm 1, in each iteration, always performs one of the steps i, ii, or iii. Besides, each step increases in at least one unit the number of cycles or in at least one unit the number of cycles and balanced cycles. This process repeats until we reach $(n + 1)$ cycles and $(n + 1)$ balanced cycles, which consequently transforms $(\pi, \breve{\pi})$ into $(\iota, \breve{\iota})$ (Remark 1). Thus, the lemma follows. $\square$

Table 1 shows the number of cycles, balanced cycles, and average per operation achieved by each step of Algorithm 1 in the worst scenario.

Table 1. Summary of number of cycles and balanced cycles increased, in the worst case, by each step of Algorithm 1.

Step	S	$\Delta c(G(\mathcal{I}), S)$	$\Delta c_b(G(\mathcal{I}), S)$	$\frac{\Delta c(G(\mathcal{I}),S)+\Delta c_b(G(\mathcal{I}),S)}{|S|}$
i	μ	0	1	1
ii	ρ	1	1	2
iii	ρ,ρ	1	1	1

Given an instance $\mathcal{I} = ((\pi, \breve{\pi}), (\iota, \breve{\iota}))$, such that the size of the permutations π and ι is n, and $G(\mathcal{I}) = (V, E, w)$, we have that $2(n + 1) = |V| = |E|$. In the worst scenario, it is possible to note that step i from Algorithm 1 takes linear time to identify the existence of a trivial negative cycle in $G(\mathcal{I})$ and to apply a move operation. Similarly, step ii also takes linear time to determine if $G(\mathcal{I})$ has a negative or balanced divergent cycle and to apply a reversal operation. Step iii requires linear time to determine that all the non-trivial cycles in $G(\mathcal{I})$ that are either negative or balanced are convergent. However, to define the parameters i, j, x, and y of the two consecutive reversals takes quadratic time, since it will depend on the interleaving cycle of the convergent cycle selected. The steps i, ii, or iii will be required at most $k \leq n$ times to sort the permutation. Thus, Algorithm 1 runs in $\mathcal{O}(n^3)$.

Lemma 7. *Given an instance $\mathcal{I} = ((\pi, \breve{\pi}), (\iota, \breve{\iota}))$, Algorithm 1 turns $(\pi, \breve{\pi})$ into $(\iota, \breve{\iota})$ using at most $2(n+1) - (c(G(\mathcal{I})) + c_b(G(\mathcal{I})))$ reversal and move operations.*

Proof. By Lemma 6, Algorithm 1 turns $(\pi, \breve{\pi})$ into $(\iota, \breve{\iota})$ using reversal and move operations. By Remark 1, to reach the target genome $(\iota, \breve{\iota})$ it is necessary to achieve $c(G(\mathcal{I})) + c_b(G(\mathcal{I})) = 2(n + 1)$. Considering the worst scenario for each step of Algorithm 1, showed in Table 1, we observe that steps i and iii increase at least by one unit, on average, per operation the value of $c(G(\mathcal{I})) + c_b(G(\mathcal{I}))$. Step ii increases the value of $c(G(\mathcal{I})) + c_b(G(\mathcal{I}))$ in two units. Thus, in the worst scenario, Algorithm 1 turns $(\pi, \breve{\pi})$ into $(\iota, \breve{\iota})$ using at most $2(n+1) - (c(G(\mathcal{I})) + c_b(G(\mathcal{I})))$ reversal and move operations, and the lemma follows. □

Theorem 2. *Algorithm 1 is a 2-approximation for the SBIRM problem.*

Proof. Considering an instance $\mathcal{I} = ((\pi, \breve{\pi}), (\iota, \breve{\iota}))$ of the SBIRM problem and the lower bound defined by Lemma 1, we have the following value: $n + 1 - \frac{c(G(\mathcal{I})) + c_b(G(\mathcal{I}))}{2}$. By Lemma 7, Algorithm 1 turns $(\pi, \breve{\pi})$ into $(\iota, \breve{\iota})$ using at most $2(n + 1) - (c(G(\mathcal{I})) + c_b(G(\mathcal{I})))$ reversal and move operations, and the theorem follows. □

3.4 Reversal, Move, and Indel Operations

In the following, we present lemmas that will be used to construct an algorithm that transforms the source genome into the target genome using reversal, move, and indel operations. Since the indel operation is a non-conservative rearrangement event, we assume that the instances of the problem may be either balanced or unbalanced.

Lemma 8. *Given an instance* $\mathcal{I} = ((\pi, \breve{\pi}), (\iota, \breve{\iota}))$ *such that there is at least one positive cycle* $C = (c^1, \ldots, c^k)$, *it is possible to increase the number of balanced cycles by one unit after applying one indel operation.*

Proof. Note that $C = (c^1, \ldots)$ needs to receive weight on its black edges in $n = \sum_{i=1}^{k} [w(e'_{c^i}) - w(e_{c^i})]$ units to become balanced. In this case, we apply an indel δ to the black edge c^1 increasing its weight by n units, and the lemma follows.   □

Lemma 9. *Given an instance* $\mathcal{I} = ((\pi, \breve{\pi}), (\iota, \breve{\iota}))$ *such that there is at least one negative trivial cycle* $C = (c^1)$, *it is possible to increase the number of balanced cycles by one unit after applying one indel operation.*

Proof. Note that C is a negative trivial cycle. We can apply an indel δ to black edge c^1 decreasing its weight by $n = |\sum_{i=1}^{k} [w(e'_{c^i}) - w(e_{c^i})]|$ units, and the lemma follows.   □

Now consider the Algorithm 2. Note that it also runs in $\mathcal{O}(n^3)$ using a similar analysis described in Algorithm 1, but considering that it takes a linear time to determine the steps where the indel operation is necessary.

Algorithm 2. The input is an instance $\mathcal{I} = ((\pi, \breve{\pi}), (\iota, \breve{\iota}))$ and the output is a sequence S of reversals, moves, and indels such that $(\pi, \breve{\pi}) \cdot S = (\iota, \breve{\iota})$.

$S \leftarrow [\,]$
while $(\pi, \breve{\pi}) \neq (\iota, \breve{\iota})$ **do**
    if there is a negative trivial cycle in $G(\mathcal{I})$ **then**
        if there is a positive cycle in $G(\mathcal{I})$ **then**
            $S \leftarrow S + [\mu_1]$ ▷ Lemma 3.
        else
            $S \leftarrow S + [\delta_1]$ ▷ Lemma 9.
        end if
    else if there is a negative or balanced divergent cycle in $G(\mathcal{I})$ **then**
        $S \leftarrow S + [\rho_1]$ ▷ Lemma 4.
    else if there is a negative or balanced convergent cycle in $G(\mathcal{I})$ **then**
        $S \leftarrow S + [\rho_1, \rho_2]$ ▷ Lemma 5.
    else
        $S \leftarrow S + [\delta_1]$ ▷ Lemma 8.
    end if
end while
return S

Lemma 10. *Given an instance* $\mathcal{I} = ((\pi, \breve{\pi}), (\iota, \breve{\iota}))$*, Algorithm 2 turns* $(\pi, \breve{\pi})$ *into* $(\iota, \breve{\iota})$ *using reversal, move, and indel operations.*

Proof. Note that initially the trivial negative cycles are turned into balanced by Lemma 3 or Lemma 9. Then, negative or balanced non-trivial cycles are treated by lemmas 4 and 5. Lastly, if Algorithm 2 reaches the last step, it means that $G(\mathcal{I})$ may have balanced trivial cycles and, at least, one positive cycle. In this case, Lemma 8 is applied. Note that, in all the cases, at least one new balanced cycle is generated. Thus, the algorithm will eventually reach the target genome since the maximum number of balanced cycles in $G(\mathcal{I})$ is $n + 1$ (Remark 1) and the lemma follows. □

Lemma 11. *Given an instance* $\mathcal{I} = ((\pi, \breve{\pi}), (\iota, \breve{\iota}))$*, Algorithm 2 turns* $(\pi, \breve{\pi})$ *into* $(\iota, \breve{\iota})$ *using at most* $2(n + 1) - (c(G(\mathcal{I})) + c_b(G(\mathcal{I})))$ *reversal, move, and indel operations.*

Proof. By Lemma 10, Algorithm 2 turns $(\pi, \breve{\pi})$ into $(\iota, \breve{\iota})$ using reversal, move, and indel operations. By Remark 1, to reach the target genome $(\iota, \breve{\iota})$, it is necessary to achieve $c(G(\mathcal{I})) + c_b(G(\mathcal{I})) = 2(n+1)$. In all cases of Algorithm 2 we have that $\frac{\Delta c(G(\mathcal{I}), \mathcal{S}) + \Delta c_b(G(\mathcal{I}), \mathcal{S})}{|\mathcal{S}|} \geq 1$, where $\mathcal{S}$ is the sequence of operations applied by the step of Algorithm 2. Then, in the worst case, $2(n+1) - (c(G(\mathcal{I})) + c_b(G(\mathcal{I})))$ reversal, move, and indel operations will be needed to turn the source genome into the target genome, and the lemma follows. □

Theorem 3. *Algorithm 2 is a 2-approximation for the* SBIRMI *problem.*

Proof. Considering an instance $\mathcal{I} = ((\pi, \breve{\pi}), (\iota, \breve{\iota}))$ of the SBIRMI problem and the lower bound defined by Lemma 2, we have the following value: $n + 1 - \frac{c(G(\mathcal{I})) + c_b(G(\mathcal{I}))}{2}$. By Lemma 11, Algorithm 2 turns $(\pi, \breve{\pi})$ into $(\iota, \breve{\iota})$ using at most $2(n + 1) - (c(G(\mathcal{I})) + c_b(G(\mathcal{I})))$ reversal, move, and indel operations, and the theorem follows. □

4   Conclusion

In this work, we investigated two genome rearrangement problems considering the scenario where the genomes share the same set of genes, and the orientation of the genes is known. The first model allows the conservative events of reversal and move, while the second also includes the non-conservative event of indel. For both cases, we showed that they belong to the NP-hard class, and we presented a 2-approximation algorithm for each problem. The results bring an advance regarding the complexity analysis considering the three studied events. It is worth mentioning that the problem considering only reversal and indel events remains with unknown complexity.

As future works, investigations focused on the complexity of the problem regarding the reversal and indel events are relevant since closing the complexity gap of those problem variants brings fundamental knowledge for the comparative genomic field.

Acknowledgment. This work was supported by the National Council of Technological and Scientific Development, CNPq (grants 140272/2020-8, 202292/2020-7 and 425340/2016-3), the Coordenação de Aperfeiçoamento de Pessoal de Nível Superior - Brasil (CAPES) - Finance Code 001, and the São Paulo Research Foundation, FAPESP (grants 2013/08293-7, 2015/11937-9, and 2019/27331-3).

References

1. Bafna, V., Pevzner, P.A.: Sorting by transpositions. SIAM J. Discr. Math. **11**(2), 224–240 (1998)
2. Berman, P., Hannenhalli, S., Karpinski, M.: 1.375-approximation algorithm for sorting by reversals. In: Möhring, R., Raman, R. (eds.) ESA 2002. LNCS, vol. 2461, pp. 200–210. Springer, Heidelberg (2002). https://doi.org/10.1007/3-540-45749-6_21
3. Biller, P., Guéguen, L., Knibbe, C., Tannier, E.: Breaking good: accounting for fragility of genomic regions in rearrangement distance estimation. Genome Biol. Evol. **8**(5), 1427–1439 (2016)
4. Biller, P., Knibbe, C., Beslon, G., Tannier, E.: Comparative genomics on artificial life. In: Beckmann, A., Bienvenu, L., Jonoska, N. (eds.) CiE 2016. LNCS, vol. 9709, pp. 35–44. Springer, Cham (2016). https://doi.org/10.1007/978-3-319-40189-8_4
5. Brito, K.L., Jean, G., Fertin, G., Oliveira, A.R., Dias, U., Dias, Z.: Sorting by genome rearrangements on both gene order and intergenic sizes. J. Comput. Biol. **27**(2), 156–174 (2020)
6. Bulteau, L., Fertin, G., Tannier, E.: Genome rearrangements with indels in intergenes restrict the scenario space. BMC Bioinform. **17**(14), 426 (2016)
7. Caprara, A.: Sorting permutations by reversals and Eulerian cycle decompositions. SIAM J. Discr. Math. **12**(1), 91–110 (1999)
8. Caprara, A., Lancia, G., Ng, S.K.: Fast Practical solution of sorting by reversals. In: Shmoys, D. (ed.) Proceedings of the 11th ACM-SIAM Annual Symposium on Discrete Algorithms (SODA 2000), pp. 12–21. Society for Industrial and Applied Mathematics, Philadelphia, PA, USA (2000)
9. Chen, X.: On sorting unsigned permutations by double-cut-and-joins. J. Comb. Optim. **25**(3), 339–351 (2013)
10. Chen, X., et al.: Assignment of orthologous genes via genome rearrangement. IEEE/ACM Trans. Comput. Biol. Bioinform. **2**(4), 302–315 (2005)
11. Christie, D.A., Irving, R.W.: Sorting strings by reversals and by transpositions. SIAM J. Discr. Math. **14**(2), 193–206 (2001)
12. Fertin, G., Jean, G., Tannier, E.: Algorithms for computing the double cut and join distance on both gene order and intergenic sizes. Algorithms Mol. Biol. **12**(1), 16 (2017)
13. Garey, M.R., Johnson, D.S.: Computers and Intractability; A Guide to the Theory of NP-Completeness. W.H. Freeman & Co., New York (1990)
14. Hannenhalli, S., Pevzner, P.A.: Transforming men into mice (Polynomial Algorithm for Genomic Distance Problem). In: Proceedings of the 36th Annual Symposium on Foundations of Computer Science (FOCS 1995), pp. 581–592 (1995)
15. Hannenhalli, S., Pevzner, P.A.: Transforming cabbage into turnip: polynomial algorithm for sorting signed permutations by reversals. J. ACM **46**(1), 1–27 (1999)
16. Oliveira, A., et al.: Sorting signed permutations by intergenic reversals. IEEE/ACM Trans. Comput. Biol. Bioinform. **18**(6), 2870–2876 (2021)

17. Oliveira, A.R., Jean, G., Fertin, G., Brito, K.L., Dias, U., Dias, Z.: Sorting permutations by intergenic operations. IEEE/ACM Trans. Comput. Biol. Bioinform. **18**(6), 2080–2093 (2021)
18. Radcliffe, A.J., Scott, A.D., Wilmer, E.L.: Reversals and transpositions over finite alphabets. SIAM J. Discr. Math. **19**(1), 224–244 (2005)
19. Siqueira, G., Alexandrino, A.O., Oliveira, A.R., Dias, Z.: Approximation algorithm for rearrangement distances considering repeated genes and intergenic regions. Algorithms Mol. Biol. **16**(1), 1–23 (2021)

Chromothripsis Rearrangements
Are Informed by 3D-Genome Organization

Natalia Petukhova[2], Alexey Zabelkin[1(✉)], Vitaly Dravgelis[1],
Sergey Aganezov[3], and Nikita Alexeev[4]

[1] ITMO University, St. Petersburg, Russia
a.zabelkin@itmo.ru
[2] Bioinformatics Research Center, Pavlov First Saint Petersburg Medical State
University, St. Petersburg, Russia
[3] Department of Computer Science, John Hopkins University, Baltimore, MD, USA
[4] Baltimore, USA

Abstract. Chromothripsis is a mutational phenomenon representing a
unique type of tremendously complex genomic structural alteration. It
was initially described and was broadly observed in cancer with lower
frequencies in other genetic disorders. Chromothripsis manifests massive
genomic structural alterations during a single catastrophic event in the
cell. It is considered to be characterized by the simultaneous shattering
of chromosomes followed by random reassembly of the DNA fragments,
ultimately resulting in newly formed, mosaic derivative chromosomes and
with a potential for a drastic oncogenic transformation. Here, we consider a question of whether the genomic locations involved in chromothripsis rearrangements' are randomly distributed in 3D genomic packing
space or have some spatial organization's predispositions. To that end,
we investigated the structural variations (SVs) observed in previously
sequenced cancer genomes via juxtaposition of involved breakpoints onto
the Hi-C contact genome map of normal tissue. We found that the average Hi-C contact score for SVs breakpoints appearing at the same chromosome (*cis*-SVs) in an individual patient is significantly higher than
the average Hi-C matrix signal, which indicates that SVs tend to involve
spatially proximal regions of the chromosome. Furthermore, we overlaid
the chromothripsis annotation of groups of SVs' breakpoints and demonstrated that the Hi-C signals for both chromothripsis breakpoint regions
as well as regular SVs breakpoints are statistically significantly higher
than random control, suggesting that chromothripsis *cis*-SVs have the
same tendency to rearrange the proximal sites in 3D-genome space. Last
but not least, our analysis revealed a statistically higher Hi-C score for
all pairwise combinations of breakpoints involved in chromothripsis event
when compared to both background Hi-C signal as well as to combination of non-chromothripsis breakpoint pairs. This observation indicates
that breakpoints could be assumed to describe a given chromothripsis
3D-cluster as a proximal bundle in genome space. These results provide

---

N. Petukhova and A. Zabelkin—These authors contributed equally.
N. Alexeev—Independent researcher

L. Jin and D. Durand (Eds.): RECOMB-CG 2022, LNBI 13234, pp. 221–231, 2022.
https://doi.org/10.1007/978-3-031-06220-9_13

valuable new insights into the spatial relationships of the SVs loci for both chromothripsis and regular genomic alterations, laying the foundation for the development of a more precise method for chromothripsis identification and annotation.

Keywords: Chromothripsis · Hi-C · Genome instability · Complex rearrangements · 3D-genome organization · Contact frequency · Oncogenesis

1   Introduction

During oncogenesis, cells acquire numerous genetic changes that confer tumor-specific properties, including immortality, escape of apoptosis and antimitotic signaling, neovascularization, invasiveness, and metastatic potential. In most cases, these oncogenic changes are considered to occur gradually, having a cumulative nature [26]. However, recent advances in sequencing and bioinformatics analysis of cancer genomes have challenged the established paradigm, as a radically new, unique mechanism of simultaneous oncogenic cell transformation - chromothripsis - has been identified, causing complex intra- and inter-chromosomal rearrangements in a short time. At the present moment, the mechanisms and the causes of chromothripsis remain unclear: it is believed that a single catastrophic event in the cell leads to simultaneous shattering of a chromosome(s), followed by random, chaotic reassembly of genomic fragments into mosaic derivative(s), along with a partial loss of genetic material [11,24]. Several nonexclusive mechanistic models have been proposed to explain the cause and high complexity of a chromothripsis event, [13,15,17,25] but the molecular mechanism of such a cellular catastrophe remains unclear and poorly understood, especially from the point of its prediction, or even accurate detection.

The phenomenon of chromothripsis was discovered relatively recently when massive genomic rearrangements were detected in patients with chronic lymphocytic leukemia [24]. Initially, the prevalence of chromothripsis was shown to be around 2–3% in all human cancer cases where the frequency of occurrence depends on the tissue origin [1,7]. For a more detailed study and analysis of the chromothripsis phenomenon, some characteristic features were proposed to distinguish it from other complex genomic rearrangements, such as: significant 1D-clustering of DNA breakpoints; copy number (CN) change not exceeding 2 or 3; chromothripsis fragments being reassembled in a chaotic way in all possible orientations [8]. Despite the high prevalence in different types of cancer along with enormous consequences and significant role in carcinogenesis, the mechanism of chromothripsis and its molecular biological pathway are still not fully understood, not to mention the preconditions for its occurrence. Some models of the chromothripsis initiation have been proposed, which reflect only a mechanistic understanding of this phenomenon [17]. Recently, the key role of chromothripsis as the main mechanism that accelerates the rearrangement and amplification of the genomic DNA, contributing to the rapid evolution of cancer cells, has been demonstrated [20]. In addition, a genome-wide analysis of cancer patients showed a significantly higher prevalence of chromothripsis compared to the initial estimates - up to 49% of all studied cases of different types of cancer had chromothripsis [30].

Recent genome-wide studies of 2,568 patients within the framework of the international PCAWG consortium proved the wide prevalence of chromothripsis among cancer patients, alongside the ambiguity of the rearrangements detection and the complexity of their interpretation [4]. To date, two algorithms for the chromothripsis detection were proposed - ShatterSeek [4] and ShatterProof [6] - both based on the statistical method for structural rearrangements assessment and the criteria being related to the chromothripsis event described by Korbel and Campbell (2013) [8]. It is worth noting that the developers of ShatterSeek themselves point to discrepancies in the assessment of chromothripsis events when comparing their results with the results obtained with the similar Shatter-Proof method - in their case, the discrepancies depend on the initial threshold of statistical sensitivity chosen in the study [4].

Since the discovery of chromothripsis as a phenomenon and its proposed distinctive properties, chromosome conformation capture (3C, 3C-derived) technology has revolutionized the analysis of sequencing data and our understanding of the 3D-genome organization: recent studies have highlighted the critical importance of genomic spatial architecture for maintaining all molecular mechanisms, gene regulation and genome stability [12,19,28]. In particular, several studies showed the influence of 3D-genome architecture on SVs occurrence [2], also its role in translocations formation has been shown [5,18,33]. We also note that the role of spatial genomic organization has been considered in comparative genomics studies, in which cross-species rearrangements are being identified and analyzed, with results showing that 3D genomic proximity correlates with the evolutionary breakpoint sites [27,29]. In this regard, it is important to investigate of the chromothripsis phenomenon and rearrangements formation in a view of the 3D-genome organization, and to identify signatures of chromothripsis in the context of genome conformation, as well as review the strictly "chaotic" nature of the rearrangements reassembly. Recently, the attempt was made to study the relationship between genome organization before and after SVs occurrence in the same population of cells [21]: the model system of human cells treated by doxorubicin (Topoisomerase II (TOP2) trapping) was used to induce chromothripsis-like rearrangements. The key criterion of SVs that occurred was the randomness of breakpoint formation [8] which was studied by juxtaposition of fragments in the Hi-C-based reconstructed derivative chromosomes of rearranged cells. It was shown that the majority of SVs (induced and spontaneous) occurred in conserved compartments whereas TOP2-linked SVs were more likely to occur in A-compartments ("active", open chromatin) and fewer within B-compartments, the spontaneous SVs distributed more uniformly with no statistical difference compared to the random background set. The similar association was found for cumulative SVs identified in cancer genomes [21]. There have also been efforts focused on methodological development for both detection and grouping of complex SVs, including those possibly originating from chromothripsis as part of the somatic evolutionary process in cancer progression [22,31].

The aim of the present study is to analyze the predispositions in the 3D-genome space for chromothripsis complex rearrangements (i.e., religated pairs and other pairwise combinations of involved breakpoints) occurring in cancer. We propose a hypothesis that chromothripsis rearrangements occur not only simultaneously and present as a clustered 1D-segment on the chromosome, but also are mutually co-localized in 3D-space in the form of a proximal bundle. We investigated the SVs determined as chromothripsis via the juxtaposition of the breakpoints on the contact genome map of normal tissue and checked the pairwise distances between all breakpoints. We hypothesize that it is the spatial co-localization of chromothripsis breakpoints in 3D-space occurring in each individual patient that could be a new parameter to include in the algorithm for more accurate chromothripsis detection.

2 Materials and Methods

2.1 Hi-C Data

Public Hi-C data (GEO accession GSE118629) was used to serve a reference of average long-range genomic interactions in the RWPE1 normal prostate ATCC CRL11609 cell line [16]. Each matrix cell H_{ij} is a contact frequency between the genome loci corresponding to the i-th and j-th bins. The initial 40Kbp resolution, Hi-C normalized matrix GSE118629 was considered. Normalized Hi-C matrix 40 Kbp [16] and hg19 annotated bins were converted into .cool format. In this study, we considered the Hi-C matrix with 400 Kbp resolution as this resolution is not as noisy as 40 Kbp Hi-C matrix but it still refined enough to detect the signal between pairs of breakpoints involved in genome rearrangements. We further note that we consider only *cis*-chromosomal intra-arm submatrices of the original Hi-C matrix as the contact signal for inter-chromosomal/arm regions is drastically lower and noisier, most likely requiring a separate analytical approach for processing and analysis.

In order to clean up the alignment bias from the Hi-C data, we normalized the values in each column and each row as described in [10]. In order to find the relative spatial proximity between two loci at the genomic space, we normalized the initial Hi-C signal by the mean value at this genomic distance. Namely, the transformation of the Hi-C matrix into the so called Observed/Expected matrix was made by standard method [10] in the following way:

$$h_{ij} := H_{ij}/M_k,$$

where H_{ij} is the score between two loci i and j, M_k – the mean Hi-C score taken over the set of bin pairs

$$D_k := \{(i,j) : |i - j| = k\}.$$

This allows us to take into account relative spatial proximity instead of absolute one. In particular this would put the Hi-C scores of short-range breakpoint pairs into the same scale as long-range ones.

2.2 SVs Data

Recent large-scale study of 2,658 cancer patients with 38 different cancer types was performed under the PanCancer project (PCAWG Consortium) [4]. WGS data was used to obtain rearrangements in patients. SVs coordinates including chromothripsis SVs data used in the present study was obtained at https:// dcc.icgc.org/releases/PCAWG. Prostate adenocarcinoma was chosen for present analysis as this type of cancer is characterized by various frequently occurring complex structural variation events.

We thus considered SVs for 175 patients with prostate adenocarcinoma. SVs' genomic coordinates were annotated with GRCh37 assembly. Initially, 14,498 SVs were collected, but we removed inter-chromosomal and inter-chromosomal-arm SVs, ultimately retaining 8,154 intra-arm cis-SVs. We further narrowed the considered SVs set by removing rearrangements with size: $|position1 - position2| < 60$ Kbp in order to exclude microinversions and other smaller variations from the analysis, as the involved breakpoints most likely do not have spatial relationship that is decoupled from their close proximity in 1D-space. Overall, we considered 5,085 cis-SVs in the present study.

2.3 Chromothripsis Rearrangements Data

We annotated each SV as "belonging" to a chromothripsis event if both of its positions (i.e., breakpoints) were located within chromothripsis region(s) identified by ShatterSeek [8]. ShatterSeek method represents a statistical approach which consideres the following key features for chromothripsis identification: 1) "clusters of breakpoint should be interleaved and have equal distributions of SVs types"; 2) "rearrangements fragments joins should follow a roughly even distribution"; 3) "oscillating CN patterns should not exceed 2 or 3 states"; 4) "interspersed loss of heterozygosity at heterozygous single-nucleotide polymorphisms". Based on these criteria, all SVs were classified into 5 groups by ShatterSeek: high-confidence and linked-to-high, low-confidence, and linked-to-low as well as no chromothripsis SVs [4]. The high-confidence annotation refers to regions which are characterized by at least 6 interleaved intrachromosomal SVs or at least 3 interleaved intrachromosomal SVs and 4 or more interchromosomal SVs, 7 adjacent segments oscillating between 2 CN states. We used the high-confidence chromothripsis group to markup SVs as belonging to chromothripsis cluster.

2.4 Breakpoints Pairwise Distances Analysis

In this paper, we study mutual configuration of the SVs breakpoints in 3D-space. To do this, we consider all possible pairwise combinations of breakpoints from the same chromosomal arm (see Fig. 1). We consider the cases when both breakpoints came from the same patient (we label such combinations as $comb\_s$) or when the first breakpoint came from one patient and the second came from another patient (we label such combinations as $comb\_d$). It is worth noting that for the intra-patient pairwise combinations $comb\_s$, we do not include pairs of

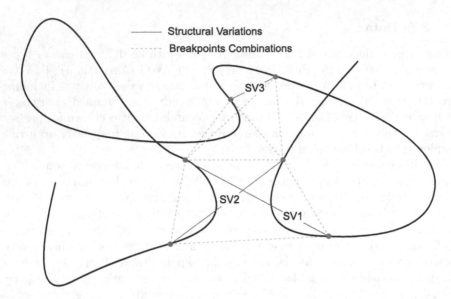

Fig. 1. Three genome rearrangements, 8 out of 12 pairwise combinations of SV1, SV2, and SV3 breakpoints are marked orange (4 pairwise combinations are omitted for better readability) (Color figure online).

breakpoints that define true SVs (i.e., a novel adjacency formed between said breakpoints) and analyze them separately. Where it is relevant, we take into account the chromothripsis labeling of involved breakpoints (as described above).

2.5 Statistical Analysis

In order to compare the H-C scores across different groups of genome loci pairs, we perform the standard Student's t-test for independent pairs. We can do so since all the compared groups Hi-C scores distributions (while not normal) belong to the domain of normal attraction by the central limit theorem. Since the performed several comparisons, we report the Bonferroni corrected p-values of the t-test.

3 Results and Discussion

We note that previous research highlighted that the breakpoints of regular translocations and *cis*-SVs have higher Hi-C contact frequencies in cancer cells [5,33]. Since chromothripsis can be represented as a collection of rearrangements occurring simultaneously, we hypothesize that breakpoints involved in all SVs from a given chromothripsis have a tendency to co-locate in 3D-space. To see if such co-localization evidence is present in the Hi-C signal, we first consider all of the pairwise combinations of breakpoints (see Fig. 1).

For these combinations, we compare the Hi-C scores between the true SVs, *comb_s*, and *comb_d* groups without any chromothripsis annotations (see Fig. 2).

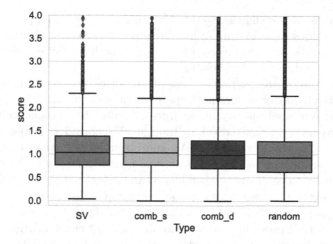

Fig. 2. The Hi-C scores for the pairwise breakpoint combinations determined by measured SVs (blue), comb_s (orange), and comb_d (green), and control set of random pairs of genome loci (red) (Color figure online).

As we expected, the Hi-C scores for the pairs of breakpoints representing actual SVs have the highest mean of 1.13. Quite surprisingly, the Hi-C scores for the both types of combinations are also higher (the mean values are 1.09 and 1.02, respectively) than one would expect to see when considering combinations of random pairs of locations with pairwise distances similar to those defined by SVs (Bonferroni corrected t-test p-values are less than 2.2e−16).

We confirmed that the signal for true rearrangements was higher than the mean Hi-C matrix score, meaning that these genomic loci have more Hi-C contact frequencies. This fact suggests that studied SVs occur at non-random pairs of loci in the genome, as they are relatively more proximal. This is consistent with previous report that showed pre-existing spatial proximity of loci involved in double-stand breaks (DSBs) [33]. Since Hi-C represents contact frequencies rather than physical nuclear distances, such an approach ensures a straightforward check-up for the "contact-first" model of rearrangements' formation [32].

We demonstrated both true *cis*-SVs and other pairwise combinations of SV-defined breakpoints originating in the same patient have higher contact frequencies than combinations of breakpoints from different patients. Therefore, we suggest that all breakpoints of each individual patient occur at some preferable loci spatially more proximal along a chromosome, and they form a mutually close bundle. However, combinations from breakpoints of different patients have higher Hi-C score than the mean matrix signal. This observation could be explained by SVs' tendency to occur in the same specific chromosome compartments, for example, in "active", open chromatin, as it was indicated in the recent study [21]. Thus, we could expect the commensurate contact frequencies for breakpoints combinations, even originated from different patients. However, we found that comb_s showed a significantly higher Hi-C score than comb_d (Bonferroni corrected p-value is less than $2.2e-16$). This suggests that cross-patient elevated signal might represent a mixture of non-breakpoint chromatin features and spatial proximity of breakpoint-involved loci, where the intra-patient signal realizes the breakpoint-induced spatial relations.

In order to perform the precise assessment of SVs in each patient and to understand the behavior of chromothripsis rearrangements in 3D-genome space, we expanded our analysis by taking into account the classification of chromothripsis events from ShatterSeek method [4]. We considered breakpoint pairs involved in chromothripsis-grouped SVs and observed that they have the same high contact frequencies as the values observed for regular SVs (the mean value is 1.12). In order to confirm the significance and relevance of the Hi-C score for breakpoints involved in regular rearrangements and, consequently, to determine the relative proximal location of chromothripsis breakpoints, additional analysis of Hi-C score of all possible pairwise combinations of chromothripsis and non-chromothripsis breakpoints was performed. We found no statistical difference between the mean signal for pairs of breakpoints involved in true chromothripsis SVs and all other combinations of chromothriptic breakpoints in the same patient (the means are 1.12 and 1.1, respectively, the t-test p-value is 0.11). On the other hand, we observed that the Hi-C score for combinations of within-chromothripsis breakpoint pairs from the same patient was significantly higher than the signal across pairwise combinations of breakpoints pairs from non-chromothripsis SVs (the mean values are 1.1 and 1.09 respectively, the Bonferroni-corrected p-value is 0.026). This means that for any pair of breakpoints of the same chromotripsis event it is as likely, from the proximal perspective, for them to form observed SVs' connections as to re-ligate in some different way, forming a distinct set of SVs on the same set of chromotriptic breakpoints. However, we observe a significant difference in average signal values when comparing regular SVs and different pairwise combinations of their breakpoints (the Bonferroni corrected t-test p-values is 0.00236). Overall, this additionally confirms non-randomness of chromothriptic breakpoints grouping and relative inter-proximity between breakpoints that comprise a given chromothripsis cluster.

Thus, chromothripsis may be promoted by the intrinsic spatial organization of the given chromosomes such that DSBs within them are rather probable to be re-ligated to another DSB on the same chromosome. Indeed, such mechanistic relevance has sense in a view of the mechanism of chromothripsis DSBs repair. Non-homologous end joining (NHEJ) is considered to be the major mechanism for chromothripsis fragments repair which ligates the DSBs in an error-prone manner [24]. Further, the detailed analysis of the breakpoints junction showed few base-pair microhomologies, if any, short deletions/duplications, and short templated or non-templated insertions. This indicates that NHEJ [9] and/or microhomology-mediated end joining (MMEJ), also known as alternative non-homologous end-joining (Alt-NHEJ) [14] are the most likely DNA repair mechanisms underlying the gluing process of the generated chromothripsis fragments in most cases. Analysis of chromothripsis junctions in the PanCancer study showed the results in accordance with these repair mechanisms [4]. Additionally, NHEJ is the predominant repair pathway in human cells and traditionally serves as a guardian of the genome, preventing genomic instability through the fast, highly efficient repair of DSBs [23]. Compared to other repair mechanisms, NHEJ surveillance machinery is consistently active and does not depend on the cell cycle phase, and involves the straightforward, though usually error-prone, ligation of DSBs by simple joining them pairwise as soon as they occur to be close to each other in space: NHEJ requires physical juxtaposition of the DSBs ends (their "synapsis") that includes complex protein machinery [34] which mechanism is still under active research. Thus, the critical requirement of sufficient NHEJ is to overcome the entropy of dissociated chromosome ends, therefore, taking into account non-specificity of NHEJ that ligate all DSBs within proximity, the 3D-genome organization is considered to be a driving force to promote joining of DSBs providing mechanical factors to facilitate NHEJ repair. Therefore, we can assume that the contact frequency between two genomic loci may correlate with the frequency of their repair, the result of which can be seen in chromothripsis rearrangements with corresponding spatial predisposition of the breakpoints.

4 Conclusions

In the present study, we analyzed chromothripsis complex rearrangements in a view of spatial organization of the genome and observed whether there are some predispositions in 3D-genome structure to form these specific structural variations. SVs in prostate cancer and corresponding large chromothripsis clusters were analyzed and mapped on the Hi-C contact matrix of normal tissue. Resulting Hi-C score for both regular and chromothripsis SVs was found to be higher than average Hi-C matrix signal, confirming previous reports that regular *cis*-SVs have a strong influence from spatial proximity to the preferred DSBs joying [33]. Obtained findings are consistent with recent work of chromothripsis-like process where indirect approach (network nodes analysis) revealed genome catastrophe joining to be non-random, showing preferential DSBs repair to local sites or inter-exchange between specific loci [3]. Moreover, our analysis revealed a statistically higher Hi-C score for all possible combinations of intra-patient chromothriptic breakpoint pairs compared to the non-chromotriptic pairwise breakpoint combinations, confirming relative inter-proximity between breakpoints that comprise a given chromothripsis 3D-cluster. In conclusion, results obtained in the present study confirm the hypothesis that 3D-genome organization plays a crucial role in complex cellular processes as well as spatial proximity promotes SVs pairing of breakage positions that has mechanistic relevance in terms of the major mechanism of chromothripsis DSBs repair (NHEJ) requiring physical juxtaposition of the DSBs ends (their "synapsis"). The obtained results present a future research perspective for a refinement of chromothripsis identification methods with a proposed additional consideration of breakpoints clustering based on their relative proximity rather than only on statistical 1D-criteria.

References

1. Cai, H., Kumar, N., Bagheri, H.C., von Mering, C., Robinson, M.D., Baudis, M.: Chromothripsis-like patterns are recurring but heterogeneously distributed features in a survey of 22, 347 cancer genome screens. BMC Genom. **15**(1), 1–3 (2014). https://doi.org/10.1186/1471-2164-15-82
2. Canela, A., et al.: Genome organization drives chromosome fragility. Cell **170**(3), 507-521.e18 (2017). https://doi.org/10.1016/j.cell.2017.06.034
3. Cleal, K., Jones, R.E., Grimstead, J.W., Hendrickson, E.A., Baird, D.M.: Chromothripsis during telomere crisis is independent of NHEJ, and consistent with a replicative origin. Genome Res. **29**(5), 737–749 (2019). https://doi.org/10.1101/gr.240705.118
4. Cortés-Ciriano, I., et al.: Comprehensive analysis of chromothripsis in 2, 658 human cancers using whole-genome sequencing. Nat. Genet. **52**(3), 331–341 (2020). https://doi.org/10.1038/s41588-019-0576-7
5. Engreitz, J.M., Agarwala, V., Mirny, L.A.: Three-dimensional genome architecture influences partner selection for chromosomal translocations in human disease. PLoS ONE **7**(9), e44196 (2012). https://doi.org/10.1371/journal.pone.0044196
6. Govind, S.K., et al.: ShatterProof: operational detection and quantification of chromothripsis. BMC Bioinform. **15**(1), 1–3 (2014). https://doi.org/10.1186/1471-2105-15-78
7. Kloosterman, W.P., Koster, J., Molenaar, J.J.: Prevalence and clinical implications of chromothripsis in cancer genomes. Current Opinion Oncol. **26**(1), 64–72 (2014)
8. Korbel, J.O., Campbell, P.J.: Criteria for inference of chromothripsis in cancer genomes. Cell **152**(6), 1226–1236 (2013). https://doi.org/10.1016/j.cell.2013.02.023
9. Lieber, M.R.: The mechanism of double-strand DNA break repair by the nonhomologous DNA end-joining pathway. Ann. Rev. Biochem. **79**(1), 181–211 (2010). https://doi.org/10.1146/annurev.biochem.052308.093131
10. Lieberman-Aiden, E., et al.: Comprehensive mapping of long-range interactions reveals folding principles of the human genome. Science **326**(5950), 289–293 (2009). https://doi.org/10.1126/science.1181369
11. Luijten, M.N.H., Lee, J.X.T., Crasta, K.C.: Mutational game changer: chromothripsis and its emerging relevance to cancer. Mutat. Res. Rev. Mutat. Res. **777**, 29–51 (2018). https://doi.org/10.1016/j.mrrev.2018.06.004
12. Lukas, J., Lukas, C., Bartek, J.: More than just a focus: the chromatin response to DNA damage and its role in genome integrity maintenance. Nat. Cell Biol. **13**(10), 1161–1169 (2011). https://doi.org/10.1038/ncb2344
13. Marcozzi, A., Pellestor, F., Kloosterman, W.P.: The genomic characteristics and origin of chromothripsis. In: Pellestor, F. (ed.) Chromothripsis. MMB, vol. 1769, pp. 3–19. Springer, New York (2018). https://doi.org/10.1007/978-1-4939-7780-2_1
14. McVey, M., Lee, S.E.: MMEJ repair of double-strand breaks (director's cut): deleted sequences and alternative endings. Trends Genet. **24**(11), 529–538 (2008). https://doi.org/10.1016/j.tig.2008.08.007
15. Morishita, M., et al.: Chromothripsis-like chromosomal rearrangements induced by ionizing radiation using proton microbeam irradiation system. Oncotarget 7(9), 10182–10192 (2016). https://doi.org/10.18632/oncotarget.7186
16. Rhie, S.K., et al.: A high-resolution 3d epigenomic map reveals insights into the creation of the prostate cancer transcriptome. Nat. Commun. **10**(1), 1–2 (2019). https://doi.org/10.1038/s41467-019-12079-8

17. Rode, A., Maass, K.K., Willmund, K.V., Lichter, P., Ernst, A.: Chromothripsis in cancer cells: an update. Int. J. Cancer **138**(10), 2322–2333 (2015). https://doi.org/10.1002/ijc.29888

18. Roukos, V., Misteli, T.: The biogenesis of chromosome translocations. Nat. Cell Biol. **16**(4), 293–300 (2014). https://doi.org/10.1038/ncb2941

19. Rowley, M.J., Corces, V.G.: Organizational principles of 3d genome architecture. Nat. Rev. Genet. **19**(12), 789–800 (2018). https://doi.org/10.1038/s41576-018-0060-8

20. Shoshani, O., et al.: Chromothripsis drives the evolution of gene amplification in cancer. Nature **591**(7848), 137–141 (2020). https://doi.org/10.1038/s41586-020-03064-z

21. Sidiropoulos, N., et al.: Somatic structural variant formation is guided by and influences genome architecture (2021). https://doi.org/10.1101/2021.05.18.444682

22. Simonaitis, P., Swenson, K.M.: Finding local genome rearrangements. Algorithms Mol. Biol. **13**(1), 1–14 (2018)

23. Sishc, B., Davis, A.: The role of the core non-homologous end joining factors in carcinogenesis and cancer. Cancers **9**(7), 81 (2017). https://doi.org/10.3390/cancers9070081

24. Stephens, P.J., et al.: Massive genomic rearrangement acquired in a single catastrophic event during cancer development. Cell **144**(1), 27–40 (2011). https://doi.org/10.1016/j.cell.2010.11.055

25. Stevens, J.B., et al.: Diverse system stresses: common mechanisms of chromosome fragmentation. Cell Death Disease **2**(6), e178–e178 (2011). https://doi.org/10.1038/cddis.2011.60

26. Stratton, M.R., Campbell, P.J., Futreal, P.A.: The cancer genome. Nature **458**(7239), 719–724 (2009). https://doi.org/10.1038/nature07943

27. Swenson, K.M., Blanchette, M.: Large-scale mammalian genome rearrangements coincide with chromatin interactions. Bioinformatics **35**(14), i117–i126 (2019)

28. Szalaj, P., Plewczynski, D.: Three-dimensional organization and dynamics of the genome. Cell Biol. Toxicol. **34**(5), 381–404 (2018). https://doi.org/10.1007/s10565-018-9428-y

29. Véron, A.S., Lemaitre, C., Gautier, C., Lacroix, V., Sagot, M.F.: Close 3d proximity of evolutionary breakpoints argues for the notion of spatial synteny. BMC Genom. **12**(1), 1–13 (2011)

30. Voronina, N., et al.: The landscape of chromothripsis across adult cancer types. Nat. Commun. **11**(1), 1–3 (2020). https://doi.org/10.1038/s41467-020-16134-7

31. Weinreb, C., Oesper, L., Raphael, B.J.: Open adjacencies and k-breaks: detecting simultaneous rearrangements in cancer genomes. BMC Genom. **15**(6), 1–11 (2014)

32. Zhang, Y.: The role of mechanistic factors in promoting chromosomal translocations found in lymphoid and other cancers. In: Advances in Immunology, pp. 93–133. Elsevier (2010). https://doi.org/10.1016/s0065-2776(10)06004-9

33. Zhang, Y., McCord, R.P., Ho, Y.J., Lajoie, B.R., Hildebrand, D.G., Simon, A.C., Becker, M.S., Alt, F.W., Dekker, J.: Spatial organization of the mouse genome and its role in recurrent chromosomal translocations. Cell **148**(5), 908–921 (2012). https://doi.org/10.1016/j.cell.2012.02.002

34. Zhao, B., Watanabe, G., Morten, M.J., Reid, D.A., Rothenberg, E., Lieber, M.R.: The essential elements for the noncovalent association of two DNA ends during NHEJ synapsis. Nat. Commun. **10**(1), 1–2 (2019). https://doi.org/10.1038/s41467-019-11507-z

Metagenomics

Using Computational Synthetic Biology Tools to Modulate Gene Expression Within a Microbiome

Liyam Chitayat Levi[1], Ido Rippin[3], Moran Ben Tulila[1], Rotem Galron[1], and Tamir Tuller[1,2(✉)]

[1] Department of Biomedical Engineering, Tel-Aviv University, Tel Aviv, Israel
tamirtul@tauex.tau.ac.il
[2] The Sagol School of Neuroscience, Tel-Aviv University, Tel Aviv, Israel
[3] Sackler Faculty of Medicine, Tel-Aviv University, Tel Aviv, Israel

Abstract. The microbiome is an interconnected network of microorganisms, which exist and influence a wide array of natural and synthetic environments. Genetic information is constantly spread across the members of the microbial community in a process called horizontal gene transfer, causing exposure of genetic alterations and modifications to all members of the community.

In order to accurately and effectively engineer microbiomes, genetic modifications must be introduced to certain species, as selectivity is a key factor in creating and fixing functional abilities within microbial environments. Moreover, introduction of genes into unwanted hosts may cause unprecedented ecological impacts, posing a major biosafety issue. Technologies in the field are usually experimentally developed for a specific host or environment, and the lack of automization and generalization limit them to a specific microbiome. Additionally, they only deal with the transformation process itself at best and do not modulate the different elements of the genetic material, neglecting considerations related to the interactions between the new genetic material and the population.

This work presents a set of computational models that automatically design a microbiome-specific plasmid that is selectively expressed in certain parts of the bacterial population. The underlying algorithm fine-tunes genetic information to be optimally expressed in the wanted hosts of the plasmid, while simultaneously impairing expression in unwanted hosts. We take into account and selectively optimize the main elements linked to gene expression and heredity. In addition, we have provided both in-silico and in-vitro analysis supporting our claim. This study was part of the TAU IGEM 2021 project (https://2021.igem.org/Team:TAU_Israel).

Keywords: Population genomics · Microbiome engineering · Gene expression · Horizontal gene transfer · Evolutionary systems biology · Synthetic biology

1 Introduction

The term "microbiome" is defined as the community of different microorganisms that coexist in an environment. Nearly every system, from natural to synthetic, is populated

© The Author(s), under exclusive license to Springer Nature Switzerland AG 2022
L. Jin and D. Durand (Eds.): RECOMB-CG 2022, LNBI 13234, pp. 235–259, 2022.
https://doi.org/10.1007/978-3-031-06220-9_14

by a unique and diverse community of organisms, which continuously interact among themselves and with their environment. Early studies of the field have shown that the animal's microbiome has a noticeable effect on key features including their host's fitness and lifespan [1]. Research regarding the human and animal microbiome in the past years has led to truly impactful results that provide new understanding of the mechanisms of host-microbiome interactions and their key influence of various physiological [2] and even psychological [3] factors. Additionally, analysis of other natural environments has yielded intriguing results as well [4]. Overall, different studies were able to demonstrate that the microbial composition is reactive to environmental changes, boosting the interest in methods for active modulation and engineering of microbiomes, causing development of various techniques.

Some techniques such as microbiome directed evolution, genomic engineering, and others that require transplant of new or otherwise altered bacteria into the environment, have succeeded in some conditions [5–8] usually for shorter time frames. The main reason for that is that the transplanted bacteria are less adapted to the new environment and therefore are disadvantaged compared to the native bacteria in the competition over the environmental niche. Other methods aim to engineer the bacterial microbes that are already present in the environment, and have mainly focused on the transformation vector itself [9]. Most of these systems have been developed specifically for a certain environment and cannot be easily applied in others, however the more crucial problem arises once the new genetic information is introduced to the environment. Bacteria constantly share genetic information through various methods of horizontal gene transfer [10], and in most cases, in order for the engineering process to be effective and accurate, it should be introduced precisely to the wanted hosts. Moreover, due to the innate variations and unpredictability of biological systems, introduction of new genetic information to an uncontrolled environment may cause unprecedented ecological impacts.

In this work, we propose a novel method that offers a different view of the biological process, in which each genetic element that is linked to gene expression is examined and synthetically altered, instead of working with genetic building blocks as given. This method is generic and computational, aiming to fit selected genetic information to a given microbiome, by modulating expression in wanted and unwanted hosts of the modification. For instance, in the case of the human gut microbiome, some bacteria are symbiotic- and others are pathogenic [11]. An effective community engineering process would likely target a subgroup of the pathogenic bacterial species which can be viewed as the wanted hosts of the modification in this case (that can include for example a gene the decrease their growth rate); however, it should probably avoid expression in the symbiotic bacteria as much as possible, which can be defined as the unwanted hosts.

This approach is designed by considering the effects of horizontal gene transfer (HGT) on the genetic construct and interactions it facilitates. Additionally, this method takes into account the various degrees of characterizations that can exist for a certain microbiome, and can function even with very minimal metagenomic information (our current implementation uses annotated genomes, and can potentially be used with metagenomically assembled genomes correspondingly). Lastly, this form is designed to modify the microbiome for longer time periods. It is relatively resistant to the environmental damage of the genetic information, as each genetic element is examined and treated individually. The design process considers the fitness effect of the modification on its proposed hosts, and modulates the burden it poses accordingly.

2 Methods

The underlying concept of this method is to tailor the different genetic components influencing the expression process using suitable models for the biophysical data regarding preferences of cellular machinery. Along evolution, small and random changes have accumulated, causing the phylogenetic differentiation which is also reflected in the changes of the innate preferences of cellular machinery between species. These preference variations between different machinery related to gene expression in different species include transcription factors (TFs), restriction enzymes, translation machinery, and many more. The combined effect of these variations causes a certain sequence to have different expression levels in different hosts (see Fig. 1).

The codon usage bias (CUB) variations can be leveraged to achieve the stated goal. The main limitation for this approach is the relatively low degree of characterization for most microbiomes, and the noisiness of metagenomic data. The described approaches are fit to predict biophysical information regarding the described variations from metagenomic sequencing and supporting databases, to ensure the generality of this method. However, additional characterization of the microbial species in the context of the environment can be used to optimize the accuracy of the engineering process.

The optimization issue could be defined as following: once a gene is selected to be used in the microbiome, two important sub-populations of any possible size are characterized- the first are the wanted hosts, which are the species that should be able to optimally express the selected gene. The second sub-population is the group of unwanted hosts, which should not be able to express the selected gene, or express it sub-optimally. The following algorithms will then distinctively deal with the main genetic elements (described in Fig. 1) related to the process of gene expression and optimize their performance in the wanted hosts while simultaneously deoptimizing it in the unwanted hosts. Therefore, the tailored elements are meant to act different in one group of hosts compared to the other, thus allowing selective expression despite the process of HGT.

An additional consideration for the engineering process, is the tradeoff between optimization of expression of the modification in wanted hosts, compared to prevention of the expression in the unwanted hosts. In many situations, there is a clear preference between these two goals, which is taken into consideration in the designing process through a predefined tuning parameter. The final goal of this method is to engineer an entire environment specific plasmid, which has improved expression for the desired hosts and impaired expression for unwanted hosts, altering the expression-related elements to be as selective as possible.

2.1 Translation Efficiency Modeling

The open reading frame (ORF), is the genetic element that codes for amino acids. Due to the redundancy of the genetic code, cellular machinery has adapted to translate certain codons more optimally than others, a bias quantified in calculated CUB scores [12].

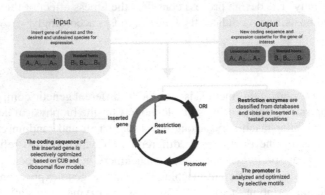

Fig. 1. Illustration of the main genetic components engineered to modulate expression of the designed plasmid in a target microbiome.

The posed cellular effect is that ribosomes are a limited resource in living organisms, and so-called "synonymous" changes in the ORF may influence the ribosomal flow, translation efficiency and fitness and can also affect other gene expression steps [13]. Optimization according to CUB, also referred to as codon harmonization, is traditionally meant to optimize expression for a single organism [14–16]. This algorithm describes the synonymous re-coding of the ORF not for a single organism, but for an entire consortium. During this process, the expression and fitness is optimized for the wanted hosts and deoptimized for the unwanted hosts.

Extraction of CUB Data: Codon usage bias preferences can be calculated under various assumptions and quantified by different indexes, according to the available data for the microbiome. In this work, we used the Codon Adaptation Index (CAI) which estimates that tendency of an ORF to include optimal/frequent codons and tRNA Adaptation Index (tAI) which measures that adaptation of a coding region to the tRNA pool for characterizing hosts' CUB scores. Their standard deviation (σ) and mean (μ) values are calculated for each gene of every organism (Fig. 2B), it is important to note that both options are available to fit the tailoring method to the available information about the species in the microbiome.

CAI Calculation: The codon adaptation index uses a reference set of genes from a specific species to assess the relative frequency of each codon with respect to its synonymous codons [12]. Each codon is given a weight based on the reference set, using RSCU (reference set codon usage) according to the following formula:

$$w_{codon} = \frac{RSCU_{codon}}{RSCU_{max}} = \frac{\frac{x_j}{\hat{x}}}{\frac{max_{1 \le j \le d} x_j}{\hat{x}}} = \frac{x_i}{max_{1 \le j \le d} x_j} \tag{1}$$

where x_i is the i-th codon of amino acid x.

The weight of the i-th codon of amino acid x is equal to the number of appearances of codon i (x_i) divided by the number of appearances of the most common codon out of the synonymous codons of amino acid x. For our model, we used a reference set of highly expressed genes from each organism. Due to the high need of the cell for such genes, those genes are under stronger evolutionary pressure to adapt to the intracellular machinery, thus expected to have more optimal codons.

The geometric mean of the codons' weights can then be used to calculate a total CAI score for a gene:

$$CAI_{seq} = \left(\prod_{i=1}^{L} w_i\right)^{\frac{1}{L}} = exp\left(\frac{1}{L}\sum_{i=1}^{L} ln(w_i)\right) \tag{2}$$

where L is the gene length in codons count.

tAI Calculation: The tRNA Adaptation Index is a measure of translational efficiency which considers the intracellular concentration of tRNA molecules and the affinity of each codon-anticodon pairing based on a reference set of genes. The first step in calculating tAI is obtaining the tRNA pool distribution. Ideally, tRNA levels should be measured directly, but since the measurement is highly inaccurate and complicated to conduct the distribution is usually calculated based on tRNA gene copy number. Due to their critical role, tRNA genes appear more than once in the genome. As a high number of occurrences in the genome likely indicates high levels of the tRNA it encodes for [17], these two values are comparable. Using the tRNA pool distribution, each codon is given a weight expressing its translational efficiency:

$$w_i = \sum_{j=1}^{n_i} (1 - s_{ij}) * tGCN_{ij} \tag{3}$$

where W_i is the weight of the i-th codon, n_i is the number of tRNA types that pair with the i-th codon, $tGCN_{ij}$ is the $tRNA_j$ gene copy number and s_{ij} is a value in the range $[0,1]$ representing the affinity of the pairing between the i-th codon and the j-th tRNA.

Codon weights are normalized by division by the maximum weight among all 61 codons and zero values are replaced with the geometric mean of the other weights, in order to avoid zeroing of the tAI gene index. The tAI gene score is calculated by the geometric mean of all the normalized codon weights, given by:

$$tAI_{seq} = \left(\prod_{i=1}^{L} w_i\right)^{\frac{1}{L}} = exp\left(\frac{1}{L}\sum_{i=1}^{l} ln(w_i)\right) \tag{4}$$

where L is the gene length in codons count.

Optimizations for In-Vitro Evaluation

Optimization is based on choosing the "most optimal" codon between the synonymous codons (which encode the same amino acid). The following CUB measurements were calculated for *E. coli* and *B. subtilis*:

CAI (codon adaptation index): as previously explained in this appendix, in this optimization, the ORF contains optimal codons that were calculated by their relative abundance in a reference set of highly expressed genes of the chosen organism.

tAI (tRNA adaptation index): as previously explained in this appendix, in this optimization, codons with higher translational efficiency are included in the ORF. It was calculated by considering the intracellular concentration of tRNA molecules and the affinity of each codon-anticodon pairing. This index was calculated based on a reference set of highly expressed genes of the chosen organism.

TDR (typical decoding rate): this optimization is based on ribosome profiling data (Ribo-Seq), which provides a snapshot of a mid-translation ribosomal position on the mRNA molecules in a cell during certain conditions.

The optimal codon is defined as the codon with the minimal loss score, which is described by the loss functions:

Ratio loss (R):

$$
loss_{codon} = \alpha \cdot \sum_{optimized\ organisms} \sum_{features} \left(1 - \frac{score_{codon}}{max(scores)}\right)
$$

$$
+ (1 - \alpha) \cdot \sum_{deoptimized\ organisms} \sum_{features} \frac{score_{codon}}{max(scores)} \tag{5}
$$

Difference loss:

$$
diff\ loss_{codon} = \alpha \cdot \sum_{optimized\ organisms} \sum_{features} 1 - (score_{codon} - max(scores))
$$

$$
+ (1 - \alpha) \cdot \sum_{deoptimized\ organisms} \sum_{features} score_{codon} - max(scores) \tag{6}
$$

The minimum of the first sum is achieved when the score of the codon in optimized organisms is close to the maximum value possible. The minimum of the second sum is achieved when the score of the same codon is distant from the maximal value (close to the minimum). So, minimization of the loss function brings an optimal solution from both points of view.

The optimization abbreviation consists of the CUB (tAI, CAI, TDR) type followed by the optimization type (R or D), i.e. tAI-D. Additionally, the reason why CAI is written without the optimization type is due to the fact that by chance, the CAI-R and CAI-D sequences are identical.

Multi-organism Optimization: The suggested model takes a holistic approach that integrates both the host preference for a certain synonymous codon based on the CUB index chosen and the significance of modifying the codon at the proteome level for every host, whether wanted or unwanted.

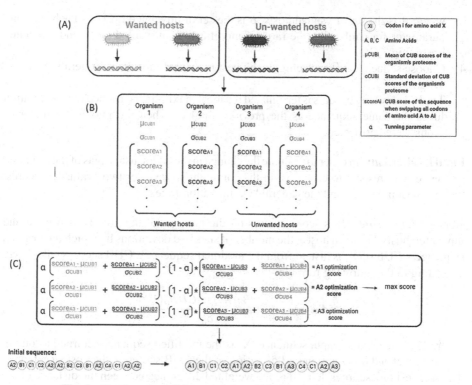

Fig. 2. One iteration of the translation (CUB) optimization algorithm.

The algorithm is based on hill climbing and consists of the following steps in each iteration: let X be the modified sequence in the current iteration. In the first iteration, X is chosen as the original input sequence. Iterate over all possible codons of all the amino acids (61 options in total), and for each:

1. Create a new sequence X' by replacing all synonymous codons of the selected codon in the current iteration in X to the selected codon.
2. Iterate over all the organisms (wanted and unwanted) and for each calculate the number of standard deviations by which the CUB score of X' of the organism is distant from the mean score of the proteome by the following formula:

$$dist(a) = \frac{\left(CUB_a\left(X'\right) - \mu_a\right)}{\sigma_a} \tag{7}$$

where σ and μ are standard deviation and mean of CUB scores of the organism's proteome, respectively.

3. Calculate the value of the target optimization function for X', defined as:

$$f\left(X'\right) = \alpha \cdot mean_{a \in A}(dist(a)) - (1 - \alpha) \cdot mean_{b \in B}(dist(b)) =$$

$$\alpha \cdot \frac{\sum_{a \in A} \frac{(CUB_a(X') - \mu_a)}{\sigma_a}}{|A|} - (1 - \alpha) \cdot \frac{\sum_{b \in B} \frac{(CUB_b(X') - \mu_b)}{\sigma_b}}{|B|} \tag{8}$$

where A is the set of wanted hosts, B is the set of unwanted hosts and α is a tuning parameter for calibrating the ratio of modification in wanted hosts and unwanted hosts (Fig. 2C).

4. Choose sequence X' producing the highest $f(x')$ score for the next iteration.

The hill climbing process is terminated if a local maximum is reached (a new iteration produces the same sequence as the previous one) or when the maximal number of iterations were performed.

Final Evaluation: In order to evaluate the change in the codon usage bias of the modified sequence with respect to each host in the community, we offer two evaluation scores: "Weakest Link Score" and the general "Optimization Index".

Weakest Link Score: This score focuses on the margins of the groups of wanted and unwanted hosts. For each host a, the number of standard deviations by which the original sequence's CUB score is distant from the modified sequence's CUB score is calculated, according to the following formula:

$$dist(a) = \frac{(CUB_a(X') - CUB_a(X))}{\sigma_a} \tag{9}$$

where X is the original input sequence, X' is the modified sequence returned as output by the model and σ is the standard deviation of the CUB scores of the host's proteome. The weakest link score is defined as the weighted difference between the distance of the host that was least optimized from the group of wanted hosts and the host that was least de-optimized from the unwanted hosts group, and is calculated as:

$$weakest\ link = \alpha \cdot min_{a \in A}(dist(a)) - (1 - \alpha) \cdot max_{b \in B}(dist(b)) \tag{10}$$

where A is the set of wanted hosts, B is the set of unwanted hosts and a is the tuning parameter. Any positive score indicates that the margins of the two groups are separated, meaning that the organism that was least optimized is still superior to all of the de-optimized organisms.

Optimization Index: This score describes the average "optimization distance" between the wanted and unwanted hosts groups. The average distance score is defined as the weighted difference of the mean optimization distances, between the group of hosts in which gene expression is optimized and the group in which it is de-optimized

$$average\ distance\ score = \alpha \cdot mean_{a \in A}(dist(a)) - (1 - \alpha) \cdot mean_{b \in B}(dist(b))$$

$$= \alpha \cdot \frac{\sum_{a \in A} \frac{(CUB_a(X') - CUB_a(X))}{\sigma_a}}{|A|} -$$

$$(1 - \alpha) \cdot \frac{\sum_{b \in B} \frac{(CUB_b(X') - CUB_b(X))}{\sigma_b}}{|B|} \tag{11}$$

A is the set of wanted hosts, B is the set of unwanted hosts and α is a tuning parameter for calibrating the ratio of modification in wanted hosts and unwanted hosts. A positive score can mean higher CUB values with respect to optimized organisms in comparison to the original values, lower CUB values with respect to deoptimized organisms compared to the original value, or both. The higher the index, the more selective is the optimization.

2.2 Transcription Optimization

During transcription of genes in prokaryotes, cellular machinery recognizes promoters, which are found upstream to a gene, and recruit supporting TFs in order to allow RNA polymerase to start transcription. The "core promoters" are defined as the exact area that the sigma factor, a component of the bacterial RNA-polymerase [18] binds to, and they are quite universal among bacteria. However, the regions which are more upstream contain additional sites recognized by TFs. In this work, we have chosen to define the promoters' sequences as the first 200 base pairs upstream to the ORF. Different organisms utilize different TFs in order to promote the transcription initiation, as each of these factors recognizes different sets of genomic sequences described as "motifs".

Motifs can have variable sites, where nucleotides can be one of several options, therefore are usually represented by a Position-Specific Scoring Matrix (PSSM) – a matrix of size $4 \times L$ (where L is the motif's length), containing the probability of each nucleotide to appear in each position in the motif. The calculation of a PSSM assumes independence of the motif's sites from one another, and prohibits insertions or deletions.

Motif Discovery: The STREME (Sensitive, Thorough, Rapid, Enriched Motif Elicitation) software tool uses a hidden Markov model (HMM) of a specified order (in this case, third order) to scan the query sequences for enriched motifs when compared to a set of control sequences, up to a certain significance threshold ($P - value$ 0.05 in this case) [19]. The relevant motifs for this purpose are motifs that are related to transcription, and are uniquely present in the promoters of the wanted hosts compared to the unwanted hosts (Fig. 3B). In order to satisfy these two requirements, two sets of motifs were searched (both defined to be 6–20 bp long):

Transcription Enhancing Motifs: To ensure that the motif is related to transcription activation, motifs were searched for each wanted host with the promoter set as the primary input (i.e. the sequence set where motifs should be enriched) and the intergenic sequences, defined as all sequences on the same strand that neither belong to the ORF nor to the promoters' sequences, as the control (background). The discovered motifs are common in sequences associated with gene expression, which likely indicates that the motifs themselves might have a desirable regulatory role.

Selective Expression Motifs: Represents signals present in the wanted hosts, compared to the unwanted hosts, used to find which of the transcription enhancing motifs will not promote transcription in unwanted hosts. To achieve this goal, motifs are searched from the third most highly expressed (inferred from expression data or CUB measurements) promoters of each wanted host compared to the promoter of each one of the unwanted hosts. For an input of n wanted hosts and m unwanted hosts, $m + 1$ sets of motifs are created for each optimized organism, resulting in $n(m + 1)$ motif sets overall.

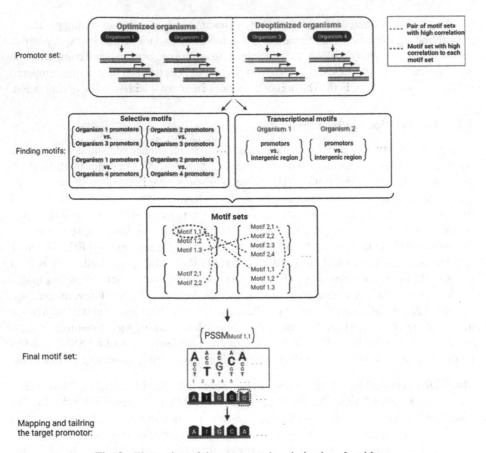

Fig. 3. Illustration of the promoters' optimization algorithm.

Construction of a Single Motif Set: In a simplified model, with only two hosts- one wanted and one unwanted, the solution is relatively instinctive: a new set of motifs that were discovered in both runs and appear in both motif sets is constructed considering the "tuning parameter" (Sect. 2). The measurement used to quantitate motif similarity is spearman correlation, which is calculated between a pair of PSSMs, and the final set is composed of selective motifs that were proven to be correlated to transcription enhancing properties, represented by their corresponding motif set. This reasoning is used in order to translate the simplified dual host algorithm into a set of analogous heuristics for a complete microbiome:

Let A be the set of wanted hosts, and B the set of unwanted hosts, S_{xy} the set of selective motifs that are enriched in organism x compared to organism y, S_x the set of transcription enhancing motifs for bacteria x, tp is the tuning parameter.

Calculate Two Thresholds: Three considered thresholds ($th1$, $th2$, $th3$) are used. $th1$ controls the basic amount of similarity between motifs to consider, $th2$ the stringency of the transcription enhancing motifs, and $th3$ the stringency of the selective motifs (results appear to be more accurate for $th1$~0.05, th2,3~0.3).

$$D_1 = th1 + tp * th2, \quad D_2 = th1 + (1 - tp) * th3 \tag{12}$$

Define a set C: union of all transcriptional motifs for the wanted hosts, $C = \bigcup_{x \in A} S_x$
Initialize set F: an empty set, used as the final motif set, $F = \varnothing$
Assign to set F: In each S_x set where x is in A, there exists a motif m' such that its Spearman correlation score with m is higher than D_1:

$$\forall x \in A, \quad \exists m' \in S_x s.t. Spearman(m, m') \geq D_1 \tag{13}$$

In each S_{xy} set where x is in A and y is in B, there exists a motif m' such that its Spearman correlation score with m is higher than D_2:

$$\forall x \in A, y \in B, \quad \exists m' \in S_{xy} s.t. Spearman(m, m') \geq D_2 \tag{14}$$

If both of the above statements are true, add m to F.

Promoter Selection and Tailoring: MAST (Motif Alignment and Search Tool) [20], is used to align the final motif set, F, to a group of promoters estimated to be the top quartile in terms of gene expression in the wanted hosts (using gene expression data, or estimations based on CUB scores of the following genes). The promoters are then ranked based on the Expect Value (E-value) of those alignments, considering the initial significance of the motif and the quality of the alignment (Fig. 3D). After identifying the best promoter candidates, which contain motifs that help transcription initiation exclusively in the wanted hosts, mismatches between a mapped motif and the promoter are then individually fixed, further tailoring and designing the promoter to fit the desired requirements (Fig. 3E).

2.3 Editing Restriction Site Presence

Restriction enzymes are part of the natural immune system of bacteria, they digest foreign DNA by cleaving specific areas determined by the nucleotide sequence. The restriction enzyme footprint of different bacteria is unique; therefore, different bacterial organisms contain distinct sets of restriction enzymes. The goal of this component is to avoid digestion of the sequence in the wanted hosts, and try to induce digestion in the unwanted hosts.

Once a sequence is digested, the flanking regions of the sequence may also rejoin in specified bacterial chromosomal repair mechanisms [21], therefore the region in which the sites are introduced has an impact on the probability of the digestion removing the unwanted modification. The other consideration to take into account is the usual tradeoff: the topological complexion and scale of effect of the genetic element, reflecting on the ability to predict and design changes in the corresponding expression process. As result, this work focuses on the modulation of restriction sites in the ORF, which has a relatively simple topology and therefore the best ability to accurately predict the effect of mutations and changes in the sequence.

Detection of Relevant Restriction Sites: The REbase version 110 [22] was utilized to find all relevant restriction sites for the wanted and unwanted hosts that can exist (allowing synonymous changes) in the ORF itself.

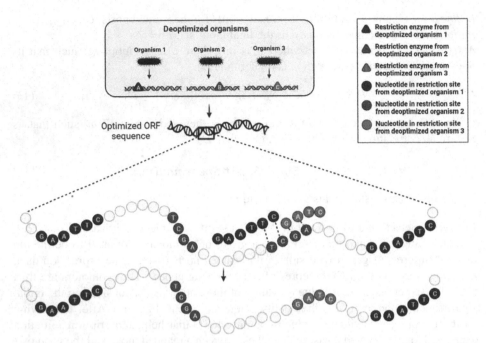

Fig. 4. Restriction sites algorithm. The restriction enzymes (triangles) are extracted from the wanted and unwanted hosts respectively; the recognition sites of the enzymes are illustrated by circles.

Insertion of Sites Originating from Unwanted Hosts: Insertion of sites that overlap can and usually does disrupt them. The main goal is to maximize the amount of sites present in the ORF, while considering the unwanted host they will be recognized by, aiming to increase the overall probability of digestion in the largest number of unwanted hosts. Therefore, the following method is used:

First, the location of potential restriction sites along the ORF is mapped, and sites that do not cause conflicts (non-overlapping) are inserted. For overlapping sites: in each overlapping position, greedily insert the site that increases the number of different organisms in the group of unwanted hosts that have a site in the sequence. If there are a couple of options, pick the site originating from the organism with the smallest amount of restriction sites (Figs. 4, 5).

Avoidance of Sites Originating from Wanted Hosts: In order to avoid sites present in the wanted hosts (again, allowing only synonymous changes), we used the corresponding module in the DnaChisel [23] software tool. It is important to note that the order of the second and third steps matters, as a possible byproduct of insertion of a site is the creation of a new restriction site that might be present in the wanted hosts, which might accidentally increase digestion in them, and should therefore be avoided.

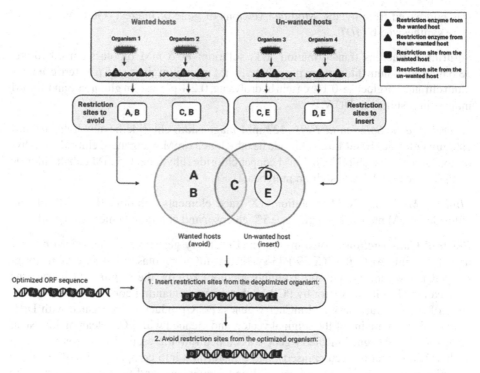

Fig. 5. Illustration of restriction site insertion.

2.4 Data Curation for *In-Silico* Analysis

The selected microbiome for model analysis is a sample of the *A. thaliana* soil micro-biome [24], which contained taxonomic lineages and 16S rRNA sequences. The annotated genomes were selected by running the 16S sequence against the BLAST rRNA software [25] (lower threshold for percent identity of the 16S rRNA sequence is 98.5%). As previously mentioned, these algorithms are designed to work with metagenomically assembled genomes in general.

Additionally, the gene used as a target for optimization is the ZorA gene, which serves as a phage resistance gene as part of the Zorya defense system, inferred to be involved with membrane polarization and infected cell death [26]. This gene can be used in a wide array of sub-populations for various different purposes, showcasing the flexibility of this framework.

2.5 In-vitro Methods

Materials and Plasmids: PCR master mix, *DpnI*, Gibson Assembly kit, PCR cleaning kit, competent *E. coli* and plasmid miniprep kit were purchased from *NEB*. *E. coli k-12, B. subtilis PY79* and AEC804-ECE59-P43-synthRBS-mCherry plasmid were kindly provided by Prof. Avigdor Eldar (Tel-Aviv University, IL). Agarose for DNA electrophoresis, Chloramphenicol, M9 minimal media and 96-well black plates were purchased from *Sigma*. LB and agar were purchased from *BD Difco*, and Ethidium Bromide solution was purchased from hylabs. Modified versions of gene of interest (GOI) and primers were synthesized by *IDT*.

Solutions: Bacillus transformation (BT) solution: 80.5 mM dipotassium dihydrate, 38.5 mM potassium dihydrogen phosphate, 3 mM trisodium citrate, 45 μM ferric ammonium citrate, 2% glucose, 0.1% casein hydrolysate, 0.2% potassium glutamate and 10mM magnesium sulfate in DDW.

Trace Elements Solution (x 100): 123 mM magnesium chloride hexahydrate, 10 mM calcium chloride, 10 mM iron chloride hexahydrate, 1 mM manganese chloride tetrahydrate, 2,4 mM zinc chloride, 0.5 mM copper chloride dihydrate, 0.5 mM cobalt chloride hexahydrate and 0.5 mM sodium molybdate.

Minimal Medium: 1X M9 solution, 1X trace elements solution, 0.1 mM calcium chloride, 1 mM magnesium sulfate, 0.5% glucose, and chloramphenicol (5 μg/mL).

Plasmid Construction: Software-designed mCherry genes were synthesized by IDT and cloned into AEC804-ECE59-P43-synthRBS-mCherry plasmid, to replace the original mCherry gene via Gibson assembly method. Briefly, the original mCherry gene was excluded from the vector by PCR, with primers containing complementary tails to each of the software-designed mCherry genes. PCR products were treated with DpnI to degrade the remains of the original vector, and cleaned with PCR cleaning kit. Next, each software-designed mCherry gene was cloned into the vector by Gibson assembly with 1:2 molar ratio (vector: insert), and transformed into competent *E. coli*. Positive colonies were confirmed by colony PCR and sequencing, and the new plasmids were extracted with miniprep kit.

Bacterial Transformation: All plasmids harboring the modified mCherry genes were separately transformed into competent *E. coli k-12* following the standard protocol, and into *B. subtilis PY79*. For the latter, one bacterial colony was suspended in BT solution (see *solutions*) and grew at 37 °C for 3.5 h. Then, the plasmid was added to the bacterial solution (1 ng/1 uL), and following 3 h incubation, bacteria was spread over pre-warmed agar plates.

Fluorescence Measurement Assay: For each tested mCherry gene, a single colony containing the modified plasmid was grown overnight in LB medium. Then, bacterial suspension was centrifuged and resuspended in PBSx1 twice. Following the second wash, the bacterial suspension was centrifuged again, and the pellet was resuspended in minimal medium (see *solutions*). The bacterial suspension was allowed to grow for 4 h. Then, bacteria were diluted with minimal medium to obtain an OD600 nm of 0.2, loaded into a 96-well plate and grew for 17 h at 37 °C with continuous shaking. Fluorescence (ex/em: 587/610 nm) and bacterial turbidity (at OD600 nm) were measured every 20 min. Each sample was tested in triplicates at three independent experiments.

Computational Log-Phase Detection: Growth curves (OD600nm) were plotted over time, and linearity that represents this phase was detected by sequential removal of the last point in the linear phase. Then, a linear trendline was fitted to the curve, and if the removal of the point increased the slope of the curve, that point was considered not part of the log phase. These iterations were conducted continuously until 1/8 of the graph is left or if two iterations did not change the calculated slope.

Statistical Analysis: We calculated P-values with a permutation test. Briefly, for every optimization, the three experiments from the same organism were averaged and a difference between *E. coli* and *B. subtilis* was calculated. Then, splitting, averaging, and distance calculations were performed, to assess if the separation between *E. coli* and *B. subtilis* is significant. The P-value is defined as the percent of splits in which the difference between the two is larger than the difference between the original split.

3   Results

The effect of the engineering process on the modulated genetic information was examined in two different scales. Firstly, the theoretical scale up of these models was questioned, understanding the applicability of the models in microbiomes of different sizes and degrees of diversity. This analysis was executed using the curated example of *A. thaliana* soil microbiome (discussed in Sect. 2.4). Secondly, the in-vivo effect of the alterations was quantified for a test case of two model bacteria as hosts.

3.1   Editing Restriction Site Presence

In order to look at generic effects and applications, the analyzed data included all the bacterial species documented in the REbase, creating synthetic sub-microbiomes of different sizes, optimizing the selected gene according to their composition, and analyzing the site presence in the final sequence (Fig. 6):

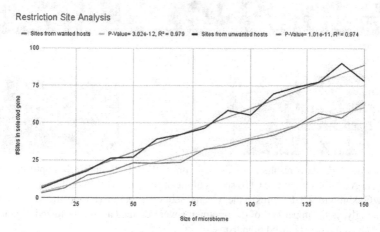

Fig. 6. Presence of different restriction sites and their origin. Random samples of different sizes (between 10 and 150), were randomly split into two equally sized groups of wanted and unwanted hosts. Afterwards, the restriction site editing algorithm (described in Sect. 2.3) was applied, and the final amount of restriction sites related to the unwanted group and the wanted group (y-axis) were tested. This process was completed 10 times for random microbiomes of each size, and the final result is the average measure of those runs.

The first, and most trivial finding is that the amount of sites present in the engineered sequence increases linearly (Pearson correlation of 0.97) with the size of the microbiome (whether if they originate from the wanted or unwanted hosts).

It is important to note that some of the sites recognized by the wanted hosts are present in the final sequence due to inability to satisfy all requirements. If a restriction enzyme is relatively promiscuous, there are many sites to avoid, the amino acid sequence has less synonymous degrees of freedom (synonymity), not all the requirements can be fulfilled and the correct translation is prioritized. Moreover, in all cases the algorithm was able to produce a sequence that has more sites originating from the unwanted hosts compared to the wanted hosts, and the gap between the number of sites from each origin increases with the microbiome size. That important discovery is the ability of this algorithm to function efficiently even (and maybe specifically) for larger and more complex microbiomes.

3.2 Translation Efficiency Modeling

The performed testing aims to test connections between the size and diversity of the curated microbiome (as described in Sect. 2.4) to the final optimization index.

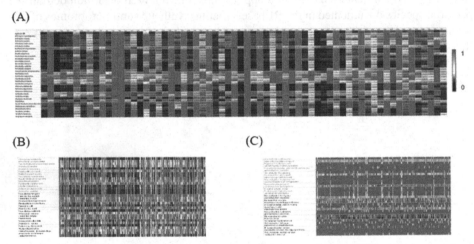

Fig. 7. Figures A–C show results from a single run of the translation (CUB) optimization algorithm. Figure A exhibits the innate CUB scores of the organisms in the microbiome, for all codons. Figures B, C exhibit the CUB scores of the selected sequence, before (B) and after (C) optimization correspondingly, as the upper half of the organisms were defined as the optimized organisms and the lower half as the deoptimized organisms.

As seen in Fig. 7A, the general CUB of the microbial species is relatively diverse, exhibiting the availability of this degree of freedom in the coding of the open reading frame. The initial comparison of the CUB scores of the sequence before (Fig. 7B) and after (Fig. 7C) optimization shows that in this case, there has been an overall global optimization of the scores. However, the improvement was much more substantial for the upper group of wanted hosts compared to the unwanted hosts, thus exhibiting the initial aim of the algorithm.

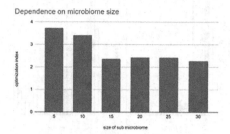
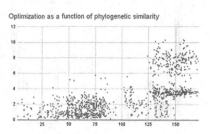

Fig. 8. A (left) tests the application of the algorithm on different sub-microbiome sizes (10 different, random splits of each size), and figure B (right) shows the correlation between phylogenetic distance (defined as the number of different characters in the 16S rRNA sequence) and optimization results.

The next interesting factor was the effect of different sub-microbiome sizes on the optimization capabilities of the ORF tailoring algorithm (Fig. 8A). In general, the optimization score remains high and relatively similar for all microbiome sizes, and the decline in the optimization score is not drastic. The standard deviation (std) becomes smaller as the microbiome has more and more species, an effect caused by two posed conditions as expected.

Comparison of the dependence of the optimization score on phylogenetic distance yielded fairly interesting results (Fig. 8B). In order to perform this analysis, every species was compared to every other species in the microbiome, when one is defined as the wanted host and the other as the unwanted host and vice versa, creating "sub-microbiome couples". On one hand, there is clear correlation between the evolutionary distance (determined by differences in the 16S rRA sequence) and the performance of the algorithm (0.737 spearman correlation). However, even for the 10% closest couples have an average optimization score of 1.215, with a std of 0.8, which is a relatively significant optimization of over 20% considering the high degree of evolutionary closeness. Overall, the analysis results indicate that the optimization process is likely significant for realistic microbial diversities and complexities.

3.3 Transcription Optimization

In order to test the significance of the transcription enhancing motifs and selective-expression motifs found by the motif discovery process, we have compared the results of a MAST run for each set of motifs.

In order to avoid bias of match results towards longer sequences, intergenic regions longer than 200 bp were normalized to 200 bp long sequences using a sliding window with a window size of 200 bp and steps of 1 bp. This modification accounts for the differences in the counts of intergenic sequences compared to promoter sequences.

As can be seen from Fig. 9A and Fig. 9B, motifs generated by both intergenic and selective runs have better match scores (indicated by lower E-value score) to the

(A)   (B)

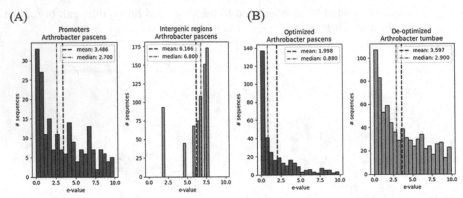

Fig. 9. E-value histogram from a single MAST run of transcription enhancing motif set and selective expression motif set. Figure A contains E-value scores of promoter sequences and sequences of intergenic regions of Arthrobacter pascens when using transcription enhancing motifs generated by a MAST run with promoter sequences defined as primary set and intergenic sequences as control set. Figure B contains E-value scores of promoter sequences of a wanted host Arthrobacter pascens and promoter sequences of an unwanted host Arthrobacter tumbae when using selective expression motifs generated by a MAST run with wanted host's promoter sequences defined as primary set and unwanted host's promoter sequences as control set. In both A and B, the mean and median E-values of the sequences in the primary set are lower than the mean and median of the control set, with a P-value of 3.17e-32 and 4.628e-23, respectively.

sequences in the primary set compared to the control set. This evidence supports the approach implemented in our transcription module, that motifs can indeed be used to find adequate promoter sequences that will promote transcription in wanted hosts, while eliminating their effect on gene expression in unwanted hosts.

3.4 In-vitro Results

In the fluorescence measurement assay, we investigated the validity of our software in altering gene expression by modifying the ORF of the mCherry gene. Several versions of ORF were designed by our software to be optimally expressed in *B. subtilis* and deoptimized for expression in *E. coli*, cloned into a plasmid, transformed into both bacteria, which were grown separately. The expression level as well as bacterial growth of each version were measured for 17 h and compared to the unmodified mCherry.

Altering gene of interest's ORF hinders the growth of deoptimized bacteria: at first, we tested the impact of software-designed GOIs (of mCherry gene) on bacterial growth, when GOI was optimized or de-optimized for expression in *B. subtilis* and *E. coli* respectively. Both bacteria that expressed five versions of the GOI were grown (as described in Sect. 2.5) and growth curves for each bacterium-expressing GOI version were plotted against time (Fig. 10A). Detection of the logarithmic phase was performed as mentioned (described in Sect. 2.1), and growth rates depicted by the slopes of linear trendline were calculated. While growth rates and maximal turbidity of *B. subtilis*, to which ORFs were optimized, were comparable to the original ORF (Fig. 10B and C), in *E. coli,* to which ORFs were deoptimized, both parameters were significantly reduced in ORF versions of tAI-D and TDR-D (Fig. 10B and C). In particular, tAI-D version largely restricted *E. coli's* growth rate by approximately 7-fold, possibly due to ribosomal traffic jams [27]. In this case, ribosomes are stalled during mRNA translation while waiting for the binding of low abundance tRNA anticodon. As a result, endogenous protein synthesis is diminished, which in turn restricts cellular growth. To assess the degree of optimization in *B. subtilis* relative to the deoptimization in E. *coli,* growth rate folds of *B. subtilis* were divided by those of *E. coli.* Higher values indicate successful optimization in *B. subtilis* and deoptimization in *E. coli.* As the graph clearly shows, tAI-D is by far optimized for *B. subtilis* and deoptimized for *E. coli* in terms of growth rates (Fig. 10D).

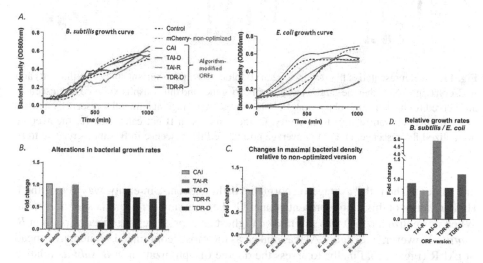

Fig. 10. ORF modification alters the growth of deoptimized bacteria (according to optimizations in appendix 1). A. Representative growth curves for B. subtilis (left) and in E. coli (right). Control (black dashed curve) stands for bacteria containing the same plasmid backbone that lacks the mCherry gene. mCherry (red dashed curve) is the original (unmodified) version of the gene, and CAI, tAI-D, tAI-R, TDR-D, and TDR-R are modified versions of mCherry gene. B. Fold change in bacterial growth rates of each ORF version relative to the growth rates in mCherry. C. Same as in B but calculated for the average maximal density. D. Fold of growth rates in B. subtilis relative to E. coli.

Correlation Between Expression Level and Model Performance: Next, we evaluated whether GOI expression levels coincided with our software predictions. The fluorescence intensity of the selected GOI mCherry, which reflects its expression levels, increased or decreased substantially in accordance to the optimized/deoptimized direction, as observed in *B. subtilis* and *E. coli* respectively (Fig. 11A). In *B. subtilis,* the tAI-D version exhibited the highest average maximal fluorescence intensity compared to the original version, while in *E. coli,* the TDR-R version exhibited the lowest (Fig. 11B).

In order to account for the alterations in fluorescence intensity to the software-designed modification in ORF sequence code and not to differences in bacterial density, it was normalized by the ratio of fluorescence intensity per bacterial density (OD600

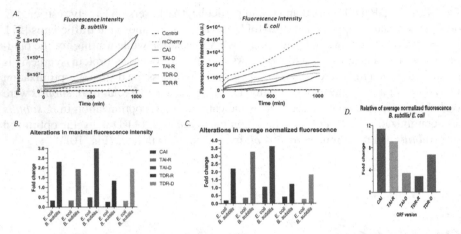

Fig. 11. A. Representative fluorescence intensity plots of all ORF variants in B. subtilis (left) and in E. coli (right) Note that the control lacked mCherry gene, and thus wasn't exhibited fluorescence, and served for background subtraction. B. Fold change in average maximal fluorescence intensity of each ORF version relative to mCherry. C. the same as in B but calculated for the average normalized fluorescence. D. Fold of average normalized fluorescence in B. subtilis relative to E. coli.

nm values). Then, the average of normalized fluorescence intensity was determined. Higher values of this ratio represent greater GOI expression per bacteria and vice versa. When normalized, tAI-D was still ranked highest as expressive-optimized version in *B. subtilis*. However, in *E. coli*, CAI was ranked as the most deoptimized version, just ahead of tAI-R (Fig. 11C). Finally, to assess the degree of optimization in *B. subtilis* relative to the deoptimization in *E. coli*, fold of the averaged normalized fluorescence of *B. subtilis* were divided by those of *E. coli* (Fig. 11D). In that way, we determined that GOI versions of CAI and tAI-R were the promising codon usage bias scores.

4   Discussion

In this work, we have introduced a novel approach to microbiome engineering, and exhibited its possibilities both computationally and experimentally. By appropriately designing the different elements related to gene expression, we were able to create a comprehensive, automatic and generic approach that can easily be applied for any given consortium of bacteria. By investing in the adjustment of each genetic element which influences gene expression, we were able to provide a comprehensive view that considers effects caused by interactions between the new genetic information and the existing population in an environment, and the new inter-microbiome interactions along with their corresponding fitness effect.

The current design approach deals with the three main processes related to gene expression: entry into the cell, transcription, and translation. First, entry into the bacterial cell is modulated by editing the presence of restriction sites, increasing chances of digestion upon entry of the plasmid into an unwanted host compared to a wanted host. Next, the transcription process is optimized by discovery of genetic motifs which are likely linked to TFs which are present explicitly in the wanted hosts and are related to transcription initiation. Lastly, the translation process includes re-coding of the ORF based on translation efficiency modulation by exploitation of the degree of freedom posed by the redundancy of the genetic code.

In order to test our models, we performed both in-silico and in-vitro tests. The in-vitro tests have shown the effect of the optimization process on the expression rate of a selected protein (Sect. 2.5), which was both higher for the wanted host (*B. subtilis*) and lower for the unwanted host (*E. coli*) simultaneously. Moreover, the attached fitness effect is just as important- while the modified sequence does not pose a significant burden compared to the initial sequence in the wanted host, this cannot be said for the presence of the plasmid in the unwanted hosts. These results can be further strengthened and examined by conducting a co-culture experiment, in order to observe the population-dynamics as well.

The clear fitness decrease in the unwanted hosts caused by the optimization process might not have that much of a significant effect in lab conditions, however the meaning of this change is that when dealing with an actual microbiome, there will be a stronger evolutionary pressure against existence of the plasmid in the unwanted hosts, thus further propagating the designed expression differentiation in the microbiome. The in-silico analysis supplied complementary views on each one of the computational methods individually, with the scale up process (from two organisms to an entire microbiome) defining the relevance of the different tests.

4.1 Future Plans

The purpose of this work is to showcase this novel approach, suggesting to individually adjust expression-related genetic elements in order to perform microbiome engineering. In order to increase the inclusivity of this approach, the following components should be added to the arsenal of existing models:

Origin of Replication (ORI): The origin of replication is the genetic element that promotes replication of the genetic material, thus it has the largest effect on the long-term interactions between the plasmid and the microbiome. The ability to fit the ORI to only some of the organisms in the microbiome should significantly improve our tool. The main challenge for this model will be understanding the innate topology of this sequence in different organisms, which is highly complex and variable.

Clustered Regulatory Interspaced Repeats (CRISPR) Memory: Another potentially very efficient approach to differentiate between organisms is based on CRISPR memory. Such an approach can be based on sequencing the regions where *CRISPR* is encoded in all the organisms in the community. At the next step, the target plasmid is designed such that it will include target sub-sequences of the *CRISPR* systems of the unwanted organisms but will not include target sub-sequences of the *CRISPR* systems of the wanted organisms.

miRNA and Other mRNA Binding RNA Genes: These types of genes bind to the mRNA and cause the down regulation of protein production rate. By estimating the levels of such RNA genes in the different organisms in the community and modeling the patterns in the mRNA that these RNA genes recognize, we can develop an additional relevant tool for differentiation. In this case, we will introduce to the target plasmid only sequences that are recognized by RNA genes that appear in the unwanted organisms and we will delete from it sub-sequences that are recognized by organisms from the wanted organisms.

Translation Initiation: The process of binding the ribosome to the ribosome binding site (RBS) on the mRNA and initiating translation requires a separate model, with consideration related to the distance between the RBS and the translation start site (TSS), and the local folding of the mRNA near it.

Multiple Optimizations: The next step to this strategy would be to generate not only one optimized sequence but an array of optional solutions, each designed to have suboptimal expression in all unwanted hosts, while simultaneously having higher expression in the wanted hosts. This addition might be a key element to enable optimal design of a plasmid meant to fit a diverse array of hosts.

Validations: In order to gain more confidence in the models and approach in general, we hope to test the functionality and effect over time in a real (or at least synthetic) microbiome. One challenge that rises as a result, is the lack of appropriate measurement to quantify the success in expression differentiation.

The OIL-PCR method by Peter J. Diebold et al. [28] allows us to assess the abundance of bacteria that contain the plasmid-encoded gene of interest (GOI) in a microbial community, which potentially reveals if our approach is able to restrict GOI abundance governed by HGT events. In the future, we intend to clean the fused product (GOI-16S rRNA) and amplify it with a specific set of forward primers targeting the fusion regions and species-specific reverse primer, targeting a variable region of the 16s rRNA gene via qPCR to identify the bacterial species. Additionally, we are planning to expand this method to combine GOI abundance and its expression by applying the same principles of GOI-abundance method but based on fusion-PCR of mRNA products of GOI and 16s rRNA genes. For this purpose, the fusion-PCR step will include primers targeting GOI's mRNA, with a reverse primer having a tail complementary to the mRNA of 16s rRNA and reverse transcriptase to establish the fused product. Potentially, this method can detect both the host identity and its GOI expression levels.

4.2 Applications

Some of the main advantages of this technology are driven by its ability to function generically and automatically for a given microbiome and selected gene. As result, it may be applicable in many different settings and environments, which have been divided into the following three categories:

Bio-sensing: Microbiomes constantly react to changes in the environment, meaning that they both "know" how to sense it and how to respond [29]. Utilization of these innate mechanisms can only be done by integrating new and targeted genetic information that has an easily readable output. More specifically, the sensing process could be targeted towards a specific bacterial species or a metabolite, as long as the gene circuit itself is designed correctly. The advantage of such selective optimization process is that expression of a biomarker in certain species could emit a readable signal for their presence, or the sensing of biomolecules could be performed by the correct species to increase signal to noise ratio (SNR) that could arise from sensing of incorrect signals by other species [30–34].

Bio Production/Bio Degradation: In addition to sensing, the functional influence of the microbiome, the functional effect of the microbiome on the environment is facilitated throughout collective synthesis and degradation of biomolecules. This additional, innate capability could be utilized in various settings including oil bioremediation, appropriate food tech scenarios, metabolic engineering, and more [35–41].

Microbiome Specific Therapy and Enhancement: The highest complexity level foreseen for the approach suggested here includes adding a missing capability to the system through the microbiome, creating "microbiome specific therapy". The advantage of selectivity and fitting here is more than clear- both in terms of functionally and in terms of safety. The main examples include human microbiome therapy [9, 42, 43], and soil microbiome therapy [44–46].

Acknowledgments. This work was continuously guided by the following researchers: Prof. Martin Kupiec, Professor Uri Gophna, and Professor Itai Benhar for valuable feedback and suggestions. Prof. Hagit Eldar-Finkelman and her lab members, which fully supported us with physical research facilities. Prof. Avigdor Eldar and PhD candidate Tasneem Bareia for both bacterial strains and plasmids, and in general advice. The authors thank the Faculties of Engineering and Life Science at Tel Aviv University for funding this research. This study was part of the TAU IGEM 2021 project (https://2021.igem.org/Team:TAU_Israel) and was also supported by Edmond J. Safra Center for Bioinformatics at Tel Aviv University.

References

1. Steinfeld, H.M.: Length of life of Drosophila melanogaster under aseptic conditions. Dissertation Thesis, 1927, University of California, Berkeley (1928)
2. Dekaboruah, E., Suryavanshi, M.V., Chettri, D., Verma, A.K.: Human microbiome: an academic update on human body site specific surveillance and its possible role. Arch. Microbiol. **202**(8), 2147–2167 (2020)
3. Appleton, J.: The Gut-Brain axis: influence of microbiota on mood and mental health. Integrat. Med. (Encinitas, Calif.), **17**(4), 28–32 (2018)
4. Reed, H.E., Martiny, J.B.H.: Testing the functional significance of microbial composition in natural communities. FEMS Microbiol. Ecol. **62**(2), 161–170 (2007)
5. Suskind, D.L., et al.: Fecal microbial transplant effect on clinical outcomes and fecal microbiome in active Crohn's disease. Inflamm. Bowel Dis. **21**(3), 556–563 (2015)

6. Bajaj, J.S., et al.: Fecal microbial transplant capsules are safe in hepatic encephalopathy: a phase 1, randomized. Placebo-Controlled Trial. Hepatology **70**(5), 1690–1703 (2019)

7. Vaughn, B.P., et al.: Increased intestinal microbial diversity following fecal microbiota transplant for active crohn's disease. Inflamm. Bowel Dis. **22**(9), 2182–2190 (2016)

8. Martin, A., Anderson, M.J., Thorn, C., Davy, S.K., Ryan, K.G.: Response of sea-ice microbial communities to environmental disturbance: an in situ transplant experiment in the Antarctic. Mar. Ecol. Prog. Ser. **424**, 25–37 (2011)

9. Foo, J.L., Ling, H., Lee, Y.S., Chang, M.W.: Microbiome engineering: current applications and its future. Biotechnol. J. **12**(3), 1600099 (2017)

10. Koonin, E.V., Makarova, K.S., Aravind, L.: Horizontal gene transfer in prokaryotes: quantification and classification. Annu. Rev. Microbiol. **55**, 709–742 (2001)

11. Khosravi, A., Mazmanian, S.K.: Disruption of the gut microbiome as a risk factor for microbial infections. Curr. Opin. Microbiol. **16**(2), 221–227 (2013)

12. Sharp, P.M., Li, W.H.: The codon adaptation index–a measure of directional synonymous codon usage bias, and its potential applications. Nucleic Acids Res. **15**(3), 1281–1295 (1987)

13. Parmley, J.L., Hurst, L.D.: How do synonymous mutations affect fitness? Bioessays: news and reviews in molecular. Cell. Dev. Biol. **29**(6), 515–519 (2007)

14. Comeron, J.M., Aguadé, M.: An evaluation of measures of synonymous codon usage bias. J. Mol. Evol. **47**(3), 268–274 (1998)

15. Victor, M.P., Acharya, D., Begum, T., Ghosh, T.C.: The optimization of mRNA expression level by its intrinsic properties-Insights from codon usage pattern and structural stability of mRNA. Genomics **111**(6), 1292–1297 (2019)

16. Bahiri-Elitzur, S., Tuller, T.: Codon-based indices for modeling gene expression and transcript evolution. Comput. Struct. Biotechnol. J. **19**, 2646–2663 (2021)

17. Sabi, R., Volvovitch Daniel, R., Tuller, T.: stAIcalc: tRNA adaptation index calculator based on species-specific weights. Bioinformatics **33**(4), 589–591 (2017)

18. Paget, M.S.: Bacterial sigma factors and anti-sigma factors: structure: function and distribution. Biomolecules **5**(3), 1245–1265 (2015)

19. Bailey, T.L.: STREME: Accurate and versatile sequence motif discovery. Bioinformatics **37**(18), 2834–2840 (2021)

20. Bailey, T.L., Gribskov, M.: Combining evidence using P-values: application to sequence homology searches. Bioinformatics **14**(1), 48–54 (1998)

21. Wigley, D.B.: Bacterial DNA repair: recent insights into the mechanism of RecBCD, AddAB and AddnAB. Nat. Rev. Microbiol. **11**(1), 9–13 (2013)

22. Roberts, R.J., Vincze, T., Posfai, J., Macelis, D.: REBASE--a database for DNA restriction and modification: enzymes, genes and genomes. Nucleic Acids Res. **38**(Database issue), D234–6 (2010)

23. Zulkower, V., Rosser, S.: DNA Chisel, a versatile sequence optimizer. Bioinformatics **36**(16), 4508–4509 (2020)

24. Bai, Y., et al.: Functional overlap of the Arabidopsis leaf and root microbiota. Nature **528**(7582), 364–369 (2015)

25. Camacho, C., et al.: BLAST+: architecture and applications. BMC Bioinf. **10**, 421 (2009)

26. Azam, A.H., Tanji, Y.: Bacteriophage-host arm race: an update on the mechanism of phage resistance in bacteria and revenge of the phage with the perspective for phage therapy. Appl. Microbiol. Biotechnol. **103**(5), 2121–2131 (2019)

27. Dana, A., Tuller, T.: The effect of tRNA levels on decoding times of mRNA codons. Nucleic Acids Res. **42**(14), 9171–9181 (2014)

28. Diebold, P.J., New, F.N., Hovan, M., Satlin, M.J., Brito, I.L.: Linking plasmid-based beta-lactamases to their bacterial hosts using single-cell fusion PCR. Elife **10**, e66834 (2021)

29. DeAngelis, K.M., Ji, P., Firestone, M.K., Lindow, S.E.: Two novel bacterial biosensors for detection of nitrate availability in the rhizosphere. Appl. Environ. Microbiol. **71**(12), 8537–8547 (2005)

30. Hynninen, A., Virta, M.: Whole-cell bioreporters for the detection of bioavailable metals. Whole Cell Sens. Syst. **II**, 31–63 (2009)

31. Kreniske, J.S., Harris, A., Safadi, W.: Landmines in the Golan Heights: A patient's perspective. Case Reports (2014)

32. Tecon, R., Van der Meer, J.: Bacterial biosensors for measuring availability of environmental pollutants. Sensors **8**(7), 4062–4080 (2008)

33. Belkin, S., et al.: Remote detection of buried landmines using a bacterial sensor. Nat. Biotechnol. **35**(4), 308–310 (2017)

34. Ramanathan, S., Ensor, M., Daunert, S.: Bacterial biosensors for monitoring toxic metals. Trends Biotechnol. **15**(12), 500–506 (1997)

35. Caplice, E.: Food fermentations: role of microorganisms in food production and preservation. Int. J. Food Microbiol. **50**(1–2), 131–149 (1999)

36. Adolfsson, O., Meydani, S.N., Russell, R.M.: Yogurt and gut function. Am. J. Clin. Nutr. **80**(2), 245–256 (2004)

37. Yu, Q., Li, Y., Wu, B., Hu, W., He, M., Hu, G.: Novel mutagenesis and screening technologies for food microorganisms: advances and prospects. Appl. Microbiol. Biotechnol. **104**(4), 1517–1531 (2020)

38. Chen, J., Zhang, W., Wan, Z., Li, S., Huang, T., Fei, Y.: Oil spills from Global Tankers: status review and future governance. J. Clean. Prod. **227**, 20–32 (2019)

39. Barron, M.G., Vivian, D.N., Heintz, R.A., Yim, U.H.: Long-term ecological impacts from oil spills: comparison of Exxon Valdez, Hebei spirit, and Deepwater Horizon. Environ. Sci. Technol. **54**(11), 6456–6467 (2020)

40. Chang, S.E., Stone, J., Demes, K., Piscitelli, M.: Consequences of oil spills: a review and framework for informing planning. Ecol. Soc. **19**(2) (2014)

41. Lawson, C.E., et al.: Common principles and best practices for engineering microbiomes. Nat. Rev. Microbiol. **17**(12), 725–741 (2019)

42. Malla, M.A., Dubey, A., Kumar, A., Yadav, S., Hashem, A., Abd_Allah, E.F.: Exploring the human microbiome: the potential future role of next-generation sequencing in disease diagnosis and treatment. Front. Immunol. **9** (2019)

43. Bakken, J.S., et al.: Treating clostridium difficile infection with fecal microbiota transplantation. Clin. Gastroenterol. Hepatol. **9**(12), 1044–1049 (2011)

44. Malusá, E., Sas-Paszt, L., Ciesielska, J.: Technologies for beneficial microorganisms Inocula used as Biofertilizers. Sci. World J. **2012**, 1–12 (2012)

45. Rodriguez, H., Gonzalez, T., Goire, I., Bashan, Y.: Gluconic acid production and phosphate solubilization by the plant growth-promoting bacterium azospirillum spp. Naturwissenschaften **91**(11), 552–555 (2004)

46. Zhu, J., Li, M., Whelan, M.: Phosphorus activators contribute to legacy phosphorus availability in agricultural soils: a review. Sci. Total Environ. **612**, 522–537 (2018)

Metagenomics Binning of Long Reads Using Read-Overlap Graphs

Anuradha Wickramarachchi[ID] and Yu Lin[✉][ID]

School of Computing, Australian National University, Canberra, Australia
{anuradha.wickramarachchi,yu.lin}@anu.edu.au

Abstract. Metagenomics sequencing enables the direct study of microbial communities revealing important information such as taxonomy and relative abundance of species. Metagenomics binning facilitates the separation of these genetic materials into different taxonomic groups. Moving from second-generation sequencing to third-generation sequencing techniques enables the binning of reads before assembly thanks to the increased read lengths. The limited number of long-read binning tools that exist, still suffer from unreliable coverage estimation for individual long reads and face challenges in recovering low-abundance species. In this paper, we present a novel binning approach to bin long reads using the read-overlap graph. The read-overlap graph (1) enables a fast and reliable estimation of the coverage of individual long reads; (2) allows to incorporate the overlapping information between reads into the binning process; (3) facilitates a more uniform sampling of long reads across species of varying abundances. Experimental results show that our new binning approach produces better binning results of long reads and results in better assemblies especially for recovering low abundant species. The source code and a functional Google Colab Notebook are available at https://www.github.com/anuradhawick/oblr.

Keywords: Metagenomics binning · Long reads · Read-overlap graph

1 Introduction

Recent advancements in sequencing technologies have accelerated microbiome research significantly. Broadly, metagenomics analysis supports the direct study of microbial genetic material from the host environments [3,32]. One fundamental problem that dominates across a wide range of research is the identification and characterization of microbial genetic material. Metagenomics binning specifically determines the species present in a given sample and further supports the downstream functional analysis of identified microorganisms. There exist two main paradigms for metagenomics binning; (1) reference-based binning (*e.g.* Kraken2 [36], Centrifuge [11], MeganLR [7], Kaiju [20], etc.) and (2) reference-free binning (*e.g.* MaxBin 2 [37], MetaBAT 2 [10], VAMB [25], etc.). Reference-free approaches are preferred when unknown species are present or

© The Author(s), under exclusive license to Springer Nature Switzerland AG 2022
L. Jin and D. Durand (Eds.): RECOMB-CG 2022, LNBI 13234, pp. 260–278, 2022.
https://doi.org/10.1007/978-3-031-06220-9_15

the reference databases are incomplete. Typically, short reads from the second-generation sequencing technologies (*e.g.* Illumina, etc.) are assembled into much longer contigs to be binned as longer contigs usually carry more pronounced genomic signals, e.g., the coverage and composition information of contigs. The coverage of an assembled contig is estimated by the aligned reads on this contig whereas the composition information is computed from the normalized ologonu-cleotide frequencies.

Long-read technologies from the third-generation sequencing is continuously gaining popularity [17], especially with the recent introduction of PacBio HiFi and Nanopore Q20+ technologies. As long reads are getting similar to contigs (assembled from short reads) in terms of length and accuracy, it is worth investigating whether long reads themselves can be binned directly before assembly. Note that the contigs binning tools cannot be directly applied to bin accurate long reads due to the absence of a coverage value for each read. MetaBCC-LR [35] and LRBinner [34] are two recent attempts to bin long reads in a reference-free manner. MetaBCC-LR and LRBinner both use k-mer coverage histograms for coverage and trinucleotide frequency vectors for composition. While MetaBCC-LR uses coverage and composition features in two subsequent steps, LRBinner combine coverage and composition features via an auto-encoder. Although these two k-mer based approaches show some promising results in binning long reads, they are highly likely to suffer from unreliable coverage estimation for individual long reads and poor sensitivity for low-abundance species due to imbalance clusters (refer to Fig. 3).

In this paper, we propose a novel binning approach (**OBLR**) to bin long reads using the read-overlap graph. In contrast with MetaBCC-LR and LRBinner, we adopt a novel coverage estimation strategy and a sampling strategy to form uniform clusters assisted by the read-overlap graph. We show that read-overlap graph assists in better estimation of read coverage and enables us to sample reads more uniformly across species with varying coverages. Moreover, the connectivity information in the read-overlap graph facilitates more accurate binning via inductive learning. Experimental results show that our new binning approach produces better binning results of long reads while reducing the extensive resources otherwise required for the assembly process.

2   Methods

Our pipeline consists of 5 steps performing the tasks, (1) building the read-overlap graph, (2) obtaining read features, (3) performing probabilistic sampling, (4) detecting clusters for sampled reads and (5) binning remaining reads by inductive learning. Figure 1 illustrates the overall pipeline of OBLR. The following sections explain each step in detail.

2.1   Step 1: Constructing Read-Overlap Graph

As the first step of the pipeline, we construct the read-overlap graph. The read-overlap graph is introduced to utilize the overlapping information between raw

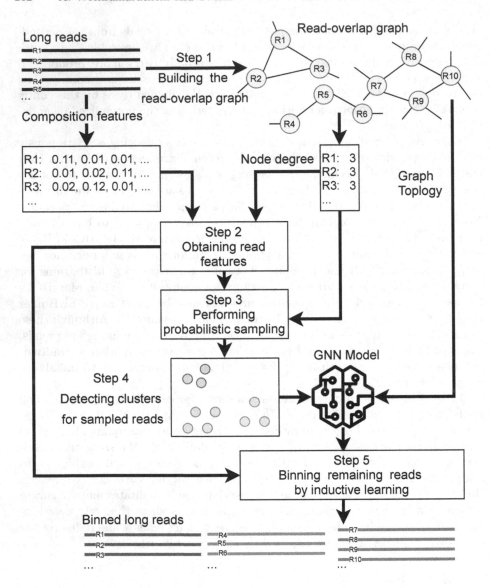

Fig. 1. An overview of the workflow of the proposed pipeline OBLR.

reads. Earlier works have demonstrated that the topology of read-overlap graph can help binning short reads [2] as well as distinguishing between genomes at the strain level [1]. As for long reads, two reads are overlapping (connected by an edge in the read-overlap graph) if and only if their overlapping length is at least $L_{overlap}$ and the overhang length is at most $L_{overhang}$ (computed according to [13]). Note that overhang refers to the region on the sequence that lies along with the aligned sequence, however, does not have matching bases to meet overlap criteria. In our pipeline, we use k-mer bin map (kbm2) program to compute the approximate overlaps between reads. We use the empirically determined values, $L_{overlap} = 2560$ and $L_{overhang} = 512$ as overlap selection criteria in default setting. Note that kbm2 is a sub-routine of the recent assembler wtdbg2 and is extremely fast to detect overlapping reads using k-mer bins without performing pairwise alignment [28]. In the read-overlap graph, each node R_i represents a read while each edge (R_i, R_j) indicates that R_i and R_j are overlapping. We also define $D(R_i)$ as the degree of R_i in this read-overlap graph.

2.2 Step 2: Obtaining Read Features

In our pipeline, we intend to derive read features that incorporate both composition and coverage information of long reads.

The composition information of long reads can be computed as their oligonucleotide frequencies which are shown to be conserved within a given species while being reasonably distinct between species [30,37]. More specifically, we compute a tetra-nucleotide frequency vector for each long read R_i, i.e., $X(R_i) \in \mathbb{R}^{136}$ as there are 136 distinct tetra-mers when combining reverse complements. This vector is used as the composition feature in our pipeline.

The coverage information of long reads usually refers to the coverage of underlying genomes from which the long reads are drawn. This is also important in metagenomics binning as long reads from the same species tend to have similar coverages [25,35]. While such coverage information is usually available for contigs assembled from short reads (as a byproduct of assembly), long reads do not come with their coverage information. However, a read from a high-coverage genome is likely to have more overlaps compared to that from a low-coverage genome. Therefore, it is a natural choice to use the node degree in the read-overlap graph to estimate the coverage of the corresponding read. This choice is supported by Fig. 2 which shows a clear correlation between the node degree in the read-overlap graph and the coverage information of the corresponding read.

In summary, we combine both tetra-nucleotide frequency vector $X(R_i)$ (for composition) and the node-degree information $D(R_i)$ (for coverage) to derive the read feature vector as $XD(R_i) = X(R_i) \times max(1, lg(D(R_i)))$ for each long read R_i. Note that the $max(\ )$ and $lg(\ )$ are introduced to dampen rigorous fluctuations in coverage, especially for low-coverage genomes. Henceforth, $XD(R_i)$ refers to the read features using degree and composition for read R_i.

2.3 Step 3: Performing Probabilistic Sampling

Clustering entire dataset at once can lead to the well-known class-imbalance problem [9] because metagenomics samples consist of species with varying coverages, i.e., imbalance clusters. However, probabilistic down sampling can effectively address this problem [16]. In order to perform such under sampling, we recall the degree information of nodes and use the Eq. 1 to compute the relative probability of sampling R_i. Note that $D(R_i) = 0$ when R_i is an isolated node and helps OBLR to discard chimeric reads. The effect of down sampling is illustrated

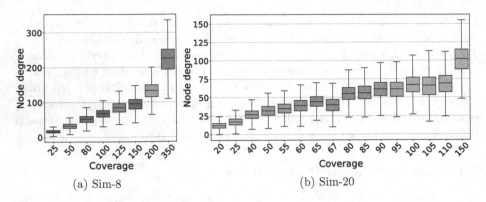

(a) Sim-8   (b) Sim-20

Fig. 2. The correlation between the node degree in read-overlap graph and the coverage information of the corresponding read for **Sim-8** and **Sim-20** datasets.

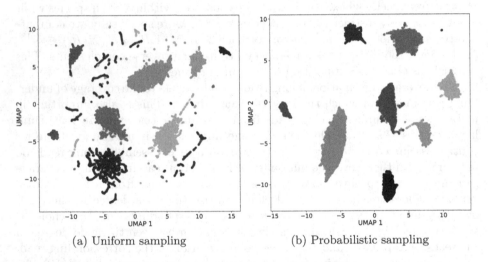

(a) Uniform sampling   (b) Probabilistic sampling

Fig. 3. Comparison of (a) uniform sampling and (2) probabilistic sampling of long reads in **Sim-8** dataset. Different colors corresponds to reads that belong to a unique species.

in Fig. 3. It is evident that clusters after down sampling are of similar sizes and with less isolated points.

$$P(R_i) = \begin{cases} \frac{1}{D(R_i)} & \text{if } D(R_i) \neq 0 \\ 0 & \text{if } D(R_i) = 0 \end{cases} \tag{1}$$

2.4   Step 4: Detecting Clusters for Sampled Reads

We use UMAP [19] to project the sampled reads into lower dimensions and HDBSCAN [18] then is applied to detect clusters for sampled reads. Note than

UMAP is a dimensionality reduction technique that is fast and scalable. HDB-SCAN is a variant of DBSCAN, however it is capable of determining clusters without fixed parameters. Thus, HDBSCAN is more robust in scenarios where the cluster densities can vary significantly. To accommodate long-read datasets with different sizes, 25,000, 50,000, 100,000, 200,000 and 400,000 are used as the sampled number of reads to detect clusters. For each instance, the Silhouette score [27] is computed. We use the sample size with the highest Silhouette score as the chosen sample size for the dataset. This enables us to determine the size to sample from a given dataset in an unsupervised manner. At the end of this step; we have a sample of reads with their bin labels $i.e.$, cluster labels.

2.5   Step 5: Binning Remaining Reads by Inductive Learning

As the read-overlap graphs may contain millions of nodes, the classic label propagation approaches face scalability issues to bin the remaining reads [15]. Therefore, OBLR employs GraphSAGE [6] to bin the remaining reads into the identified clusters in the previous step. GraphSAGE is a Graph Neural Network (GNN) architecture and has been designed to perform inductive learning using large-scale graphs [6]. GraphSAGE can be represented as a layer in a GNN that aggregates the neighborhood features to represent the features of a node itself. Formally, the l-th layer can be formulated according to Eqs. 2 and 3 [38].

$$a_i^{(l)} = \text{Mean}(h_j^{(l-1)} : j \in \mathcal{N}(R_i)) \tag{2}$$

$$h_v^{(l)} = \text{Concatenation}(h_v^{(l-1)}, a_v^{(l)}) \tag{3}$$

where $h_i^{(l)}$ is the feature vector of node R_i at layer l. Note that $h_v^{(0)}=XD(R_i)$ and $\mathcal{N}(R_i)$ represent neighbors of node R_i. While GraphSAGE supports arbitrary aggregate functions, we choose the $Mean(\ )$ as the aggregation operator to be tolerant towards noise and false connections in the read-overlap graph due to repeats. Furthermore, we use $Concatenation(\ )$ as the layer-wise feature combination strategy to retain features from both the node itself and its neighborhood.

We use two GraphSAGE layers followed by a fully-connected layer with K outputs, where K is the number of bins estimated in Step 4. Two GraphSAGE layers use $LeakyRELU$ activation while the final layer uses $log\ softmax$ activation resulting in the output probabilities for K bins. We train the GNN using 200 epochs using sampled reads binned in Step 4 and use negative log-likelihood (cross-entropy) as the loss function. During the training phase, we use a neighbor sampler that samples up to 20 neighbours in GraphSAGE layers. We use Adam optimizer for gradient descent. The trained GNN on sampled reads provides assignment probabilities for remaining reads to the bins derived in Step 4. The remaining reads are thus assigned to the bins with the highest probabilities.

3   Experimental Setup

We evaluate our pipeline using two simulated and three publicly available real datasets. Note that, all the experiments are conducted and evaluated using the same set of parameters. Detailed information about the datasets are available in Appendix A.

3.1   Simulated Datasets

We simulate four PacBio datasets using SimLoRD [29] containing 8, 20, 50 and 100 [35] species with average read length 5,000 bp and the default PacBio error profiles in SimLoRD (insertion = 0.11, deletion = 0.04 and substitution = 0.01). These two datasets are named as **Sim-8**, **Sim-20**, **Sim-50** and **Sim-100** respectively.

3.2   Real Datasets

Three real datasets with known reference genomes are also used to evaluate read-level binning performance. Long reads from these datasets were aligned to references using Minimap2 [14] to obtain the ground truth.

- **ZymoEVEN**: Oxford Nanopore reads sequenced from GridION device from NCBI Accession Number *ERR3152364* [24]. The dataset consists of 10 species with average read length 4,119.
- **SRR9202034**: PacBio CCS reads of the ATCC MSA-1003 Mock Microbial Community from NCBI BioProject number *PRJNA546278* Accession Number *SRR9202034*. The dataset contains 15 species with more than 0.1% relative abundance and average read length 8,263.
- **SRX9569057**: PacBio-HiFi reads of the NCBI BioSample *SAMN16885726* Accession Number *SRX9569057*. This dataset contains 13 species (18 strains) with more than 0.1% relative abundance and average read length 9,093.

Table 1 summarizes the information about the datasets including the number of species, dataset sizes and the number of nodes (reads) and edges of the read-overlap graphs.

3.3   Baselines and Evaluation Criteria

We benchmark our approach against two recent long-read binners, MetaBCC-LR [35] and LRBiner [34]. For a binning result of K bins against the ground truth of S species, we populate a matrix a of size $K \times S$, where a_{ks} denotes the number of reads assigned to bin k from the species s of the sample. The binning results are evaluated using precision Eq. 4, recall Eq. 5 and F1-score Eq. 6 [37]. We used AMBER [21] to compute completeness and purity, genome fractions using MetaQUAST [22], assembly CPU-time and memory usage to evaluate performance.

Table 1. Summary of the datasets

Dataset	No. of species	Dataset size (GB)	No. of nodes	No. of edges
Sim-8	8	3.5	432,333	47,984,545
Sim-20	20	5.3	666,735	42,642,457
Sim-50	50	9.5	1,119,439	86,245,400
Sim-100	100	24.6	2,991,815	1,198,753,181
ZymoEVEN	10	8.2	1,688,672	611,447,694
SRR9202034	15	19.5	2,358,257	2,105,962,083
SRX9569057	13	18.0	1,978,852	1,421,138,836

$$precision = \frac{\sum_k max_s\{a_{ks}\}}{\sum_k \sum_s a_{ks}} \tag{4}$$

$$recall = \frac{\sum_s max_k\{a_{ks}\}}{\sum_k \sum_s a_{ks}} \tag{5}$$

$$F1-score = 2 \times \frac{Precision \times Recall}{Precision + Recall} \tag{6}$$

4 Results and Discussion

We evaluate binning results at the read-level using precision, recall, F1 score, the number of bins produced. We further evaluate each bin using *per-bin F1-scores* using AMBER [21]. Moreover, we conducted a quantitative evaluation of assemblies using MetaQuast [22] before and after binning.

4.1 Binning Results

We benchmark OBLR against MetaBCC-LR [35] and LRBiner [34] which are two recent long-read binning tools as presented in Table 2. We observed that OBLR results in the highest F1-scores across all the datasets with the overall best performance. OBLR also produces more accurate estimates of the number of bins in most datasets.

These observations are further supported by the AMBER [21] evaluation summarized in Fig. 4 and 5 where OBLR produces the best per bin F1-scores among all three long-read binners. Per bin F1-score evaluates each bin separately using their purity and completeness while penalizing false bin splits and merged bins. Note that MetaBCC-LR and LRBinner suffer from fragmented binning results (*i.e.*, overestimated number of bins) because bins with low completeness are significantly penalized by AMBER. This observation is explained further in Appendix B.

Table 2. Comparison of binning results of MetaBCC-LR, LRBinner and OBLR.

Dataset	No. of bins	Criteria	MetaBCC-LR	LRBinner	OBLR
Sim-8	8	Precision	90.78%	99.14%	**99.33%**
		Recall	96.18%	99.14%	**99.33%**
		F1 score	93.40%	99.14%	**99.33%**
		Bins detected	13	**8**	**8**
Sim-20	20	Precision	82.97%	90.53%	**97.88%**
		Recall	81.95%	88.23%	**97.88%**
		F1 score	82.46%	89.36%	**97.88%**
		Bins detected	29	18	**20**
Sim-50	50	Precision	82.23%	91.92%	**92.94%**
		Recall	70.56%	77.03%	**97.81%**
		F1 score	75.95%	83.82%	**95.32%**
		Bins detected	32	31	**45**
Sim-100	100	Precision	**90.50%**	82.60%	87.61%
		Recall	84.54%	92.78%	**95.00%**
		F1 score	88.54%	87.39%	**91.16%**
		Bins detected	**89**	63	74
ZymoEVEN	10	Precision	**93.09%**	72.41%	75.44%
		Recall	73.84%	92.97%	**95.33%**
		F1 score	82.36%	81.41%	**84.23%**
		Bins detected	8	**9**	8
SRR9202034	15†	Precision	91.30%	93.16%	**98.48%**
		Recall	69.59%	91.94%	**98.52%**
		F1 score	78.98%	92.55%	**98.50%**
		Bins detected	11	10	**15**
SRX9569057	13†	Precision	80.94%	80.47%	**95.03%**
		Recall	85.82%	90.68%	**97.70%**
		F1 score	83.31%	85.27%	**96.35%**
		Bins detected	23	16	**14**

† Species with at least 0.1% abundance.

4.2   Assembly Results

We perform assembly using the long-read metagenomic assembler metaFlye [12]. Table 3 demonstrates the genome fraction and resource usage for assembling raw reads (termed *Raw*) and assembling reads binned by OBLR (termed *Binned*), respectively. Gains in genome fraction are relatively higher for simulated datasets with more species (*e.g.*, **Sim-50** and **Sim-100**) while the most significant benefit being the drastic reduction of the resources consumed. Binning long reads from real datasets before assembly in general maintains the genome fraction

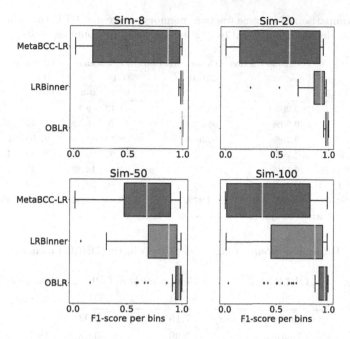

Fig. 4. Per bin F1-score comparison between MetaBCC-LR, LRBinner and OBLR computed by AMBER [21] for simulated datasets.

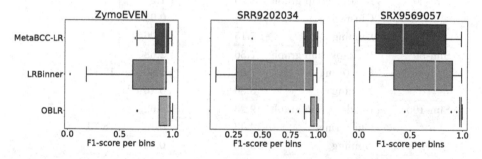

Fig. 5. Per bin F1-score comparison between MetaBCC-LR, LRBinner and OBLR computed by AMBER [21] for real datasets.

(94.13% to 94.27% in **SRX9569057**, 86.51% to 86.67% in **ZymoEVEN** and 90.30% to 90.39% in **SRR9202034**) while significantly saving on the resources. The saving on peak memory varies from 40% to 80%. However, CPU hours consumed remains comparable due to re-indexing of reads and the near-linear time complexity of the assembly process.

5   Implementation

We use multiple optimizations in the OBLR pipeline. In Step 1, we chunk the reads in to blocks of 250,000 reads, and each block of reads is mapped with

Table 3. Comparison of genome fraction, memory usage and CPU time consumed for assemblies conducted using metaFlye [12] before and after binning reads.

Dataset	Genome fraction raw	Genone fraction OBLR	Peak memory (GB)		CPU time (Hours)	
			Raw	Binned	Raw	Binned
Sim-8	99.90%	99.90%	44.12	9.14	7.98	7.76
Sim-20	99.80%	99.85%	71.70	8.52	14.75	12.84
Sim-50	99.25%	99.32%	58.12	14.36	22.95	20.09
Sim-100	97.70%	97.77%	51.95	25.91	76.62	43.78
ZymoEVEN	86.51%	86.67%	31.67	14.82	15.17	13.22
SRR9202034[†]	90.30%	90.39%	52.80	28.48	173.20	140.60
SRX9569057[†]	94.13%	94.27%	49.43	25.84	112.32	98.82

[†] Genome fraction computed from species with at least 0.1% abundance.

Table 4. Resource usage of each step in the OBLR pipeline.

Dataset	OBLR step	Peak memory (GB)	CPU time (H)	Peak GPU memory (GB)
Sim-8	Read-overlap graph	7.15	0.93	–
	Clustering	5.39	0.06	18.70
	Binning	6.38	0.01	
Sim-20	Read-overlap graph	7.26	1.27	–
	Clustering	5.94	0.05	11.56
	Binning	7.15	0.06	
Sim-50	Read-overlap graph	7.36	2.68	–
	Clustering	7.08	0.06	18.75
	Binning	8.80	0.16	
Sim-100	Read-overlap graph	19.23	26.62	–
	Clustering	14.43	0.13	18.73
	Binning	19.03	0.65	
ZymoEVEN	Read-overlap graph	4.78	4.8	–
	Clustering	22.4	0.22	11.57
	Binning	28.98	0.11	
SRR9202034	Read-overlap graph	36.54	118.73	–
	Clustering	15.09	0.15	18.89
	Binning	18.48	0.08	
SRX9569057	Read-overlap graph	4.97	82.81	–
	Clustering	30.20	0.29	18.96
	Binning	38.75	0.15	

entire dataset resulting in several mapping files. Finally, the mapping files are merged into a single file containing edges between reads and degree. For Step 2 we use seq2vec [33] to compute the composition vectors. In Steps 3 and 4 we use Rapids.AI [31] GPU libraries [26] (on NVIDIA RTX 3090 with 24 GB VRAM) for UMAP and HDBSCAN. Finally, we use PyTorch Geometric [5] in Step 5 for GraphSAGE. We conducted our experiments on an Ubuntu 20.04.3 LTS system running on AMD Ryzen 9 5950X with 16-core Processor with 32 threads and 128 GB of RAM.

Table 4 tabulates the resource usage of each step in OBLR pipeline. In addition we also present the resource utilization of kbm2 [28]. Note that the GPU memory utilization is fixed as we perform clustering and silhouette scores only upto 400,000 data points. However, resource usage for other steps vary depending on the read size distribution on each read block and the number of data points.

6 Conclusion

We presented a novel long-read binner, OBLR, which utilizes the read-overlap graph to bin long reads in metagenomic samples. Recent advances such as the k-mer bins mapping (kbm2) [28] enables extremely fast detection of overlapping reads and construction of the read-overlap graph before assembly. OBLR thus makes use of the read-overlap graph to improve the state-of-the-art long-read binning approaches. The read-overlap graph not only helps to estimate the coverage of single long read, but also allow us to sample the long-reads more uniformly across species of varying abundances. The connectivity information in the read-overlap graph further incorporates the overlapping information between reads into the binning process as overlapped reads are more likely to be in the same species. As a result, OBLR demonstrated promising results in producing more accurate bins for long-read datasets and has the potential to improve on metagenomics assemblies in terms of computing resources and genome fraction, especially for low-abundance species. In the future, we plan to investigate how OBLR can be adapted to take the advantage of the high-accuracy long reads including PacBio HiFi and Nanopore Q20+, how to incorporate the binning process into long-read metagenomics assemblies [4,12], and how to connect metagenomics binning to comparative genomics and phylogenetic studies [8,23].

A   Dataset Information

Tables 5 and 7 demonstrate the simulated and real dataset information respectively. Note that Table 5 and 6 tabulate the coverages used for simulation using SimLoRD [29] while Table 7 indicate abundances from the dataset sources.

Table 5. Information of simulated datasets.

Dataset	Number of reads	Total size	Species	Coverage
Sim-8	432,333	3.5 GB	Acetobacter pasteurianus	25
			Bacillus cereus	50
			Chlamydophila psittaci	80
			Escherichia coli	125
			Haemophilus parainfluenzae	350
			Lactobacillus casei	200
			Thermococcus sibiricus	150
			Streptomyces scabiei	100
Sim-20	666,735	5.3 GB	Amycolatopsis mediterranei	25
			Arthrobacter arilaitensis	65
			Brachyspira intermedia	20
			Corynebacterium ulcerans	40
			Erysipelothrix rhusiopathiae	55
			Enterococcus faecium	50
			Mycobacterium bovis	80
			Photobacterium profundum	85
			Streptococcus pyogenes	100
			Xanthobacter autotrophicus	150
			Rhizobium leguminosarum	100
			Francisella novicida	150
			Candidatus Pelagibacter ubique	67
			Halobacterium sp	65
			Lactobacillus delbrueckii	60
			Paenibacillus mucilaginosus	90
			Rickettsia prowazekii	100
			Thermoanaerobacter brockii	110
			Yersinia pestis	105
			Nitrosococcus watsonii	95

Table 6. Information of simulated dataset containing 50 species.

Dataset	Number of reads	Total size	Species	Coverage
Sim-50	1,119,439	9.5 GB	Azorhizobium caulinodans	25
			Bacillus cereus	35
			Bdellovibrio bacteriovorus	21
			Bifidobacterium adolescentis	44
			Bifidobacterium animalis	31
			Campylobacter jejuni	11
			Clostridium tetani	36
			Clostridium thermocellum	31
			Corynebacterium diphtheriae	42
			Corynebacterium ulcerans	33
			Ehrlichia ruminantium	26
			Enterococcus faecium	24
			Erysipelothrix rhusiopathiae	44
			Escherichia coli	20
			Fervidicoccus fontis	49
			Francisella novicida	42
			Francisella tularensis	49
			Fusobacterium nucleatum	39
			Haemophilus influenzae	12
			Haemophilus parainfluenzae	11
			Haemophilus somnus	44
			Helicobacter pylori	47
			Hyphomicrobium sp	44
			Lawsonia intracellularis	46
			Metallosphaera cuprina	33
			Methanosarcina barkeri	44
			Micrococcus luteus	46
			Mycobacterium bovis	42
			Mycoplasma gallisepticum	29
			Neisseria meningitidis	38
			Nitrosococcus watsonii	42
			Paenibacillus mucilaginosus	14
			Paenibacillus sp	31
			Photobacterium profundum	45
			Pseudogulbenkiania sp	25
			Pseudomonas putida	10
			Rhizobium leguminosarum	20
			Rickettsia prowazekii	38

(*continued*)

Table 6. (*continued*)

Dataset	Number of reads	Total size	Species	Coverage
			Rickettsia rickettsii	100
			Ruegeria sp	200
			Shewanella sp	90
			Sodalis glossinidius	120
			Staphylococcus aureus	220
			Streptococcus pyogenes	110
			Streptococcus suis	100
			Streptomyces scabiei	110
			Symbiobacterium thermophilum	250
			Thermoanaerobacter sp	220
			Thermococcus sibiricus	210
			Variovorax paradoxus	100

Table 7. Information of real datasets.

Dataset	Number of reads	Total size	Species	Abundance
ZymoEVEN	1,688,672	8.2 GB	*P. aeruginosa*	9.7%
			Escherichia coli	9.9%
			Salmonella enterica	10.0%
			Lactobacillus fermentum	9.3%
			Enterococcus faecalis	12.2%
			Staphylococcus aureus	11.2%
			Listeria monocytogenes	14.5%
			Bacillus subtilis	19.3%
			Saccharomyces cerevisiae	2.1%
			Cryptococcus neoformans	1.8%
SRR9202034	2,358,257	19.5 GB	Acinetobacter baumannii	0.18%
			Bacillus pacificus	1.80%
			Bacteroides vulgatus	0.02%
			Bifidobacterium adolescentis	0.02%
			Clostridium beijerinckii	1.80%
			Cutibacterium acnes	0.18%
			Deinococcus radiodurans	0.02%
			Enterococcus faecalis	0.02%
			Escherichia coli	18.0%
			Helicobacter pylori	0.18%
			Lactobacillus gasseri	0.18%

(*continued*)

Table 7. (*continued*)

Dataset	Number of reads	Total size	Species	Abundance
			Neisseria meningitidis	0.18%
			Porphyromonas gingivalis	18.0%
			Pseudomonas aeruginosa	1.80%
			Rhodobacter sphaeroides	18.0%
			Schaalia odontolytica	0.02%
			Staphylococcus aureus	1.80%
			Staphylococcus epidermidis	18.0%
			Streptococcus agalactiae	1.80%
			Streptococcus mutans	18.0%
SRX9569057	1,978,852	18 GB	Faecalibacterium prausnitzii	14.82%
			Veillonella rogosae	20.01%
			Roseburia hominis	12.47%
			Bacteroides fragilis	8.36%
			Prevotella corporis	6.28%
			Bifidobacterium adolescentis	8.86%
			Fusobacterium nucleatum	7.56%
			Lactobacillus fermentum	9.71%
			Clostridioides difficile	1.10%
			Akkermansia muciniphila	1.62%
			Methanobrevibacter smithii	0.17%
			Salmonella enterica	0.0065%
			Enterococcus faecalis	0.0011%
			Clostridium perfringens	0.00009%
			Escherichia coli (JM109)	1.83%
			Escherichia coli (B-3008)	1.82%
			Escherichia coli (B-2207)	1.65%
			Escherichia coli (B-766)	1.66%
			Escherichia coli (B-1109)	1.77%
			Candida albicans	0.16%
			Saccharomyces cerevisiae	0.16%

B   Interpretation of AMBER Per-bin F1-Score

The binning evaluations are presented using Precision, Recall and F1 score. Furthermore, stricter evaluations are presented using AMBER [21] for MetaBCC-LR, LRBinner and OBLR. This section explains the evaluation metrics in detail and discuss as to why AMBER evaluations are poor in some cases where number of bins predicted is further away from the actual number of species in the dataset. Note that the bin assignment matrix a can be presented as $M \times N$, illustrated in Table 8. Note that $N = 5$ and $M = 7$.

Table 8. Binning matrix

	Bin 1	Bin 2	Bin 3	Bin 4	Bin 5	Bin 5	Bin 7
Species 1	$a_{11} = 99$	$a_{12} = 0$	$a_{13} = 0$	$a_{14} = 0$	$a_{15} = 0$	$a_{16} = 1$	$a_{17} = 0$
Species 2	$a_{21} = 0$	$a_{22} = 100$	$a_{23} = 0$	$a_{24} = 0$	$a_{25} = 0$	$a_{26} = 0$	$a_{27} = 0$
Species 3	$a_{31} = 0$	$a_{32} = 20$	$a_{33} = 0$	$a_{34} = 0$	$a_{35} = 0$	$a_{36} = 0$	$a_{37} = 0$
Species 4	$a_{41} = 0$	$a_{42} = 0$	$a_{43} = 1000$	$a_{44} = 50$	$a_{45} = 0$	$a_{46} = 0$	$a_{47} = 0$
Species 5	$a_{51} = 0$	$a_{52} = 0$	$a_{53} = 0$	$a_{54} = 0$	$a_{55} = 200$	$a_{56} = 0$	$a_{57} = 300$

Recall is computed for each species, by taking the largest assignment to a bin. Precision is computed per bin taking the largest assignment of the bin to a given species. In contrast, AMBER uses purity and completeness to compute the per-bin F1 score using the following equations, for each bin b.

The true positives are computed using the majority species in a given bin. Because of this, if a bin appears as a result of a false bin split (1% reads), the Completeness of the bin will be very low as the majority of it (approximately 1%) according to AMBER evaluation. In comparison, the recall of the species using Eq. 5 will report 99% since 99% of the reads are in a single bin despite having the false bin split. Similarly, the false split of the bin will report a greater precision as long as the bin has no other species mixed according to Eq. 4. Consider the following running example.

Example 1. Suppose Species 1 has $a_{11} = 99$ and $a_{16} = 1$ with rest of the row having no reads and Bin 1 and 6 has no reads from another species. Purity in this case will be 100% for both bins 1 and 6 while completeness will be 99% and 1% respectively. F1-score will be 99.5% and 1.98% with average being very low at 50.7%. Recall will be 99% for Species 1 with 100% precision on both bins 1 and 6 since there are no impurities in each bin, thus, F1-score is 99.5% for each bin.

Example 2. Suppose Bin 2 has $a_{22} = 100$ and $a_{32} = 20$, with two species 2 and 3, with no other contaminants and species 2 and 3 are fully contained in the bin. Now, the purity of the bin is 83.33% and completeness is 83.33%, hence, F1-score is 83.33%. Recall for species 2 and 3 will be 100% since it is not broken into multiple bins. However, the precision for Bin 2 will be 83.33%, hence a F1 score of 90.91%.

This means, AMBER penalize whenever a species is broken into pieces across bins while not significantly penalizing bin mergers between large bins and smaller bins. This is because, dominant species in the bin will determine the purity and completeness.

References

1. Baaijens, J.A., El Aabidine, A.Z., Rivals, E., Schönhuth, A.: De novo assembly of viral quasispecies using overlap graphs. Genome Res. **27**(5), 835–848 (2017)
2. Balvert, M., Luo, X., Hauptfeld, E., Schönhuth, A., Dutilh, B.E.: Ogre: overlap graph-based metagenomic read clustering. Bioinformatics **37**(7), 905–912 (2021)
3. Chen, K., Pachter, L.: Bioinformatics for whole-genome shotgun sequencing of microbial communities. PLOS Comput. Biol. **1**(2) (2005)
4. Feng, X., Cheng, H., Portik, D., Li, H.: Metagenome assembly of high-fidelity long reads with hifiasm-meta. arXiv:2110.08457 (2021)
5. Fey, M., Lenssen, J.E.: Fast graph representation learning with PyTorch Geometric. In: ICLR Workshop on Representation Learning on Graphs and Manifolds (2019)
6. Hamilton, W.L., Ying, R., Leskovec, J.: Inductive representation learning on large graphs. In: Proceedings of the 31st International Conference on Neural Information Processing Systems, pp. 1025–1035 (2017)
7. Huson, D.H., et al.: Megan-LR: new algorithms allow accurate binning and easy interactive exploration of metagenomic long reads and contigs. Biol. Direct **13**(1), 1–17 (2018)
8. Huson, D.H., Richter, D.C., Mitra, S., Auch, A.F., Schuster, S.C.: Methods for comparative metagenomics. BMC Bioinf. **10**(1), 1–10 (2009)
9. Japkowicz, N., Stephen, S.: The class imbalance problem: a systematic study. Intell. Data Analysis **6**(5), 429–449 (2002)
10. Kang, D.D., et a.: Metabat 2: an adaptive binning algorithm for robust and efficient genome reconstruction from metagenome assemblies. PeerJ **7**, e7359 (2019)
11. Kim, D., Song, L., Breitwieser, F.P., Salzberg, S.L.: Centrifuge: rapid and sensitive classification of metagenomic sequences. Genome Res. **26**(12), 1721–1729 (2016)
12. Kolmogorov, M., et al.: metaflye: scalable long-read metagenome assembly using repeat graphs. Nat. Methods **17**(11), 1103–1110 (2020)
13. Li, H.: Minimap and miniasm: fast mapping and de novo assembly for noisy long sequences. Bioinformatics **32**(14), 2103–2110 (2016)
14. Li, H.: Minimap2: pairwise alignment for nucleotide sequences. Bioinformatics **34**(18), 3094–3100 (2018)
15. Liang, D.M., Li, Y.F.: Lightweight label propagation for large-scale network data. In: IJCAI, pp. 3421–3427 (2018)
16. Liu, X.Y., Wu, J., Zhou, Z.H.: Exploratory undersampling for class-imbalance learning. IEEE Trans. Syst. Man Cybern. Part B (Cybernetics) **39**(2), 539–550 (2009). https://doi.org/10.1109/TSMCB.2008.2007853
17. Logsdon, G.A., Vollger, M.R., Eichler, E.E.: Long-read human genome sequencing and its applications. Nat. Rev. Genet. **21**(10), 597–614 (2020)
18. McInnes, L., Healy, J., Astels, S.: HDBSCAN: hierarchical density based clustering. J. Open Source Softw. **2**(11), 205, e7359 (2017)
19. McInnes, L., Healy, J., Melville, J.: Umap: Uniform manifold approximation and projection for dimension reduction (2020)
20. Menzel, P., Ng, K.L., Krogh, A.: Fast and sensitive taxonomic classification for metagenomics with Kaiju. Nat. Commun. **7**, 11257 (2016)
21. Meyer, F., et al.: Amber: assessment of metagenome binners. Gigascience **7**(6), giy069 (2018)
22. Mikheenko, A., Saveliev, V., Gurevich, A.: Metaquast: evaluation of metagenome assemblies. Bioinformatics **32**(7), 1088–1090 (2016)

23. Nayfach, S., Pollard, K.S.: Toward accurate and quantitative comparative metagenomics. Cell **166**(5), 1103–1116 (2016)
24. Nicholls, S.M., Quick, J.C., Tang, S., Loman, N.J.: Ultra-deep, long-read nanopore sequencing of mock microbial community standards. Gigascience **8**(5), giz043 (2019)
25. Nissen, J.N., et al.: Improved metagenome binning and assembly using deep variational autoencoders. Nat. Biotechnol. **39**(5), 555–560 (2021)
26. Nolet, C.J., et al.: Bringing UMAP closer to the speed of light with GPU acceleration (2020)
27. Rousseeuw, P.J.: Silhouettes: a graphical aid to the interpretation and validation of cluster analysis. J. Comput. Appl. Math. **20**, 53–65, e7359 (1987)
28. Ruan, J., Li, H.: Fast and accurate long-read assembly with WTDBG2. Nat. Methods **17**(2), 155–158, e7359 (2020)
29. Stöcker, B.K., Köster, J., Rahmann, S.: Simlord: simulation of long read data. Bioinformatics **32**(17), 2704–2706 (2016)
30. Strous, M., Kraft, B., Bisdorf, R., Tegetmeyer, H.: The binning of metagenomic contigs for microbial physiology of mixed cultures. Front. Microbiol. **3**, 410 (2012)
31. Team, R.D.: RAPIDS: Collection of Libraries for End to End GPU Data Science (2018). https://rapids.ai
32. Tyson, G.W., et al.: Community structure and metabolism through reconstruction of microbial genomes from the environment. Nature **428**(6978), 37–43 (2004)
33. Wickramarachchi, A.: anuradhawick/seq2vec: release v1.0 (2021). https://doi.org/10.5281/zenodo.5515743, https://doi.org/10.5281/zenodo.5515743
34. Wickramarachchi, A., Lin, Y.: Lrbinner: binning long reads in metagenomics datasets. In: 21st International Workshop on Algorithms in Bioinformatics (WABI 2021). Schloss Dagstuhl-Leibniz-Zentrum für Informatik (2021)
35. Wickramarachchi, A., Mallawaarachchi, V., Rajan, V., Lin, Y.: Metabcc-LR: metagenomics binning by coverage and composition for long reads. Bioinformatics **36**(Supplement_1), i3–i11 (2020)
36. Wood, D.E., Lu, J., Langmead, B.: Improved metagenomic analysis with kraken 2. Genome Biol. **20**(1), 1–13 (2019)
37. Wu, Y.W., Simmons, B.A., Singer, S.W.: Maxbin 2.0: an automated binning algorithm to recover genomes from multiple metagenomic datasets. Bioinformatics **32**(4), 605–607 (2016)
38. Xu, K., Hu, W., Leskovec, J., Jegelka, S.: How powerful are graph neural networks? arXiv:1810.00826 (2018)

A Mixed Integer Linear Programming Algorithm for Plasmid Binning

Aniket Mane[✉], Mahsa Faizrahnemoon, and Cedric Chauve

Department of Mathematics, Simon Fraser University, Burnaby, BC, Canada
{amane,mfaizrah,cedric.chauve}@sfu.ca

Abstract. The problem of analysing bacterial isolates in order to detect plasmids has been widely studied. With the development of Whole Genome Sequencing (WGS) technologies, several approaches have been proposed to bin contigs into putative plasmids. Reference-based approaches aim to bin contigs by mapping or comparing their sequences against databases of previously identified plasmids or plasmid genes. On the other hand, de novo approaches use contig features such as read coverage and length for plasmid binning. Hybrid approaches that combine both strategies have also been proposed recently.

We present PlasBin a mixed integer linear programming based hybrid approach for plasmid binning. We evaluate the performance of several binning methods on a real data set of bacterial samples.

1 Introduction

Antimicrobial resistance (AMR) has recently emerged as a major public health concern. AMR is developed in bacteria through the acquisition of antimicrobial genes. This is often facilitated by mobile genetic elements (MGEs). Mobile genetic elements enabling antimicrobial resistance can be exchanged between bacteria via, among other mechanisms, evolutionary events known as horizontal gene transfer (HGT) [8]. The family of plasmids falls under the umbrella of MGEs. Plasmids are circular, double-stranded segments of DNA that are separate from chromosomes and are known to carry AMR genes. Thus, the identification and detection of plasmids from whole-genome sequencing (WGS) data is important for understanding the propagation of AMR genes.

With the advent of DNA sequencing technologies, it is now possible to obtain WGS data at a relatively low cost. For bacterial genomes, reads sequenced using WGS technologies are generally processed by a genome assembler, such as Unicycler [19] or SPAdes [4], to produce an assembly, often in the form of contigs; recent assemblers such as SPAdes and Unicycler also generate an assembly graph. The problem of detecting plasmids from the output of an assembler can be studied at three different levels: classification, binning and assembly.

At the *classification* level, the aim is to classify whether a contig originates from a plasmid or a chromosome. Erstwhile tools such PlasmidFinder [9] map query contigs against a reference database of plasmid markers. Several approaches

L. Jin and D. Durand (Eds.): RECOMB-CG 2022, LNBI 13234, pp. 279–292, 2022.
https://doi.org/10.1007/978-3-031-06220-9_16

based on machine learning have also been utilized for contig classification. These approaches attempt to learn the genomic signal, often in the form of k-mer composition, from a training data set of known plasmid sequences. Current state-of-the-art tools include mlplasmids [3], PlasClass [16], PlasFlow [11] and RFPlasmid [6].

At the *binning* level, the aim is to group contigs into bins, with the expectation that the contigs in a bin likely originate from the same plasmid. Various approaches have been proposed for plasmid binning. MOB-recon – part of the recently developed MOB-suite [17] – is a reference-based method that maps contigs against a database of known plasmid sequences and a database of plasmid marker sequences known as replicons and relaxases. Plasmids from the reference database are clustered according to their sequence similarity. Contigs are then binned together if they match with plasmids belonging to the same reference cluster. Putative plasmid bins thus constructed are then filtered using the presence/absence of known replicons or relaxases. The reliance on a reference database can potentially hinder the ability of reference-based methods to identify novel plasmids. De-novo approaches do not depend on reference sequences. Instead, they use contig features as well as, for some methods, additional information from the assembly graph. Plasmids often occur in the genome in several copies. Thus, the coverage of plasmids is expected to be significantly different than that of chromosomes. Recycler [18] peels off cycles from the assembly graph assuming uniform coverage of sequenced plasmids and using length thresholds. PlasmidSPAdes [1] also relies on coverage features: it estimates the chromosomal coverage from the assembly graph, removes contigs with similar coverage as that of the chromosome and then computes putative plasmid bins from the connected components of the remaining graph. However, this strategy may fail to identify plasmids that have a low copy number. Lastly some plasmid binning methods employ a hybrid strategy by combining ideas from both of the above approaches. Such methods first use a reference databases to assign weights to the contigs. They then use contig features such as coverage and topography of the assembly graph to separate contigs into putative plasmid bins. Gplas [2] uses the signal from known plasmidic sequences to assign a probability that the contig is plasmidic in origin; gplas uses mlplasmids [3] to compute this probability. Contigs classified as chromosomal are removed from the assembly and the remaining contigs are then separated into plasmid bins, using a graph-theoretical method. HyAsP [14] is another hybrid plasmid binning tool. It first computes the plasmid genes density of each contig by mapping the contigs against a database of known plasmid genes. It then uses a greedy heuristic to extract cycles or paths from the WGS assembly graph as putative plasmid bins, with the goal to maximize the plasmid gene density over the selected contigs while maintaining a uniform read coverage.

Finally, at the *assembly* level, contigs that are in the same plasmid bin are linearly or circularly ordered. Note that plasmid assembly methods by default are also useful for contig classification and plasmid binning. From the plasmid binning approaches mentioned above, Recycler and HyAsP have actually been designed as assembly methods.

In the present work, we focus on the plasmid binning problem, motivated by the fact that, for downstream analysis, the most useful information is in the form of groups of genes that belong to a given plasmid, more than in the order of these genes along the plasmid sequence; for example, plasmid typing can be done from the gene content and does not consider gene order [10,12]. Hence, although our method does perform plasmid assembly, we focus on plasmid binning and evaluate the results accordingly. We propose a hybrid approach that uses a mixed integer linear programming (MILP) formulation for plasmid binning. We compare our method, PlasBin, against HyAsP, plasmidSPAdes, MOB-suite and gplas on a data set of 133 bacterial samples and evaluate their performance using precision, recall and F1 statistics. Our experiments indicate that the hybrid methods generally perform better than other approaches, and that our method PlasBin obtains the best average F1 score, slightly outperforming HyAsP.

2 Hybrid Approach for Plasmid Binning Using Mixed Integer Linear Programming

We present PlasBin, an optimization based hybrid approach for the plasmid binning problem. PlasBin is based on the same principles as HyAsP, which utilizes a greedy approach to extract paths or cycles from an assembly graph. The evaluation of HyAsP demonstrated that, at the time of its design, it outperformed existing methods. The rationale of PlasBin is that by relying on a global optimization approach, using an objective function based on the HyAsP greedy path extension score, it will be able to avoid the pitfall of locally optimal solutions, while improving some of HyAsP limitations.

2.1 Input: Contigs and the Assembly Graph

The two main input for PlasBin are the contigs and the assembly graph, obtained using Unicycler [19][1]. Every contig has two extremeties, a head and a tail, denoted by c_h and c_t for a contig c. Pairs of contigs that are potentially adjacent in the genome sequence are connected to each other via edges of the assembly graph. Although, we do not consider the order of contigs for the purpose of plasmid binning, the edges help in ensuring that the contigs chosen to belong to a plasmid bin form a path or a cycle in the graph. We represent the edges of the assembly graph as pairs of contig extremeties: an edge between c_h and d_h is represented as (c_h, d_h). PlasBin also takes as input a database of genes known to occur in plasmids, called as the reference database from now.

Contig Features. PlasBin considers several features associated to a contig, namely its %GC content, sequencing coverage, length and plasmid gene density.

---

[1] We rely on Unicycler as it is a widely used bacterial genome assembler, but any assembler providing an assembly graph can be used.

- The %GC content of a sequence is the proportion of the sequence consisting of either G or C bases. It is a useful feature as the %GC content of plasmids is generally expected to be lower than that of chromosomes [15]. For contig c, we denote by gc_c its %GC content.
- Plasmids might occur in multiple copies in a cell, as opposed to the chromosome, which is single-copy. Thus, plasmidic contigs are likely to exhibit higher sequencing coverage, while contigs originating from the same plasmid are expected to have similar coverage. We define the coverage of a contig as the normalized coverage computed by Unicycler, *i.e.* the average base coverage of the contig normalized by the median coverage of the contigs in the assembly graph; thus chromosomal contigs are expected to have a normalized coverage close to 1. The coverage of contig c is denoted by rc_c.
- We denote by ℓ_c the length of contig c.
- Taking a cue from HyAsP, we associate each contig with its *plasmid gene density*, which is the proportion of the sequence matching with a set of genes known to occur in plasmids. We compute the plasmid gene density of a contig by mapping genes from the reference database to contigs using blastn (version 2.6.0) [7], considering matches that show an identity of at least 95% and cover at least 95% of the gene. The plasmid gene density of contig c is denoted by gd_c.

Seed Contigs. Based on the features described above, some contigs are more likely to be part of a plasmid than others. We refer to these contigs as seeds, a notion introduced in HyAsP. Seeds are contigs that pass certain thresholds on the features length, coverage and gene density. The parameters defining seed contigs can be modified by the user, but the default values are taken from HyAsP. The seed eligibility of contig c, a boolean feature, is denoted by s_c.

Preprocessing: Duplicating High-Coverage Contigs. PlasBin assumes that a plasmid has been sequenced with near-uniform coverage; thus all contigs from a given plasmids, provided they are not repeated elsewhere in the genome. In order to handle the possibility of a contig to be repeated within a plasmid bin, or in different plasmid bins, we duplicate contigs based on their coverage as follows: if a contig c has normalized coverage k we create $\lceil k \rceil$ nodes $c_1, c_2, ..., c_{\lceil k \rceil}$ and any edge incident to c in the original graph is duplicated into edges incident to every c_i in the modified graph. The other attributes (gene density, GC content, seed eligibility and length) remain the same as for the original contig c. As a result of this preprocessing, every node of the modified assembly graph has normalized coverage 1; this is a major difference between PlasBin and HyAsP, where the Unicycler normalized coverage was used as is in the greedy path extension heuristic.

2.2   PlasBin Workflow

PlasBin works iteratively. During each iteration an MILP computes a path in the preprocessed assembly graph that optimizes an objective function defined in terms of contig features; by the way we preprocessed the original assembly graph, this path can be a walk in the original assembly graph, thus allowing for repeated contigs. Contigs on this path represent a plasmid bin. At the end of each iteration, we obtain as output both the list of contigs belonging to a plasmid bin and a list of edges defining in the corresponding path. The MILP also enforces that at least one seed belongs to the path.

Once a plasmid bin is generated by the MILP, we update the assembly graph by removing contigs from the bin as well as the edges incident to these contigs. PlasBin then moves on to the next iteration with the updated assembly graph as input. This process continues until no seed contig is present.

Next, similarly to HyAsP, we classify every plasmid bin as putative or questionable based on certain thresholds on the overall plasmid gene density and the cumulative length of the contigs in the bin.

2.3   MILP Formulation

We now describe the MILP used in every iteration to compute a path in the provided assembly graph. This is the central part of PlasBin, aimed at replacing the greedy heuristic of HyAsP.

Decision Variables: To each contig, contig extremity and edge in the assembly graph, we associate a decision binary variables to indicate if it is part of the computed optimal path. For each contig, we also associate decision variables pertaining to %GC content, gene density and seed eligibility. These variables take the value of the respective feature if and only if the contig in question is part of the solution ($x_c = 1$), otherwise they take value 0. Finally, we introduce a continuous variable MGC that records the mean %GC content of contigs in the computed path. For a graph $AG = (V, E)$, the total number of variables is $O(|E|)$. Table 1 below describes the decision variables. We discuss later in this section how we handle variables whose formulation is a priori non-linear.

Table 1. Decision variables used in the MILP; p refers to the computed path.

Variable	Type	Description
x_c	Binary	$x_c = 1$ if node $c \in p$
$x_{c,ext}$	Binary	$x_{c,ext} = 1$ if extremity ext of contig $c \in p$
y_e	Binary	$y_e = 1$ if edge $e \in p$
GC_c	Continuous	$GC_c = gc_c$ if $c \in p$, 0 otherwise
S_c	Boolean	$S_c = s_c$ if $c \in p$, False otherwise
L_c	Integer	$L_c = \ell_c$ if $c \in p$, 0 otherwise
L_p	Integer	$L_p = \sum_{c \in p} \ell_c$
MGC	Continuous	$\sum_{c \in p}(gc_c * \ell_c)/L_p$

Objective Function: We formulate the MILP as a minimization problem, that aims to find a path in the assembly graph that minimizes a linear combination of the (negated) plasmid gene density and of a term measuring the deviation of %GC content across the contigs forming the path.

- Gene density: The aim is to maximize the plasmid gene density over the whole path. The gene density of a contig is a number between 0 and 1 so, to account for the high variability of contigs length, we weight the gene density of selected contigs by their lengths and integrate the term $-\sum_c(gd_c)*L_c$ in the objective function (the sum is over all contigs).
- %GC content deviation: We aim to minimize the sum of the deviations between the %GC content of selected contigs and the mean %GC content of the whole path. This feature was shown in HyAsP to contribute greatly to the method accuracy. Similarly to the gene density, we weight the %GC of each contig in the path by its length and incorporate the term $\sum_c(|MGC-gc_c|)*L_c$ in the objective function. Here, MGC is the mean %GC content of the path introduced in Table 1.

The objective function is thus, a linear combination of the two terms above, with each term being weighted equally. Note that the optimality criteria for HyAsP also contains a term pertaining to coverage uniformity. However, we circumvent the need for computing the deviation in contig coverage through the modifications made to the assembly graph as described in Sect. 2.1.

An important remark is that, if we were to exclude the %GC content term from the objective function, the problem to address would be a shortest path problem, with the additional constraint that the optimal path should contain a seed. This would be amenable to efficient shortest paths algorithms. However, as mentioned above, the %GC content term is important toward selecting paths that are more likely to originate from plasmids, which makes the optimization problem we address more complicated.

Constraints: We provide now an overview of the constraints required to solve the problem of finding a path in the assembly graph that is optimal with regard to the objective function.

1. Constraint to ensure the presence of a seed in the path: This is done by requiring that the sum of the binary decision variable x_c over all seed contigs is at least 1.
2. Constraints to handle variables in the denominator: In order to minimize %GC content deviation, we also need to compute the mean %GC content of the path, MGC. The mean %GC content of a plasmid is given by:
$$MGC = \frac{\sum_c gc_c*\ell_c*x_c}{L_p}$$

This is not a linear constraint as L_p is a variable in the denominator. However, it is possible to obtain a linear relaxation for the above constraint through the use of McCormick envelopes [13]. This is possible since x_c is a binary variable and thus, has definite upper and lower bounds.

3. Constraints to handle absolute values: The objective function contains the absolute value of the difference between the mean %GC of the path (MGC) and individual %GC (GC_c). This in itself is not linear. So, we introduce a difference variable d_c which is the absolute value of the difference between the mean %GC and contig c %GC for a specific path. For each contig we add the following constraints: $d_c >= MGC - GC_c$, $d_c >= GC_c - MGC$.
4. Constraints to define a path: In every iteration, we want the MILP to output a path. This is the most challenging aspect of our formulation, as it is notorious that modelling the search of an arbitrary path in a graph through a polynomial number of linear constraints is not possible. To handle this issue, we start from degree constraints that enforce that every contig extremity belongs to at most one selected edge. However, this leads to the possibility that the generated solution consists of disjoint components, all but one of which are cycles. Thus, we would like to add constraints to eliminate potential cycles. To address this issue, we use the delayed constraint generation method [5]: whenever the solution found by the MILP contains cycles, we add constraints explicitly forbidding these cycles and repeat. This results into an iterative process that goes on until we obtain a singular component defining a path as our solution. A possible scenario in this approach is an exponential number of iterations. To avoid this, we bound the number of iterations.

The number of constraints, excluding cycles-forbidding constraints added in the delayed constraint generation step, is in $O(|E|)$.

Implementation. The PlasBin MILP is solved using Gurobi solver (version 9.1.2). The code was written in python, using the Gurobi API. Each individual iteration of the MILP was allowed to run for a maximum of 4 h until the optimality gap threshold of 5% is reached. If the time threshold was reached before the optimality gap threshold, the MILP output the feasible solution at the time of stopping. Any solution thus obtained would consist of a path and possibly some cycles. Constraints to prevent these cycles were then added for the next iteration. The code is available from https://github.com/cchauve/PlasBin.

3   Experimental Results

We evaluated the results of the PlasBin on a dataset of 133 bacterial genomes and 377 plasmids from a collection of real plasmids compiled in [17]. This follows the experimental evaluation of HyAsP. Some of the methods such as HyAsP, MOB-suite and PlasBin are partly or entirely reference-based. To simulate the use of a realistic reference database, we split our data into a reference set and a test set. Samples released before 19 December 2015 were used to build the reference database and those released after that date formed the test set. The test set consisted of 66 samples, with a total of 147 plasmids, for which Illumina sequencing data was re-assembled using Unicycler to provide contigs and assembly graphs (we refer to [17]). We evaluated the following methods: PlasBin, HyAsP, MOB-recon, plasmidSPAdes and gplas.

3.1 Performance Comparison of Plasmid Binning Tools

The results of PlasBin were evaluated using precision, recall and F1-score. Putative plasmid bins (PlasBin, HyAsP) or plasmid contigs (MOB-suite, plasmidSPAdes, gplas) were mapped against the reference plasmids database using BLAST+ [7]. The precision (resp. recall) was defined as the proportion of predicted plasmid lengths matched with the reference plasmids (resp. the proportion of reference plasmid lengths covered by alignments to the predicted plasmids). Although PlasBin provides questionable plasmid bins in its output, they are not to be considered as part of the predicted bins. Hence, we focus on the statistics for putative plasmid bins.

Gplas requires the use of mlplasmids to compute the probability of contigs to originate from a plasmid. Since mlplasmids currently supports only 4 species (*Escherichia coli, Enterococcus faecium, Klebsiella pneumoniae* and *Acinetobacter baumannii*), gplas was evaluated only over the subset of samples belonging to these species.

The performance of PlasBin was first compared against HyAsP, MOB-suite and plasmidSPAdes for all the 66 test samples. The mean statistics for all the tools are given in Table 2, while full statistics are shown in Fig 1.

Table 2. Mean statistics for various plasmid assembly tools

Tool	Precision	Recall	F1
PlasBin	0.859	0.884	0.860
HyAsP	0.873	0.855	0.849
plasmidSPAdes	0.743	0.821	0.723
MOB-suite	0.869	0.644	0.686

We observe that the hybrid methods HyAsP and PlasBin have comparable performances. HyAsP and MOB-suite show the best precision, attaining a precision of over 0.9 for many samples, while PlasBin consistently yields a precision close to 0.8. On the other hand, plasmidSPAdes yields lower precision than the other methods. In terms of recall, we observe that plasmidSPAdes, PlasBin and HyAsP can account for most of the plasmid contigs with a recall of 0.8 or more. PlasBin performed the best, consistently identifying 80% of true plasmid contigs, with HyAsP and plasmidSPADes only slightly lagging behind PlasBin. MOB-suite on the other hand showed weaker recall than all the other tools.

Overall, considering the F1-score, we observe that PlasBin and HyAsP generally perform better than the other tools, with a slight advantage for PlasBin. Both methods consistently have an F1 score over 0.8. The performances of plasmidSPAdes and MOB-suite varied across the samples. MOB-suite sometimes outperformed PlasBin in terms of precision but not in terms of recall. On the other hand, plasmidSPAdes showed slightly poorer precision and recall than the hybrid methods. This suggests that the two-pronged strategy of using reference-based information as well as known plasmid characteristics tends to yield better results.

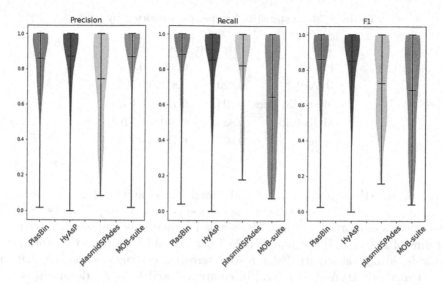

Fig. 1. Precision, recall and F1-score statistics for all tools (except gplas)

We then used gplas on the subset of samples belonging to *Escherichia coli, Enterococcus faecium or Klebsiella pneumoniae*[2] (Table 3 and Fig. 2).

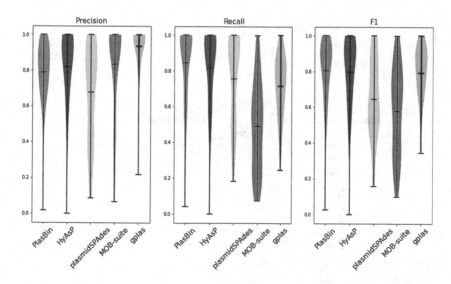

Fig. 2. Precision, recall and F1-score statistics for all tools for three species supported by gplas

---

[2] Our data set did not contain any samples from *Acinetobacter baumannii*.

Table 3. Mean statistics for species supported by gplas

Tool	Precision	Recall	F1
PlasBin	0.788	0.843	0.804
HyAsP	0.820	0.817	0.807
plasmidSPAdes	0.623	0.784	0.632
MOB-suite	0.835	0.491	0.576
gplas	0.935	0.714	0.791

We observe that gplas tends to avoid false positives and has the best precision, which is expected given it relies on species-specific training datasets. For the three species supported by gplas, PlasBin had median precision over 0.8, Overall, in terms of F1-score, PlasBin and HyAsP performed better than the other three methods, with gplas slightly behind. An interesting comment is the fact that the hybrid methods HyAsP and PlasBin compared well to gplas despite this tool using species-specific training data.

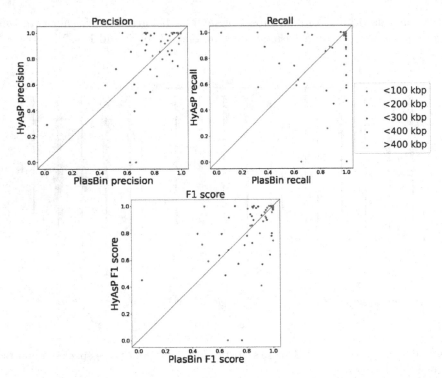

Fig. 3. Precision, recall and F1-score for PlasBin and HyAsP. Each point in the scatter-plot represents a sample. Points have been assigned a color according to the cumulative length of true plasmids in the sample.

3.2   Comparison of PlasBin and HyAsP

Motivated by the fact that we expected PlasBin, by design, to outperform HyAsP, which we did not observe to a significant point, we compared the results of PlasBin and HyAsP for each sample, with a special focus on the total length of the true and predicted plasmids. The scatterplots of the precision, recall and F1-score per sample of both tools are shown alongside in Fig 3.

The precision and recall were computed while accounting for the length of the plasmids in the set of predicted and reference plasmids respectively. From Fig 4, it can be seen that in most cases, the sum of the lengths of incorrectly predicted contigs is small, often less than 5000 bp and consistently less than 20000 bp. However, considering the size of plasmids in the data set, a relatively small absolute error often translates to a large relative error thereby resulting in low precision.

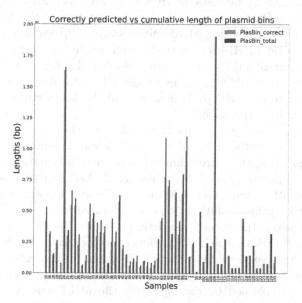

Fig. 4. Cumulative length of correctly predicted contigs (red) and of all the contigs in bins predicted by PlasBin (blue), for each sample (Color figure online)

3.3   Computational Footprint

PlasBin relies on an MILP, that can require large computing resources (time and memory), while HyAsP, being a greedy heuristic, induces a low computational footprint. We ran our experiments on a standard laptop computer, with a quadcore processor and 16 GB of memory. On average, over all considered samples, PlasBin required a little more than an hour of computation to complete (maximum of 8 h), with a memory footprint close to 6 GB (maximum 10 GB). So, while being more costly than HyAsP, PlasBin has a reasonable computational footprint. The details have provided in Table 4.

Table 4. Runtime statistics for various tools (in minutes)

Tool	Min	Max	Median	Mean
PlasBin	5	473	22	117
HyAsP	1	19	3	4
plasmidSPAdes	4	363	16	43
MOB-suite	3	15	5	8
gplas	2	12	3	4

4   Discussion

In this paper, we presented a hybrid method for plasmid binning using, based on replacing the greedy heuristic of HyAsP by an exact MILP approach. The performance of PlasBin was compared against other state-of-the-art plasmid binning methods, namely, HyAsP, MOB-suite, plasmidSPAdes and gplas. In some instances, gplas and MOB-suite showed better precision, however, PlasBin as well as HyAsP consistently showed better recall and a better F1-score, with a slight advantage to PlasBin in terms of the F1-score. As a result, these two methods had the best overall performance of all the five methods used, lending support to the hybrid approach.

Despite having an objective function similar to the extension criteria used in HyAsP, the two methods function in a different manner. At any point in the HyAsP algorithm, there are a limited number of alternatives to extend the plasmid chain being assembled. As a result, HyAsP can only search locally. PlasBin on the other hand uses a branch-and-bound algorithm and performs the search for the optimal solution globally. In many instances, particularly with regards to precision, the greedy heuristic performs better than PlasBin. For recall, however, PlasBin consistently performs better than the other approaches. These observations suggests that the MILP might be lenient in its choice of plasmid contigs. The presence of chromosomal contigs in plasmids predicted by PlasBin may be explained by the following: the MILP tries to maximize the gene density within a plasmid bin, which may result in a plasmid prediction with two contigs with high plasmid gene density joined by a chain of low gene density contigs. If contigs from this intermediate chain indeed originate from chromosomes, it might result in poor precision.

The objective function of PlasBin is a linear combination of two terms, one each pertaining to the gene density and the %GC content deviation of selected contigs. Currently, we consider each term to be equally important. However, it may be worthwhile to explore different linear combinations of the two terms.

Gplas consistently showed a high level of precision. This can likely be credited to the the novel strategy of identifying the plasmid contigs before binning. Once the plasmid contigs are identified using mlplasmids, gplas only considers those contigs in the next step of separating contigs into plasmid bins. This allows gplas to avoid false positives and also significantly reduces the size of the graph needed to be traversed.

Our MILP formulation, especially its objective function, is amenable to incorporate other features of the contigs. The comparison with gplas suggests that incorporating the probability of a contig to originate from a plasmid could be beneficial, without the need to reduce drastically the assembly graph to the subgraph of likely plasmidic contigs as is done by gplas. While gplas is a species-specific tool for contigs classification, that requires a species-specific training datasets, there are other plasmid contig classification tools such as PlasClass [16] that are more general. Integrating contigs classification results and lenient graph pruning is a promising research avenue.

References

1. Antipov, D., Hartwick, N., Shen, M.W., Raiko, M., Lapidus, A.L., Pevzner, P.A.: plasmidSPAdes: assembling plasmids from whole genome sequencing data. Bioinformatics **32**(22), 3380–3387 (2016). https://doi.org/10.1093/bioinformatics/btw493

2. Arredondo-Alonso, S., et al.: gplas: a comprehensive tool for plasmid analysis using short-read graphs. Bioinformatics **36**(12), 3874–3876 (2020). https://doi.org/10.1093/bioinformatics/btaa233

3. Arredondo-Alonso, S., et al.: mlplasmids: a user-friendly tool to predict plasmid- and chromosome-derived sequences for single species. Microb. Genom. **4**(11), e000224 (2018). https://doi.org/10.1099/mgen.0.000224

4. Bankevich, A., et al.: Spades: a new genome assembly algorithm and its applications to single-cell sequencing. J. Comput. Biol. **19**(5), 455–477 (2012). https://doi.org/10.1089/cmb.2012.0021

5. Bertsimas, D., Tsitsiklis, J.: Introduction to Linear Optimization. Athena Scientific, 1st edn. (1997)

6. van der Graaf-van Bloois, L., Wagenaar, J.A., Zomer, A.L.: RFPlasmid: predicting plasmid sequences from short-read assembly data using machine learning. Microb. Genom. **7**(11) (2021). https://doi.org/10.1099/mgen.0.000683

7. Camacho, C., et al.: BLAST+: architecture and applications. BMC Bioinf. **10**, 421 (2009). https://doi.org/10.1186/1471-2105-10-421

8. Carattoli, A.: Plasmids and the spread of resistance. Int. J. Med. Microbiol. **303**(6), 298–304 (2013). https://doi.org/10.1016/j.ijmm.2013.02.001

9. Carattoli, A., et al.: In silico detection and typing of plasmids using plasmidfinder and plasmid multilocus sequence typing. Antimicrob. Agents Chemother. **58**(7), 3895–3903 (2014). https://doi.org/10.1128/AAC.02412-14

10. Dewar, A., et al.: Plasmids do not consistently stabilize cooperation across bacteria but may promote broad pathogen host-range. Nat. Ecol. Evol. **5**(12), 1624–1636 (2021). https://doi.org/10.1038/s41559-021-01573-2

11. Krawczyk, P., Lipinski, L., Dziembowski, A.: Plasflow: predicting plasmid sequences in metagenomic data using genome signatures. Nucleic Acids Res. **46**(6), e35 (2018). https://doi.org/10.1093/nar/gkx1321

12. Luo, L., et al.: Comparative genomics of Chinese and international isolates of Escherichia albertii: population structure and evolution of virulence and antimicrobial resistance. Microb. Genom. **7**(12) (2021). https://doi.org/10.1099/mgen.0.000710

13. McCormick, G.P.: Computability of global solutions to factorable nonconvex programs: Part I—convex underestimating problems. Math. Program. **10**(1), 147–175, e000224 (1976). https://doi.org/10.1007/BF01580665
14. Müller, R., Chauve, C.: HyAsP, a greedy tool for plasmids identification. Bioinformatics **35**(21), 4436–4439 (2019). https://doi.org/10.1093/bioinformatics/btz413
15. Nishida, H.: Comparative analyses of base compositions, DNA sizes, and dinucleotide frequency profiles in archaeal and bacterial chromosomes and plasmids. Int. J. Evol. Biol. **2012**, 342482 (2012). https://doi.org/10.1155/2012/342482
16. Pellow, D., Mizrahi, I., Shamir, R.: Plasclass improves plasmid sequence classification. PLoS Comput. Biol. **16**(4), 1–9 (2020). https://doi.org/10.1371/journal.pcbi.1007781
17. Robertson, J., Nash, J.: MOB-suite: software tools for clustering, reconstruction and typing of plasmids from draft assemblies. Microb. Genom. **4**(8), e000206 (2018). https://doi.org/10.1099/mgen.0.000206
18. Rozov, R., et al.: Recycler: an algorithm for detecting plasmids from de novo assembly graphs. Bioinformatics **33**(4), 475–482 (2016). https://doi.org/10.1093/bioinformatics/btw651
19. Wick, R.R., Judd, L.M., Gorrie, C.L., Holt, K.E.: Unicycler: Resolving bacterial genome assemblies from short and long sequencing reads. PLoS Comput. Biol. **13**(6), 1–22 (2017). https://doi.org/10.1371/journal.pcbi.1005595

Genomic Sequencing

Benchmarking Penalized Regression Methods in Machine Learning for Single Cell RNA Sequencing Data

Bhavithry Sen Puliparambil$^{(\boxtimes)}$ (iD), Jabed Tomal (iD), and Yan Yan (iD)

Thompson Rivers University, 805 TRU Way, Kamloops, BC V2C 0C8, Canada
bsen20@mytru.ca, {jtomal,yyan}@tru.ca

Abstract. Single Cell RNA Sequencing (scRNA-seq) technology has enabled the biological research community to explore gene expression at a single-cell resolution. By studying differences in gene expression, it is possible to differentiate cell clusters and types within tissues. One of the major challenges in a scRNA-seq study is feature selection in high dimensional data. Several statistical and machine learning algorithms are available to solve this problem, but their performances across data sets lack systematic comparison. In this research, we benchmark different penalized regression methods, which are suitable for scRNA-seq data. Results from four different scRNA-seq data sets show that Sparse Group Lasso (SGL) implemented by the SGL package in R performs better than other methods in terms of area under the receiver operating curve (AUC). The computation time for different algorithms varies between data sets with SGL having the least average computation time. Based on our findings, we propose a new method that applies SGL on a smaller pre-selected subset of genes to select the differentially expressed genes in scRNA-seq data. The reduction in the number of genes before SGL reduce the computation hardware requirement from 32 GB RAM to 8 GB RAM. The proposed method also demonstrates a consistent improvement in AUC over SGL.

Keywords: Single cell RNA sequencing · Machine learning · LASSO · Feature selection · High dimensional data · R

1 Introduction

Single cell RNA sequencing (scRNA-seq) technology is gaining popularity in current biological research. With scRNA-seq technology, researchers can simultaneously explore thousands of cells in a tissue and their average gene expression levels, as well as the gene expression profile of each individual cell in that tissue [1]. One of the many applications of scRNA-seq technology is differentiating tumor cells from normal healthy cells by comparing their molecular signatures. However, the scRNA-seq data itself is not without its challenges [2]. One could say it is the very definition of the curse of dimensionality ($p >> n$, where p is the

L. Jin and D. Durand (Eds.): RECOMB-CG 2022, LNBI 13234, pp. 295–310, 2022.
https://doi.org/10.1007/978-3-031-06220-9_17

number of variables and n is the number of observations) problem in machine learning (ML). For scRNA-seq data, the number of genes (variables) far exceeds the number of cells (observations). One way to solve the $p >> n$ problem is to employ feature selection.

Feature selection is the method of selecting variables that are more useful for predicting the target variable. Random forests, Recursive Feature Elimination (RFE), and penalized regression are some of the commonly used feature selection methods in machine learning. Multiple studies have been published on the application of random forests for scRNA-seq data [3,4]. For very high dimensional data such as scRNA-seq, using RFE alone tends to be computationally expensive [5]. Furthermore, some of the penalized regression algorithms are developed specifically for scRNA-seq data showed varied success.

Penalized regression in machine learning has several versions such as ridge regression, least absolute shrinkage and selection operator (LASSO) regression [6] and a combination of ridge and lasso known as elastic net regression [7]. Each of these is useful for a different problem when dealing with scRNA-seq data. For instance, ridge regression is useful for bringing some of the coefficients of the model features closer to 0. To reduce the dimensions, the features close to zero are then forced to be excluded while others are retained in the model. This notion is also known as hard thresholding. This issue is better solved by lasso regression proposed by [6], which can make coefficients equal to absolute zero (soft thresholding). There are many variants of lasso such as Drop lasso [8], Group lasso [9], Sparse Group lasso [10], and Big lasso [11]. One of our objectives is to see which of these algorithms are better suited for different kinds of scRNA-seq data.

There exist several other variants of lasso such as fused lasso [12], adaptive lasso [13], and prior lasso [14]. Fused lasso was proposed for time series or image-based data; adaptive lasso was proposed for proportional hazards regression; and prior lasso was proposed for biological data which requires prior information to be incorporated [14]. Since these algorithms are not suitable for scRNA-seq data, they are not included in this study.

Currently, there are some studies on penalized regression methods for high-dimensional data [15,16]. However, to our knowledge, there is no comprehensive study of how the 7 methods selected for our research perform in comparison with each other, specifically when dealing with scRNA-seq data. Our research aims to fill this knowledge gap and thereby provide a comprehensive guideline as to the performance of these methods. In addition, a new algorithm is proposed which uses less number of genes while improving the AUC of the top algorithm.

The rest of this article is organized as follows. Section 2 explains the different algorithms and metrics used; Sect. 3 introduces the scRNA-seq data sets and the design of this research. Sections 4 and 5 present the results, discuss the findings and biological interpretations, and propose future directions for this research.

2   Methods

The penalized regression algorithms and the performance metrics in this study are described below.

2.1   Penalized Regression

Penalized regression allows feature selection for high dimensional data such as scRNA-seq data by producing sparse solutions which are predictive models based on the expression of a limited number of genes. Penalized regression has two parts, a loss function and a penalty. Let the regression equation be,

$$Y = \mathbf{X}\beta + \epsilon, \tag{1}$$

where Y is a $n \times 1$ vector for response variable, $\mathbf{X}$ is $n \times p$ matrix for predictor variables, β is $p \times 1$ vector of coefficients and $\epsilon \sim N(0, \sigma^2 I)$ is the error term. We consider that both $\mathbf{X}$ and Y are centered and scaled in Eq. 2. The estimated penalized regression coefficients is given by,

$$\hat{\beta} = \underset{\beta}{\mathrm{argmin}}(\frac{1}{n}||Y - \mathbf{X}\beta||^2 + \lambda||\beta||), \tag{2}$$

where $\lambda \geq 0$ is the tuning parameter to be estimated using cross-validation and $||\beta||$ is the norm of coefficient vector β. The first term $\frac{1}{n}||Y - \mathbf{X}\beta||^2$ is the loss function and the second term $\lambda||\beta||$ is the penalty. The main difference between different penalized regressions algorithms is how they apply the penalty. L1 norm and L2 norm are popular choices for penalty [8], which are defined as,

$$L1\ norm = ||\beta||_1 = \sum_{i=1}^{p} |\beta_i|, \tag{3}$$

$$L2\ norm = ||\beta||_2^2 = \sum_{i=1}^{p} \beta_i^2. \tag{4}$$

Ridge Regression is a penalized regression where penalty term is the sum of squared coefficients (Eq. 4). It is especially useful when predictor variables are highly correlated.

The Least Absolute Shrinkage and Selection Operator (LASSO) [6] minimizes the residual sum of squares subject to the sum of the absolute coefficients being less than the tuning parameter. LASSO uses L1 norm (Eq. 3) as penalty. Compared to the ridge regression, which can only shrink coefficients towards 0, lasso can make some coefficients exactly equal to zero thereby producing a more interpretable model.

Elastic net regression proposed by [7] is a combination of L1 and L2 penalties. Therefore, the elastic net regression enjoys the properties of both ridge and lasso regression. The L1 norm part of the penalty produces a sparse model and the L2 norm part of the penalty shrinks very large coefficients [17].

Group LASSO: Yuan and Lin [9] proposed Group LASSO for selecting subsets of important variables. Compared to the lasso which selects variables, group lasso selects groups of variables. This notion is particularly useful in processing scRNA-Seq data because we would like to include or exclude the group of genes that lie in a pathway related to the outcome rather than individual genes. Assume that there are $j = 1, 2, \cdots, J$ groups of variables and n observations. For each group, let $\mathbf{X}_j$ be $n \times p_j$ submatrix of $\mathbf{X}$ with columns corresponding to predictor variables in group j and $\boldsymbol{\beta}_j$ be the corresponding coefficient vector of length p_j. Then the regression equation for group lasso can be written as,

$$Y = \sum_{j=1}^{J} \mathbf{X}_j \boldsymbol{\beta}_j + \epsilon. \tag{5}$$

Note that for $\boldsymbol{\beta} = (\boldsymbol{\beta}_1', \boldsymbol{\beta}_2', \cdots, \boldsymbol{\beta}_j')'$, $\mathbf{X} = (\mathbf{X}_1, \mathbf{X}_2, \cdots .\mathbf{X}_j)$, and $\mathbf{X}_j' \mathbf{X}_j = I_{p_j}$, the above regression equation simplifies to Eq. (1). For a symmetric and positive definite kernel matrix $K_j = p_j I_{p_j}$, the group lasso estimate is

$$\hat{\boldsymbol{\beta}} = \operatorname*{argmin}_{\boldsymbol{\beta}} (\frac{1}{n} \|Y - \sum_{j=1}^{J} \mathbf{X}_j \boldsymbol{\beta}_j\|^2 + \lambda \sum_{j=1}^{J} \|\boldsymbol{\beta}_j' K_j \boldsymbol{\beta}_j\|^{\frac{1}{2}}), \tag{6}$$

where $\lambda \geq 0$ is the tuning parameter.

Sparse Group LASSO: The shortcoming of group lasso is that while it gives a sparse set of groups, all the coefficients in a selected group will be nonzero. But sometimes both sparsity of groups and variables within each group are desired. In scRNA-seq data, identifying important genes within the biological pathways is often of interest. Simon et al. [10] proposed sparse group lasso as a solution to this specific problem.

Drop LASSO: scRNA-seq data often become intricated by noise features that need to be dropped out. Dropout noise occurs when scRNA-seq fails to detect some genes even though they are expressed in the cell [8]. Consequently, those genes will cause zeros in the data set. Khalfaoui and Vert [8] proposed Drop Lasso as a better-adapted solution to data complication by dropout noise. It is a combination of the dropout regularisation technique proposed by [18] and lasso proposed by [6]. It creates a sparse linear model robust to the noise by artificially augmenting the training set with new examples corrupted by dropout.

Big LASSO: Zeng and Breheny [11] proposed big lasso algorithm in R for handling ultra-high dimensional and large scale data. Their approach handles out-of-core computation seamlessly by loading data into memory only when necessary. This is done with the help of memory-mapped files which store massive data on the disk. The big lasso algorithm also possesses efficient feature screening rules which can accelerate the computation. The major differences between Big lasso and lasso are in out-of-core computation and parallel processing.

2.2 Clustering

Clustering is the process of grouping data into clusters so that objects of the same cluster are similar to each other [19]. Two popular clustering methods are hierarchical clustering and K-Means clustering which are described below.

In **hierarchical clustering**, objects are grouped into clusters by partitioning data hierarchically [19]. These clusters are then graphically represented as a tree-like diagram known as a dendrogram. The dendrogram is very useful in deciding the optimal number of clusters. In this study, we use hierarchical clustering to group genes into clusters prior to applying into group lasso and SGL.

K-Means clustering algorithm starts by randomly placing k centroids in p data points scattered in the n-dimensional space [20]. Clusters are formed by assigning data points to the nearest centroid. The algorithm progress iteratively by moving the centroids at each step such that the clustering error is minimized. K-Means clustering when applied in conjunction with Lasso improved prediction accuracy [21]. K-means clustering is employed to cluster cells at the final step of the algorithm proposed in this research.

2.3 K-Fold Cross-validation

Cross-validation is a data partitioning method used to estimate the prediction error of models and to tune model parameters [22]. We will use K-Fold Cross-Validation for tuning the hyper parameters for all the algorithms in this study. In K-Fold Cross-Validation, the data are first divided into k subsets of cells. One of the k subsets is used as the test set and the remaining $k - 1$ subsets are used as the training set. Then prediction error is calculated for k repeats by selecting a different test set each time. The average error for k repeats is used as the prediction performance of the model. In this study, we use 10-fold cross-validation for measuring performance.

2.4 ROC AUC

A receiver operating characteristic curve (ROC) is a graph that plots the true positive rate (TPR) on the y-axis and the false positive rate (FPR) on the x-axis [23]. ROC is used to evaluate the prediction ability of a binary classification model. The area under the ROC curve is known as ROC AUC. The AUC value reflects the overall prediction performance of a classifier. The AUC is theoretically and empirically better than the accuracy metric for evaluating the classifier performance and discriminating an optimal solution especially for unbalanced classification. However, the computational cost of AUC is higher than accuracy and misclassification error [24].

3 Research Design and Data

3.1 Experimental Data

In this study, 4 scRNA-seq data from 3 different species (Human, mouse, and plant) are used. The first two data sets (GSE60749 and GSE71585) are selected from a collection of 40 curated scRNA-seq data from conquer website (http://imlspenticton.uzh.ch:3838/conquer/) created by [25]. Conquer website is a collection of consistently processed, analysis-ready, and well-documented publicly available scRNA-seq data. There are currently 40 data sets on conquer website, each having count and transcripts per million (TPM) estimates for genes, as well as quality control and exploratory analysis reports. The other two data sets (GSE81861 and GSE123818) are downloaded from Gene Expression Omnibus (GEO). Table 1 shows the data sets with GEO accession numbers, number of genes, number of cells in each class, species, technology used for scRNA-seq and the source from where data were accessed. The cell groups are selected as per the label from the original experiment. No additional batch effect correction is performed during preprocessing in this research.

Table 1. Experimental data sets

Data set	Genes	Cells	Organism	Technology	Source
GSE60749	224444	183:84	Mus Musculus	Illumina HiSeq	conquer
GSE71585	24058	79:57	Mus Musculus	Fluidigm BioMark	conquer
GSE81861	57241	272:160	Homo Sapiens	Fluidigm based single cell RNA-seq protocol	GEO
GSE123818	27629	1099:1099	Arabidopsis Thaliana	Illumina NextSeq	GEO

The first data set GSE60749 is of species Mus musculus. This data set was generated in the study of gene expression variability in pluripotent stem cells (PSCs) by single-cell expression profiling of PSCs under different chemical and genetic perturbations conducted by [26]. Gene expression levels are quantified as transcripts per million reads (TPM). For our research, we selected 183 individual v6.5 mouse embryonic stem cells (mESCs) and 84 Dgcr8 -/- mESCs that lack mature miRNAs (knockout of a miRNA processing factor). The 183 individual mESCs are assigned to class 1 and 84 Dgcr8 -/- mESCs are assigned to class 0. The data included 22443 genes initially which was reduced to 15508 after preprocessing in which all the genes with no variance in expression across all the cells were removed.

The second data set GSE71585 is also of the species Mus musculus. Our research included mouse species data set because numerous clinical trials are conducted on mice prior to human trials. GSE71585 data set was generated by scRNA-Seq of adult mouse primary visual cortex in a study [27] conducted to understand cell type diversity in the nervous system. There are 1809 cells and 24057 genes in this data set. Gene expression levels are quantified as transcripts per million reads (TPM). Out of all the cells, 79 Ntsr1_tdTpositive_cell are assigned to class 0 and 57 Ntsr1_tdTnegative_cell are assigned to class 1.

After removing the genes which did not vary in expression across all cells, there are 17870 genes in total.

The third data set GSE81861 is of species homo sapiens and is from the analysis of transcriptional heterogeneity in colorectal tumors [28]. Intratumoral heterogeneity is a major obstacle to cancer treatment and a significant confounding factor in bulk-tumor profiling. Therefore [28] conducted an analysis of transcriptional heterogeneity in colorectal tumors and their microenvironments using scRNA-seq. There are 272 primary colorectal tumor cells and 160 matched normal mucosa cells. Gene expression levels are quantified as fragments per kilobase per million reads (FPKM). A binary classification problem was created by assigning 272 primary colorectal tumor cells to class 1 and 160 matched normal mucosa cells to class 0. The data was then transposed to form a matrix of 432 rows (cells) and 57242 columns (genes). Standardization and normalization were not carried out for this data set because it negatively affected the performance of Lasso algorithms in preliminary analysis. All the genes which were not expressed (0 values) or equally expressed among all cells were removed, thereby reducing the number of genes to 38090.

Our last data set GSE123818 belong to the plant species Arabidopsis Thaliana and it contains thousands of cells. This data is obtained from the study of Spatiotemporal Developmental Trajectories in the Arabidopsis Root [29]. The study generated mRNA profiles of 6-day-old wild-type (wt) and shortroot-knockout (shr) Arabidopsis thaliana roots by deep sequencing of single cell and bulk RNA libraries (wild type only), in duplicate (bulk & wild-type single cell) and singlicate (shr-3), using Illumina NextSeq. There are 4727 wt cells and 1099 shr cells in this dataset. All 1099 shr cells are selected for our research. To have balanced classes, 1099 samples are selected at random from 4727 wt cells. Class 0 was assigned to wt cells and class 1 was assigned to shr cells to create the binary classification problem. There are 27629 genes in the data set which are reduced to 24075 after removing the genes which do not vary in expression (same value) across all cells.

3.2   Research Design

As the first step, each data set is pre-processed to be compatible for use in different R packages. Two cell groups, as labeled by the individual experiment in which the data set was created, are assigned 0 or 1 for the binary classification problem. Note that in the future we may expand the study to include more cell groups and verify algorithms for the multi-classification problem. With the processed data, we verify how each algorithm performs in terms of AUC and computation time when dealing with scRNA-seq data from different species. We use the same performance measures (AUC, and computation time) and cross-validation for all data sets to ensure a fair comparison. In this step, a 10-fold Stratified Cross-Validation is conducted to fine-tune the hyperparameters for each algorithm, and then the performance metrics for all the algorithms are calculated. We used hierarchical clustering for grouping variables prior to Group Lasso and Sparse Group Lasso. After comparing the performance metrics, best-performing algorithms are selected and combined to form a new algorithm. Finally, the performance metrics of the new algorithm is compared with that of the top-performing algorithm. Figure 1 illustrates the proposed algorithm. The computation has been done using Ubuntu 20.04.4 LTS (GNU/Linux 5.4.0-100-generic x86_64) with 32 GB RAM hosted by Compute Canada. R version 4.1.2 was used for software programming.

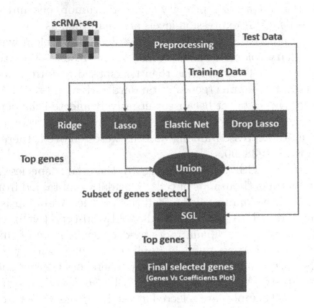

Fig. 1. Schematic diagram of the proposed algorithm. In this algorithm, there is a significant reduction in the number of genes prior to the execution SGL. Once the final set of genes are selected, they are used to cluster cell groups in the data set.

4   Results

Our **first** objective is to compare the performance of the methods. The resulting average cross-validated AUC (CV-AUC) and computation time are shown in Table 2 and Table 3, respectively. Figure 2 shows the average CV-AUC across all 4 data sets for each algorithm. From Table 2 and Fig. 2 we observe that the top 5 algorithms are SGL, grplasso, droplasso, biglasso, and lasso. Notice that SGL and grplasso outperform all other methods in terms of average CV-AUC, whereas ridge regression algorithm in glmnet package has the least average CV-AUC. This could be because grplasso and SGL incorporate grouping of genes information into model, whereas ridge regression treats all the genes equally.

Algorithm 1. Steps to implement the proposed algorithm

1. Load data set into R and assign classes 1 and 0 to the two selected group of cells to form a binary classification problem.
2. Shuffle cells within each class to randomize data points.
3. Remove genes with no variability in expression across all cells from the data set.
4. Split the data set into training (90%) and testing data (10%) for 10-fold cross validation.
5. Repeat the steps for a 10-fold CV
   (a) Fit ridge, lasso, elastic net and drop lasso.
   (b) Select the top genes from each algorithm. The top genes are the genes which have coefficients above a cut off (mean of absolute value of coefficients).
   (c) Form a gene pool by taking union of only the top genes from all 4 models.
   (d) Fit SGL with the new gene pool pre-grouped by hierarchical clustering.
   (e) Save the coefficients of SGL.
6. Find the average of coefficients for each gene across 10 folds and then sort the genes.
7. Visualize the gene Vs coefficients plot and select the final set of genes which are above the elbow of the curve.

Across data sets, the average computation time is the least for SGL and biglasso, while the most time-consuming algorithm is ridge. Ridge regression uses all genes making computation complex. On the other hand, SGL can make an entire group of genes, as well as some of the genes within selected groups, zero resulting in a sparse matrix and lesser computation. It is notable that computation time for GSE81861 data set is higher compared to that of GSE60749 for most of the algorithms due to more number of non-zero coefficients for the former data set.

The **second** objective of this study is to combine lasso algorithms in order to improve AUC and gene selection. From the discussion of the results of the first objective, we see that SGL and grplasso are good candidates for forming a new algorithm. In terms of gene selection, SGL performs better than grplasso. SGL is therefore chosen over grplasso for the new computational algorithm. SGL achieves better AUC than biglasso in comparable time for data sets of size 20 MB to 250 MB when tested on a computer with a 32 GB processor. Therefore biglasso is not included in the new algorithm.

For the new algorithm, we select the ridge, lasso, elastic net, and droplasso to form a filter which creates a gene pool with the number of genes being significantly reduced. The gene pool is formed by taking a union of top genes from 4 algorithms because we observed that different methods might give different set of top genes. A union of top genes is, therefore, more likely to capture the important differentially expressed genes. The gene pool is used as input to SGL to calculate the AUC.

In the new algorithm, hierarchical clustering and SGL are executed with a gene pool that has significantly less number of genes. This reduction in number of genes enables us to do the computation with an 8 GB processor for two data

sets (GSEGSE60749, GSE71585) and get the same AUC as obtained with the
32 GB processor. This new computational algorithm can be used for other high
dimensional data sets as well for feature selection. We note that the AUC of the
proposed algorithm shown in Table 4 is equal to or better than that of SGL. The
final selection of genes is found using a genes Vs coefficients plot of the SGL
fit. In the last step, we use K-Means clustering for cell clustering with the final
selection of genes for each data set. The clustering results are shown in Fig. 3.

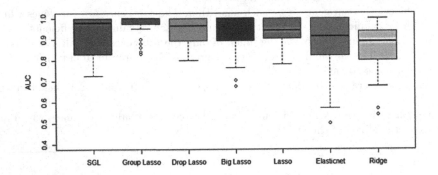

Fig. 2. Average Cross-validation AUC across all 4 data sets. Even though group lasso
has better AUC than SGL, SGL is better in terms of gene selection. Selecting the
differentially expressed genes is of more importance for a scRNA-seq data set compared
to prediction AUC.

5   Discussion

In this section, we discuss the final subset of genes selected with the new algorithm, and the cell clustering for each data set.

As shown in Fig. 3, the data set GSE60749 clustered with two classes is well
separated. The top genes of this data set identified by the new algorithm are
44441, 44260, 44454, 44446, 44450, 44440, Pbld, Lifr, Hist2h4, and AK203176.
One interesting finding is that 44441, 44260, 44454, and 44440 genes are non-
coding RNAs called piRNA. This indicates a possibility of an association between
the knockout of miRNA processing factor and piRNAs which has not been studied so far. piRNAs are found in humans and rats. Although their function must
still be resolved, the abundance of piRNAs in germline cells and the male sterility
of Miwi mutants suggest a role in gametogenesis [30].

The new algorithm can cluster primary visual cortex cell groups in GSE71585
well. The top 2 differentially expressed genes in Ntsr1 (neurotensin receptor 1)
tdT (tdTomato - an exceptionally bright red fluorescent protein) positive cells
and Ntsr1 tdT negative cells are Calm2 and Snap25. Calm2 is active in the
pathways of Alzheimer's disease and Glycogen Metabolism [31]. Snap25 gene

Table 2. Algorithm performance comparison (Metric = Average CV-AUC).

Algorithm	R Package	GSE60749	GSE71585	GSE81861	GSE123818
Sparse Group Lasso	SGL	1	**0.98**	**0.92**	0.83
Group Lasso	grplasso	1	**0.98**	0.87	**0.99**
Drop Lasso	droplasso	0.99	0.94	0.87	**0.97**
Big Lasso	biglasso	1	1	0.80	0.95
Lasso	glmnet	1	0.96	0.85	0.94
Elastic net	glmnet	1	0.63	0.86	0.93
Ridge	glmnet	0.99	0.84	0.71	0.90

Table 3. Algorithm performance comparison (Metric = Average computation time in seconds).

Algorithm	R Package	GSE60749	GSE71585	GSE81861	GSE123818
Sparse Group Lasso	SGL	6.53	**1.73**	**2.97**	5.66
Group Lasso	grplasso	**1.12**	2.51	29.66	3.78
Drop Lasso	droplasso	13.57	7.18	59.33	**3.36**
Big Lasso	biglasso	3.11	4.77	7.23	20.30
Lasso	glmnet	3.18	2.76	13.39	48.54
Elastic net	glmnet	3.57	2.66	13.59	51.51
Ridge	glmnet	58.07	26.71	3.77	17.58

enables syntaxin-1 binding activity. It is present in 10 different biological pathways. NCBI records [32] show that it is used to study attention deficit hyperactivity disorder, obesity, schizophrenia, and type 2 diabetes mellitus. The human ortholog of this gene is implicated in Down syndrome and congenital myasthenic syndrome 18. The strong association between the lack of tdT protein in Ntsr1 cells and these genes identified by our method shows merit for further study.

For GSE81861, cell groups are clustered with some overlap in classes. The top genes are FABP1, SAT1, PHGR1, LGALS4, FRYL, MT1E, HSP90AA1, and HNRNPH1. FABP1 gene enables long-chain fatty acids binding activity. It is involved in 13 biological pathways including metabolism, and Peroxisome proliferator-activated receptor (PPAR) signaling pathway [33]. The second gene SAT1 is also involved in 13 biological pathways including metabolism. Defects in this gene are associated with keratosis follicularis spinulosa decalvans (KFSD) [34]. LGALS4 gene is underexpressed in colorectal cancer [35]. Similarly, HSP90AA1 gene is also an important gene found in 115 pathways such as signaling by EGFR, EGFRvIII, and ERBB2 in Cancer [36]. HNRNPH1 gene found in 12 pathways may be associated with hereditary lymphedema type I. Knockdown of heterogeneous nuclear ribonucleoprotein H1 (HNRNPH1) by siRNA inhibits the early stages of HIV-1 replication in 293T cells infected with VSV-G

Table 4. Comparison of performance (AUC) between SGL with all genes and SGL using new algorithm which reduces number of genes by union of top genes from ridge, lasso, elastic net and droplasso. The new algorithm is consistently improving over SGL.

Data set	All genes	Gene pool	SGL	New algorithm
GSE60749	224444	5965	1	1
GSE71585	24058	5448	0.98	1
GSE81861	57241	5823	0.92	0.94
GSE123818	27629	10857	0.83	0.85

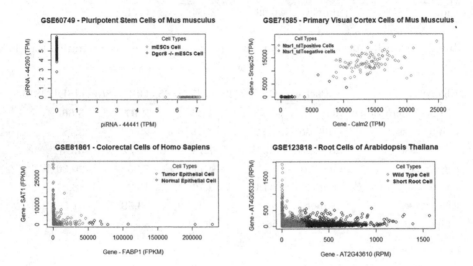

Fig. 3. Cell clustering (K-Means) with final selection of genes for all 4 data sets. The top gene (piRNA 44441) alone can perfectly differentiate two cell groups in GSE60479 data set. There is some overlap of cell clusters within the other 3 data sets.

pseudotyped HIV-1 [37]. 4 genes (SAT1, MT1E, HSP90AA1, and HNRNPH1) out of 8 top genes from a colorectal tumor have strong interaction with HIV-1 proteins. The association between cancer and HIV-1 has been widely studied [38–41] by medical researchers. Evidently, the new algorithm can select a highly relevant subset of genes from the given samples of human colorectal cancer cells.

GSE123818 data set have more overlap between cell clusters compared to that of the other data sets. The top genes of this data set as found with the new algorithm are AT2G43610, AT4G05320, AT2G07698, and AT3G51750. One of five polyubiquitin genes in A. thaliana, AT2G43610 gene is found in growth and developmental stages such as root development [42]. AT4G05320 gene encodes the highly conserved 76-amino acid protein ubiquitin which is attached to proteins targeting degradation [43]. AT2G07698 gene is expressed during seed development stage [44]. AT3G51750 codes a hypothetical protein which is involved in root and seed development [45]. All of the selected genes are related to the growth and developmental stages in Arabidopsis thaliana. We recommend further study of these genes in relation to root development and degradation.

6 Conclusion and Future Work

As evident from the results and detailed discussion above, the sparse group lasso algorithm with a reduced set of genes can select a highly relevant subset of genes that are strongly associated with the cell clusters in scRNA-seq samples. Here we recognize that lasso algorithms have many hyperparameters which can be customized to arrive at different results. The output of grplasso and SGL packages can also change depending on the number of groups in the input data and the type of grouping used. The new algorithm proposed in this research is a combination of lasso algorithms that showed consistently better AUC than the top-performing lasso algorithm (SGL). The proposed algorithm does not require deep knowledge in the grouping of genes in scRNA-seq data and yet identified a highly relevant set of differentially expressed genes. A 32 GB RAM Linux remote server was used to execute SGL on the whole scRNA-seq data set, whereas, the proposed algorithm can be executed on a small computer of 8 GB RAM because it uses a small subset of genes to run SGL instead of all of the genes. This research can expand in the future to include more algorithms and related R packages, such as nuisance penalized regression [46] or Seagull [47] which also implement lasso, group lasso, and sparse group lasso algorithms. scRNA-seq data sets may have more than one cell group. Therefore, another direction worth exploring is verifying the R packages such as msgl [48] which can implement multinomial classification.

Acknowledgements. The authors would like to acknowledge the funding for this research from,

1. TRU Internal Research Fund (IRF) awarded to Dr. Jabed Tomal, Department of Mathematics and Statistics, Thompson Rivers University and Dr. Yan Yan, Department of Computing Science, Thompson Rivers University.

2. Natural Sciences and Engineering Research Council of Canada (NSERC) awarded to Dr. Jabed Tomal, Department of Mathematics and Statistics, Thompson Rivers University and Dr. Yan Yan, Department of Computing Science, Thompson Rivers University.

The authors also acknowledge Compute Canada for hosting the 32 GB Linux remote server which is used for computation in this research.

References

1. Slovin, S., et al.: Single-cell RNA sequencing analysis: a step-by-step overview. RNA Bioinform. 343–365 (2021). https://doi.org/10.1007/978-1-0716-1307-8_19
2. Kiselev, V.Y., Andrews, T.S., Hemberg, M.: Challenges in unsupervised clustering of single-cell RNA-seq data. Nat. Rev. Genet. **20**(5), 273–282 (2019)
3. Kaymaz, Y., Ganglberger, F., Tang, M., Fernandez-Albert, F., Lawless, N., Sackton, T.B.: HieRFIT: Hierarchical Random Forest for Information Transfer. bioRxiv (2020). https://doi.org/10.1101/2020.09.16.300822

4. Pouyan, M.B., Kostka, D.: Random forest based similarity learning for single cell RNA sequencing data. Bioinformatics **34**(13), i79–i88 (2018)
5. Chen, X.W., Jeong, J.C.: Enhanced recursive feature elimination. In: Sixth International Conference on Machine Learning and Applications (ICMLA 2007), pp. 429–435. IEEE (2007)
6. Tibshirani, R.: Regression shrinkage and selection via the lasso: a retrospective. J. Roy. Stat. Soc. Ser. B (Stat. Methodol.) **73**(3), 273–282 (2011)
7. Zou, H., Hastie, T.: Regularization and variable selection via the elastic net. J. Roy. Stat. Soc. Ser. B (Stat. Methodol.) **67**(2), 301–320 (2005)
8. Khalfaoui, B., Vert, J.P.: DropLasso: a robust variant of Lasso for single cell RNA-seq data. arXiv preprint arXiv:1802.09381 (2018)
9. Yuan, M., Lin, Y.: Model selection and estimation in regression with grouped variables. J. Roy. Stat. Soc. Ser. B (Stat. Methodol.) **68**(1), 49–67 (2006)
10. Simon, N., Friedman, J., Hastie, T., Tibshirani, R.: A sparse-group lasso. J. Comput. Graph. Stat. **22**(2), 231–245 (2013)
11. Zeng, Y., Breheny, P.: The biglasso package: a memory-and computation-efficient solver for lasso model fitting with big data in R. arXiv preprint arXiv:1701.05936 (2017)
12. Tibshirani, R., Saunders, M., Rosset, S., Zhu, J., Knight, K.: Sparsity and smoothness via the fused lasso. J. Roy. Stat. Soc. Ser. B (Stat. Methodol.) **67**(1), 91–108 (2005)
13. Zou, H.: The adaptive lasso and its oracle properties. J. Am. Stat. Assoc. **101**(476), 1418–1429 (2006)
14. Jiang, Y., He, Y., Zhang, H.: Variable selection with prior information for generalized linear models via the prior lasso method. J. Am. Stat. Assoc. **111**(513), 355–376 (2016)
15. Scialdone, A., et al.: Computational assignment of cell-cycle stage from single-cell transcriptome data. Methods **85**, 54–61 (2015)
16. Cao, X., Xing, L., Majd, E., He, H., Gu, J., Zhang, X.: A systematic evaluation of methods for cell phenotype classification using single-cell RNA sequencing data. arXiv preprint arXiv:2110.00681 (2021)
17. Zou, H., Hastie, T.: Regression shrinkage and selection via the elastic net, with applications to microarrays. JR Stat. Soc. Ser. B **67**, 301–20 (2003)
18. Srivastava, N., Hinton, G., Krizhevsky, A., Sutskever, I., Salakhutdinov, R.: Dropout: a simple way to prevent neural networks from overfitting. J. Mach. Learn. Res. **15**(1), 1929–1958 (2014)
19. Rani, Y., Rohil, H.: A study of hierarchical clustering algorithm. ter S on Te SIT **2**, 113 (2013)
20. Hartigan, J.A., Wong, M.A.: Algorithm AS 136: a K-means clustering algorithm. J. Roy. Stat. Soc. Ser. C (Appl. Stat.) **28**(1), 100–108 (1979). https://doi.org/10.2307/2346830
21. Hua, J., Liu, H., Zhang, B., Jin, S.: Lak: lasso and K-means based single-cell RNA-seq data clustering analysis. IEEE Access **8**, 129679–129688 (2020)
22. Bates, S., Hastie, T., Tibshirani, R.: Cross-validation: what does it estimate and how well does it do it? arXiv preprint arXiv:2104.00673 (2021)
23. Park, S.H., Goo, J.M., Jo, C.H.: Receiver operating characteristic (ROC) curve: practical review for radiologists. Korean J. Radiol. **5**(1), 11–18 (2004). https://doi.org/10.3348/kjr.2004.5.1.11
24. Hossin, M., Sulaiman, M.N.: A review on evaluation metrics for data classification evaluations. Int. J. Data Mining Knowl. Manag. Process **5**(2), 1 (2015)

25. Soneson, C., Robinson, M.D.: Bias, robustness and scalability in differential expression analysis of single-cell RNA-Seq data. bioRxiv, 143289 (2017)
26. Kumar, R.M., et al.: Deconstructing transcriptional heterogeneity in pluripotent stem cells. Nature **516**(7529), 56–61 (2014). https://doi.org/10.1038/nature13920
27. Tasic, B., et al.: Adult mouse cortical cell taxonomy revealed by single cell transcriptomics. Nat. Neurosci. **19**(2), 335–346 (2016). https://doi.org/10.1038/nn.4216
28. Li, H., et al.: Reference component analysis of single-cell transcriptomes elucidates cellular heterogeneity in human colorectal tumors. Nat. Genet. **49**(5), 708–718 (2017). https://doi.org/10.1038/ng.3818
29. Denyer, T., Ma, X., Klesen, S., Scacchi, E., Nieselt, K., Timmermans, M.C.: Spatiotemporal developmental trajectories in the Arabidopsis root revealed using high-throughput single-cell RNA sequencing. Dev. Cell **48**(6), 840–852 (2019)
30. Girard, A., Sachidanandam, R., Hannon, G., et al.: A germline-specific class of small RNAs binds mammalian Piwi proteins. Nature **442**, 199–202 (2006). https://doi.org/10.1038/nature04917
31. Calm2 calmodulin 2 [Mus musculus (house mouse)] [Internet]. Bethesda (MD): National Library of Medicine (US), National Center for Biotechnology Information (2022). https://www.ncbi.nlm.nih.gov/gene/12314. Accessed 17 Jan 2022
32. Snap25 synaptosomal-associated protein 25 [Mus musculus (house mouse)] [Internet]. Bethesda (MD): National Library of Medicine (US), National Center for Biotechnology Information (2022). https://www.ncbi.nlm.nih.gov/gene/20614. Accessed 17 Jan 2022
33. Fabp1 fatty acid binding protein 1, liver [Mus musculus (house mouse)] [Internet]. Bethesda (MD): National Library of Medicine (US), National Center for Biotechnology Information (2022). https://www.ncbi.nlm.nih.gov/gene/14080. Accessed 17 Jan 2022
34. SAT1 spermidine/spermine N1-acetyltransferase 1 [Homo sapiens (human)] [Internet]. Bethesda (MD): National Library of Medicine (US), National Center for Biotechnology Information (2022). https://www.ncbi.nlm.nih.gov/gene/6303. Accessed 17 Jan 2022
35. LGALS4 galectin 4 [Homo sapiens (human)] [Internet]. Bethesda (MD): National Library of Medicine (US), National Center for Biotechnology Information (2022). https://www.ncbi.nlm.nih.gov/gene/3960. Accessed 17 Jan 2022
36. HSP90AA1 heat shock protein 90 alpha family class A member 1 [Homo sapiens (human)] [Internet]. Bethesda (MD): National Library of Medicine (US), National Center for Biotechnology Information (2022). https://www.ncbi.nlm.nih.gov/gene/3320. Accessed 17 Jan 2022
37. HNRNPH1 heterogeneous nuclear ribonucleoprotein H1 [Homo sapiens (human)] [Internet]. Bethesda (MD): National Library of Medicine (US), National Center for Biotechnology Information (2022). https://www.ncbi.nlm.nih.gov/gene/3187. Accessed 17 Jan 2022
38. König, R., et al.: Global analysis of host-pathogen interactions that regulate early-stage HIV-1 replication. Cell **135**(1), 49–60 (2008). https://doi.org/10.1016/j.cell.2008.07.032
39. Nunnari, G., Smith, J.A., Daniel, R.: HIV-1 Tat and AIDS-associated cancer: targeting the cellular anti-cancer barrier? J. Exp. Clin. Cancer Res. **27**(1), 1–8 (2008)
40. Corbeil, J., et al.: Productive in vitro infection of human umbilical vein endothelial cells and three colon carcinoma cell lines with HIV-1. Immunol. Cell Biol. **73**(2), 140–145 (1995)

41. Alfano, M., Graziano, F., Genovese, L., Poli, G.: Macrophage polarization at the crossroad between HIV-1 infection and cancer development. Arterioscler. Thromb. Vasc. Biol. **33**(6), 1145–1152 (2013)
42. The Arabidopsis Information Resource (TAIR). https://www.arabidopsis.org/servlets/TairObject?type=locus&name=At2g43610. www.arabidopsis.org. Accessed 17 Jan 2022
43. The Arabidopsis Information Resource (TAIR). https://www.arabidopsis.org/servlets/TairObject?type=locus&id=126703. www.arabidopsis.org. Accessed 17 Jan 2022
44. The Arabidopsis Information Resource (TAIR). https://www.arabidopsis.org/servlets/TairObject?type=locus&name=At2g07698. www.arabidopsis.org. Accessed 17 Jan 2022
45. The Arabidopsis Information Resource (TAIR). https://www.arabidopsis.org/servlets/TairObject?type=locus&name=At3g51750. www.arabidopsis.org. Accessed 17 Jan 2022
46. Sun, Q., Zhang, H.: Targeted inference involving high-dimensional data using nuisance penalized regression. J. Am. Stat. Assoc. **116**(535), 1472–1486 (2021)
47. Klosa, J., Simon, N., Westermark, P.O., Liebscher, V., Wittenburg, D.: Seagull: lasso, group lasso and sparse-group lasso regularization for linear regression models via proximal gradient descent. BMC Bioinform. **21**(1), 1–8 (2020)
48. Vincent, M., Hansen, N.R.: Sparse group lasso and high dimensional multinomial classification. Computat. Stat. Data Anal. **71**, 771–786 (2014)

Deciphering the Tissue-Specific Regulatory Role of Intronless Genes Across Cancers

Katia Aviña-Padilla[1] , José Antonio Ramírez-Rafael[1] ,
Octavio Zambada-Moreno[1] , Gabriel Emilio Herrera-Oropeza[2] ,
Guillermo Romero[3] , Ishaan Gupta[4] , and Maribel Hernández-Rosales[1(✉)]

[1] CINVESTAV-Irapuato, Libramiento Norte Carretera Irapuato León Kilómetro 9.6,
36821 Irapuato, Guanajuato, Mexico
maribel.hr@cinvestav.mx
[2] King's College London, Strand, London WC2R 2LS, UK
[3] Data-Pop Alliance, 99 Madison Avenue, New York, NY 11211, USA
[4] Indian Institute of Technology - Delhi, Main Road, IIT Campus,
Hauz Khas, New Delhi 110016, India

Abstract. Intronless genes (IGs) or single-exon genes lacking introns
are found across Eukaryotes. IGs are not regulated by the splicing machinery and may be subject to lower post-transcriptional gene
expression variability. Therefore, IGs might be potential candidates for
biomarkers with better predictability and easier regulation as targets for
therapy. Cancer is a complex disease that relies on progressive uncontrolled cell division linked with multiple dysfunctional biological processes. Tumor heterogeneity remains the most challenging feature in cancer diagnosis and treatment. Given the clinical relevance of IGs, we aim
to identify their unique expression profiles and interactome, that may
act as functional signatures across eight different cancers. We identified 940 protein-coding IGs in the human genome, of which about 35%
were differentially expressed across the analyzed cancer datasets. Specifically, 78% of differentially expressed IGs were undergoing transcriptional reprogramming with elevated expression in tumor cells. Remarkably, in all the studied tumors, a highly conserved induction of a group
of deacetylase-histones located in a region of chromosome 6 enriched
in nucleosome and chromatin condensation processes. This study highlights that differentially expressed human intronless genes across cancer
types are prevalent in epigenetic regulatory roles participating in specific protein-protein interaction (PPI) networks for ESCA, GBM, and
LUAD tumors. We determine that IGs play a key role in the tumor phenotype at transcriptional and post-transcriptional levels, with important
mechanisms such as interactomics rewiring.

Keywords: Single-exon genes · Comparative genomics · Cancer
genomics · Epigenetics · Tumor heterogeneity · Cancer evolutionary
genomics

Supported by CONACYT.

1   Introduction

Most eukaryotic gene structures contain exons interrupted by non-coding introns that are removed by RNA splicing to generate the mature mRNA. Although the most prevalent class of genes in the human genome contain Multiple Exon Genes (MEGs), about 5% of the genes are Intronless (IGs) or Single-Exon (SEGs) that lack introns. Due to the absence of introns and associated post-transcriptional splicing, IGs may be subject to lower post-transcriptional gene expression variability suggesting a potential role as clinical biomarkers and drug targets that deserve careful consideration in diseases such as cancer [13,24,27]. Several previous studies have identified the role of IGs in cancer [2,5,13,17,36]. For example the *RPRM* gene increased cell proliferation and tumor suppression activity in gastric cancer [2]; *CLDN8* gene is associated with colorectal carcinoma and renal cell tumors [5] while *ARLTS1* is upregulated in melanoma; *PURA* & *TAL2* are upregulated in leukemia [13] and protein kinase *CK2α* gene is up-regulated in all human cancers [17].

A remarkable instance of IGs acting in a clinical role is *SOX11*, a member of *SOXC* (SRY-related HMG-box) gene family of transcription factors involved in embryonic development and tissue remodeling by participating in cell fate determination [4]. *SOX11* has been associated with tumorigenesis, with aberrant nuclear protein expression in Mantle Cell Lymphoma (MCL) patients [8,30,35,37]. This TF is not expressed in normal lymphoid cells or other mature B cell lymphomas (except Burkitt lymphoma), but it is highly expressed in conventional MCL, including the cyclin D1-MCL subtype [36]. Hence, *SOX11* represents a widely used marker in the differential diagnosis of MCL and other types of small B-cell neoplasias in the clinical hemato-oncology practice [18,31].

However, to date, a comparative analysis integrating transcriptomic and interactomics profiles of IGs across diverse types of cancer is missing. Hence, this work aims to identify and characterize their expression, functional role and interactomics profiles across different selected cancers.

2   Results

In this study we use RNA-sequencing data from 3880 tumor samples belonging to 8 cancer types from the cancer genome atlas or TCGA (Appendix Table 1). We selected the four most prevalent cancers, namely Breast Invasive Carcinoma (BRCA), Colon Adenocarcinoma (COAD), Lung Adenocarcinoma (LUAD) and Prostate Adenocarcinoma (PRAD), along with the four most aggressive cancers with high intrinsic heterogeneity, namely Bladder Urothelial Carcinoma (BLCA), Esophageal Carcinoma (ESCA), Glioblastoma Multiforme (GBM) and Kidney Renal Clear Cell Carcinoma (KIRC). We found that 338 out of the 940 genes identified as IG-encoded proteins are undergoing differential regulation in tumors compared to the normal tissue. GBM had the most number of differentially expressed (DE)-IGs at 168, followed by KIRC at 130, and both PRAD and COAD at 116. While in BRCA 104, LUAD 87, BLCA 97 and ESCA 86 were determined.

2.1   Functional Assignment and Gene Expression of IGs in Normal Tissue

First, we performed a functional assignment analysis to characterize the biological role and the expression profiles of IG-encoded proteins in normal tissue. Our results determined that 940 proteins in the human genome are classified as IGs. Secondly, when performing a functional enrichment analysis, it stands out their relevance in chromosome and genetic material organization as well as their response to sensory stimulus in biological processes (Fig. 1).

Furthermore, we examined their constitutive expression on healthy tissue samples corresponding to those of the analyzed tumors. This analysis based on the GTex database, is depicted in Fig. 2. We observe that the identified tissue samples have more similar transcriptional profiles when belonging to the same tissue. There is a clear separation of clusters belonging to the brain, with very different profiles in comparison to the rest of tissues. Kidney cortex, lung, prostate and breast, esophagus mucosa, colon transverse clusters are very well distinguished from the rest; while all the gastrointestinal tissues are clustered in the same expression area. In contrast, bladder samples do not seem to have a tissue-specific expression profile. Hence, IGs have tissue-specific expression in most of the tissues corresponding to the cancer diseases/tumors studied here.

2.2   IGs Tend to Have a More Induced Gene Expression Pattern When Compared to MEGs

In order to characterize the distinct biological behavior of IGs, we studied the overall gene expression patterns of IGs compared to those of MEGs in all the tumors. Notably, a greater percentage of upregulated genes are found among DE-IGs than DE-MEGs in all the different types of cancers analyzed.

For instance, when comparing both populations of genes, we identified statistical significance for an over-representation of the up-regulated IGs over the MEGs group among DEGs in PRAD ($p\text{-}value = 2.8108\,e{-}07$), ESCA ($p\text{-}value = 7.0858\,e{-}05$), LUAD ($p\text{-}value = 0.0015$) and BRCA ($p\text{-}value = 0.0276$) cancers. Moreover, we aimed to compare if the upregulation levels are higher in IGs than in MEGs transcripts. Our analysis revealed that in BRCA, BLCA, ESCA and GBM IGs tend to express in higher levels than MEGs, as shown in Appendix Fig. 8.

2.3   Upregulated IGs Across Cancer Types Encode for Highly Conserved HDAC Deacetylate Histones Involved in Negative Gene Regulation

To dig insight into the role of the prevalent upregulation mechanism identified for the DE-IGs, we aimed to characterize the groups of induced IGs across the analyzed cancer types. 338 out of 940 (35%) IG encoded proteins in the human genome are deregulated in at least one of the eight analyzed tumor types, where 106 (30%) are found to be up-regulated in one cancer type and down-regulated

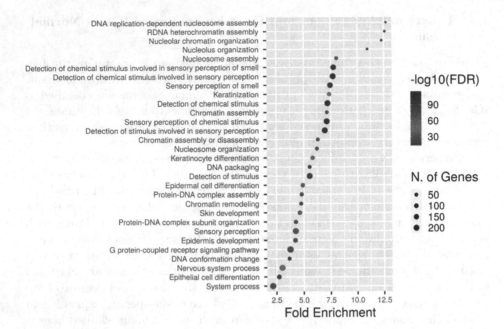

Fig. 1. Functional enrichment of IG-encoded proteins in the human genome.
Genes are depicted by a dot, the size of the dot represents the number of genes. FDR is
calculated based on a nominal *p-value* from the hypergeometric test. Fold Enrichment
is defined as the percentage of *intronless genes* belonging to a pathway, divided by the
corresponding percentage in the background. FDR reports how likely the enrichment
is by chance, higher values are colored on a scale of red to blue. In the x-axis, Fold
Enrichment indicates how drastically genes of a certain pathway are overrepresented.

in another type. In a higher number of them, 222 (65%), upregulation is conserved in two or more cancers, which suggests they are undergoing transcriptional reprogramming with higher rates of upregulated levels of their mRNAs as
an outcome. Moreover, this upregulation mechanism is highly shared among the
different cancers (Fig. 3), for instance, most of the genes upregulated in BLCA
and ESCA are shared with the other tumors, while tumors with a significant but
less shared upregulation are GBM, PRAD and KIRC, which could be expected
given their remarkable heterogeneity (Fig. 3).

To delve insights into the conservation of the activation of this negative gene
expression mechanism in the cancer genomes, we identified the most conserved
upregulated histones in all the diseases. We identified a group of 13 histone-
related genes whose expression is highly conserved among the cancers (Appendix
Fig. 10 and Appendix Fig. 11). Moreover, this group of upregulated genes is
statistically enriched in a proximal region (26.1968-27.8927 position in Mbps) in
chromosome 6 in the human genome (Appendix Fig. 12).

HASPIN was found to be the only DE-IG shared among all the tumor
genomes, highlighting its relevance in tumorigenesis. This protein kinase is

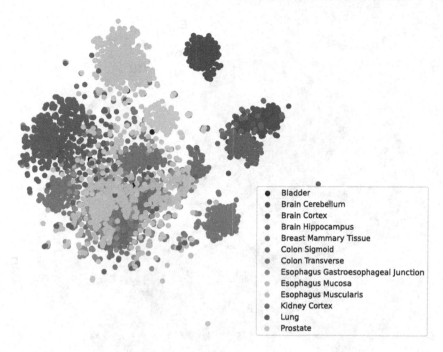

Fig. 2. IG expression profile across healthy tissues. t-SNE algorithm summarises the expression pattern of different samples in a given tissue. Each sample is represented as a dot; color indicates the tissue. Dots distance represents transcriptional profile similarity: the closer they are in the plane, the more similar is their transcriptional profile.

known to be required for histone H3 phosphorylation, necessary for chromosomal passenger complex accumulation at centromeres of mitotic cells [10,34]. In addition to its chromosomal association, it is associated with centrosomes and spindles during mitosis. The overexpression of this gene is related to a delayed progression through early mitosis [10].

Functional enrichment analysis was carried out for the deregulated IGs in all the studied diseases (Appendix Table 2). GBM is the disease with the most DE-IGs and with the most diverse functional roles for the induced genes. The GBM-specific enriched terms are pathways of neurodegeneration, cell-cell adhesion, and gland development. The cancer-specific enriched terms chaperone-mediated protein folding and regulation of neuron apoptotic processes were found for esophagus and colon cancers, respectively. Further, the Reactome pathways *R-HSA-321481* deacetylases histones (HDACs) and *R-HSA-3214858* RMTs methylate histone arginines were also enriched in IGs while the GO:0006335 term *DNA replication-dependent chromatin assembly* was also enriched suggesting an essential role in cancer biology.

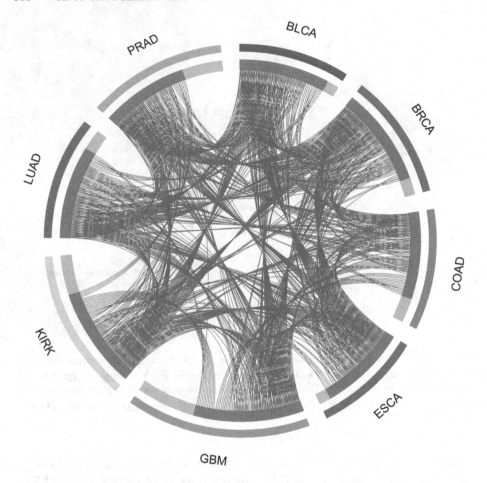

Fig. 3. Upregulated IGs across the analyzed cancer genomes. The Circos plot depicts how upregulated genes for each cancer overlap. On the outside shell, each arc represents a cancer genome: BLCA in red, BRCA in blue, COAD in green, ESCA in purple, GBM in orange, KIRC in yellow, LUAD in brown, and PRAD in pink. On the inside shell, the dark orange color represents the proportion of upregulated genes shared with other cancers, while the light orange color represents the proportion of genes that are uniquely upregulated in a cancer. Purple arcs represent shared upregulated genes, the greater the number of arcs and the longer the dark orange area imply a greater overlap among the upregulated IGs across cancers. (Color figure online)

2.4 IG Downregulation Is Conserved in Breast and Colon Cancers and Is Involved in Signaling and Cell-Specific Functions

When comparing the downregulated IGs and their enriched functional terms among the diseases, it was pointed out that the breast and colon cancers share all the repressed IGs with their functional pattern. Downregulated genes in those cancers are involved in the regulation of cellular localization, sensory organ development, regulation of membrane potential, adipogenesis, epithelial to mesenchymal transition in colorectal cancer, growth, actin filament base process, positive regulation of cell death, class A/1 (rhodopsin-like receptors) and the regulation of secretion by cell biological processes (Appendix Table 2). Notably, the most shared downregulated pathway among cancer genomes is the regulation of anatomical structural size, shared among BRCA, PRAD, GBM, LUAD and BLCA tumors. Additionally, unique roles are found in lung and prostate cancers. They have downregulated genes involved in the vitamin D receptor pathway and the wounding response, respectively.

2.5 Cancer-Specific Differentially Expressed IGs

To study the specificity in the regulation of human IGs, we quantified the number of cancer-specific DE-IGs (Fig. 4). We also built a bipartite network to identify cancer-specific and shared deregulated IGs, where upper nodes represent DE-IGs and bottom nodes represent cancer types. We link an IG to a cancer if it was found to be deregulated in that cancer (Appendix Fig. 9). Among the 338 DE-IGs, we identified that 35% were specific for a cancer type (Appendix Table 3). GBM followed by KIRC and PRAD with 35, 18, and 16 DE-IGs, respectively, displayed the highest specificity for IG expression.

GBM. Among the specific DE-IGs found in GBM, an evident functional enriched first group includes mainly cellular proliferation and cytoskeleton-related genes. An enrichment of microtubule-based movement, transport along the microtubule, cytoskeleton-dependent intracellular transport and microtubule-based transport is identified (FDR = 0.0132). Moreover, the second group of specific deregulated genes is enriched in functions related to tumor suppression, a negative regulatory mechanism of cell growth control, which usually inhibits tumor development. We identified GBM-specific deregulated genes with crucial involvement in the *p53* pathway, such as *KLLN*, *INSM2* and *SFN*.

KIRC. The cancer-specific deregulated IGs in KIRC include the tissue-specific *KAAG1* gene (kidney-associated antigen 1) along with the transcription factors *POU5*, *TAL* and *bHLH*. Moreover, transmembrane proteins such as the proto-cadherin beta 10 (*PCDH10*), related to the *Wnt* signaling, were also upregulated. *PCDH10* has been classified as a tumor suppressor in multiple cancers. It induces cell cycle retardation and increases apoptosis by regulating the *p53/p21/Rb* axis and *Bcl-2* expression [19].

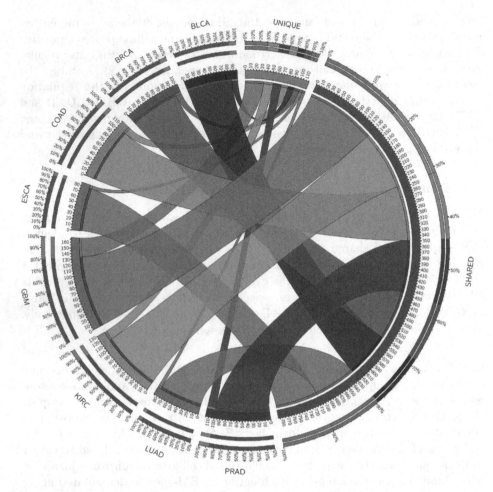

Fig. 4. Shared and unique differentially expressed IGs across cancers. Circa plot depicts the number of shared DE-IGs across the 8 types of tumors under study. In the left hemisphere, the total number of DE-IGs in each cancer is represented by an arc: BLCA (red), BRCA (orange), COAD (yellow), ESCA (light green), GBM (green), KIRC (turquoise), LUAD (light blue), and PRAD (dark blue). The unique (fuchsia) and shared (purple) groups are depicted in the right hemisphere. The inner arcs represent the distribution of the DE-IGs in each cancer, while the outside arcs represent the intersection of both hemispheres on a scale of 0–100%. Data was obtained from the TCGA database. (Color figure online)

PRAD. IGs specifically downregulated in PRAD include *RAB6D* (member RAS oncogene) and *MARS2* (methionyl-TRNA Synthetase 2) related to amino acid metabolism, while *PBOV1*, a 135 amino acid protein with a transmembrane domain was highly upregulated (logFC = 3.1699). This gene was mapped to 6q23-q24, a region associated with loss of androgen dependence in prostate tumors [3]. Other upregulated IGs include transcription factors such as *FOXB2* (Forkhead Box B2) and *ATOH1* a *bHLH*, which is associated with goblet cell carcinoid and Merkel cell carcinoma diseases [33]. Moreover, *FRAT2*, an essential positive regulator of the *Wnt* signaling and *LENG9*, a member of the conserved cluster of receptors of the leukocyte-receptor complex (*LRC*) were also upregulated [26].

BRCA. Few cancer-specific differentially expressed IGs that code for TFs were identified for BRCA, while specific gene expression is more abundant for transmembrane proteins in this cancer, as is the case of a group of glycoproteins and transmembrane receptors involved in signal transduction. For instance, upregulation of *PIGM* (phosphatidylinositol glycan anchor biosynthesis class M), *GP5* (glycoprotein V platelet), *CALML5* (calmodulin-like 5), coiled-coil domain-containing proteins *CCDC96* and *CCDC87* were identified specifically for this cancer.

LUAD. Regarding the specific differential expression found for lung tumors, the following TFs are significantly upregulated: *TAF1L*, a transcription factor that is related to chronic inflammatory diseases by regulating apoptotic pathways including regulation of *TP53* activity, and *FOXE1*, a forkhead TF previously related to thyroid cancer [11]. Interestingly, specific transmembrane signaling receptors are upregulated. *MAP10* is a microtubule-associated protein and *THBD*, a thrombomodulin endothelial-specific type I membrane receptor that binds thrombin. Among *THBD* related pathways are collagen- and fibrin clot formation as well as calcium ion binding and cell surface interactions at the vascular wall.

BLCA. In BLCA, only *ZXDB* (Zinc Finger X-Linked Duplicated B) TF is identified as specifically deregulated. Transcripts for signal transduction proteins are specifically upregulated for *OR52E8* (Signaling by GPCR), *LRRC10B* (Leucine-Rich Repeat Containing 10B) and *MAS1L*, a proto-oncogene which is a peptide ligand-binding receptor linked to sensor neurons for pain stimuli detection.

ESCA and COAD. No specific TFs are specifically differentially expressed for colon and esophageal cancer. Gastrointestinal adenocarcinomas of the tubular gastrointestinal tract, including esophagus, stomach, colon, and rectum, share a spectrum of genomic features, including TF-guided genetic regulation [28]. In agreement with this, we found that transmembrane proteins related to GPCR signaling pathways are specifically undergoing differential expression for both cancers; for instance, in colon, *GPR25* (G protein-coupled receptor 25), *CCDC85B* (coiled-coil domain containing 85B), *CCDC184* (coiled-coil domain containing 184), *SPRR1A* (small proline-rich protein 1A), *PROB1* (proline-rich essential protein 1). Meanwhile, for esophagus tumors, *OR51B4* related to signaling by GPCRs and *LDHAL6B* involved in glucose metabolism and respiratory electron transport were identified.

Remarkably, when comparing IGs deregulation across the eight cancer types, we identified that ESCA and LUAD shared 93% of the DE-IGs, while BRCA, BLCA, COAD, PRAD and KIRC shared more than 86% at least with another cancer. Finally, and presumably given its "*multiforme*" nature, GBM is the cancer type identified with a significant but lesser percentage of shared DE-IGs (79%). Notwithstanding, our results show that GBM represents the tumor where the IGs have more differential expression, specificity and concerted functional roles. This could be explained due to their gene expression tissue-specific relevance in the brain.

2.6   Proteins Encoded by Cancer-Specific Deregulated IGs Interact with Distinct Groups of Proteins in PPI Networks

Gene expression is a phenomenon where coupled biochemical interactions take place to transcribe mRNA for protein production. There is a high regulation in the balance of this event. Hence differences in protein composition, production or abundance are a consequence of disruptions in cell phenotypes. The differential expression pattern of the mRNA is key in determining a cell state at the molecular and physiological level [1]. It has been shown that genes involved in "*similar diseases*" share protein-protein interactions (PPI) and higher expression profiling similarity [12].

To determine if the DE-IGs play a crucial role in inhibiting or exacerbating biological reactions at a physical level, we obtained the Protein-Protein Interactions (PPIs) with highest confidence (according to STRING database) where DE-IGs for each cancer are involved. Our results show that a considerable proportion of proteins encoded by DE-IGs in each cancer interacts with specific groups of proteins, due to the low percentage of PPIs that are shared between two or more cancers (Fig. 6). In this analysis, we will consider that an interaction is shared between cancers if a DE-IG in one cancer is also deregulated in another cancer, and the interaction of the IG-encoded protein with another protein is reported by STRING, otherwise, we classify that interaction as cancer-specific.

For instance, 100% of the interactions found in LUAD are cancer-specific, followed by the high specificity of interactions found in ESCA (99.3865%) with only a unique shared PPI of centromere complex with BLCA (*HIST1H2BJ, CENPA*). GBM has a similar pattern with 95.8333% of unique interactions and only 19 shared ones. In BLCA, 59.16% of the analyzed PPI are specific to this cancer type. PRAD is a condition that possesses 53.1339% cancer-specific interactions and COAD has 42.1455%. For BRCA, 30.62% of its interactions are only found in this cancer, the lowest fraction of unique interactions. KIRC has the second-lowest with only 31.55% of unique interactions.

If two cancers share PPIs related to the cancer-specific DE-IGs, there could be an underlying affected process characteristic for such diseases. To determine those specific processes, protein complexes involved in shared interactions were examined, finding that most of the identified proteins belong to families of core histones. Overall, the analyzed PPIs indicate a very distinct pattern of interactomics for each cancer. The tumors that share the greatest proportion of PPIs are breast and colon, exhibiting 145 common interactions which represents 44.61% of all interactions identified for such diseases. The second higher similarity found is for prostate and kidney tumors, sharing 227 interactions (28.64% of all). The rest of the tumors share at most 21.32% interactions (Appendix Fig. 13 and Appendix Table 4).

These cancer-specific interactions were analyzed deeper to delve into their functional role in each cancer (Appendix Fig. 14). As it could be expected due to their high specificity at the PPI level, this network approach shows specific and defined clusters for glioblastoma (Fig. 5b) and lung (Fig. 5d). In less-defined clusters, esophageal and bladder cancer can be observed in panels e) and f). In contrast, in the case of prostate, bladder and kidney tumors, the interactions are linked by common DE-IGs in a cluster. On the other hand, we identified that the communities in the network are defined primarily by the type of proteins. Functional enrichment shows specific biological processes intrinsic to the physical interactions implied for the genes in each cancer (Appendix Fig. 14). For instance, proteins involved in BLCA-specific PPIs are conducting mainly chromatin organization and DNA repair reactions, while in colon tumors, proteins play a concerted role in the regulation of transcriptional processes. In keeping with this, in esophageal tissue, the interactors are involved in the regulation of different classes of non-coding RNAs. Meanwhile, in glioblastoma and lung tumors, proteins are specifically interacting for splicing activity and transcription initiation processes, respectively. Kidney and prostate tumors show specific protein-protein interactions for DNA-replication and protein-protein complex formation, and DNA organization and packaging. In contrast, proteins with specific interactions in breast tumors, have an important activity linked to cellular growth and apoptotic processes.

Intrinsic of the PPI networks is the topological information capable of characterizing cancer proteins [16]. Surprisingly, our comparison of IGs with their physical interactors at a protein-protein level shows a tendency to lower betweenness centrality, lower degree and a few interactions with other IGs. Therefore, even though IGs are not hubs in the network of interactions where they participate, changes in their regulatory role can cause a cascade of disruptions in interactions that might lead to malignancy.

Altogether, these results indicate opposite patterns to those reported for cancer genes, which suggest that the behavior tendency of IG proteins is to interact with *oncogenic genes* [16].

For a closer approach, the nearest fifty interactors of the DE-IG encoded proteins in each cancer were analyzed. In Fig. 6 we can observe that cancer-specific DE-IG encoded proteins interact with specific groups of proteins in distinct cancers, suggesting that DE-IGs affect particular post-transcriptional processes.

The most important protein complexes for each type of cancer were determined for cancer-specific deregulated IGs and their interactors (Appendix Table 5). In total, we detected 27 protein complexes where cancer-specific DE-IGs and

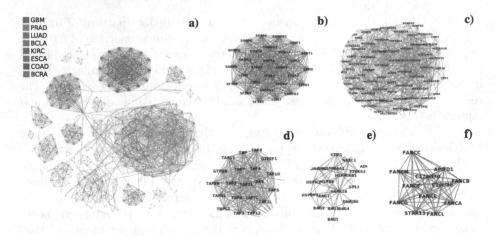

Fig. 5. Protein-protein interaction network of DE-IGs and their 50 closest interactors across cancers. a) Unique and shared PPI across cancer are shown, each cancer has a specific color for its interactions. Close-ups of the biggest and more isolated clusters in the network are presented: b) GBM, c) PRAD, d) LUAD, e) ESCA, and f) BLCA. PPI data was retrieved from the String database.

their interactors are found. Colon cancer is the disease containing more DE-IGs in the same complex: 9 out of 13 corresponding to the disassembly of the beta-catenin Wnt signaling pathway. Bladder DE-IGs are interacting mainly with the proteins of the CD40 signaling pathway and are involved in multi-multicellular organism processes. Likewise colon, breast cancer-specific IGs are related to Wnt signaling pathway, but also to follicle-stimulating hormone signaling.

Glioblastoma-specific proteins have a major participation in splicing and vesicle transport. Kidney's are more related to regulation of feeding behavior. Esophageal's IGs are involved in protein folding, and tRNA and ncRNA processing. Lung cancer proteins participate in transcription preinitiation complex assembly, as well as in the BMP signaling pathway. Prostate cancer DE-IGs are found mainly related to chromatid sister cohesion and translation and postranslation processes. The protein complex related to Wnt signaling pathway is associated with a large group of deregulated genes among Bladder, Breast, Colon, Kidney and Glioblastoma, being a total of 17 IGs with disrupted expression.

2.7   DE-IGs Participate in the Genetic *"rewiring"* of Cancer Cells

Cancer cells undergo significant genetic *"rewiring"* as they acquire metastatic traits and adapt to survival in multiple environments with varying nutrient availability, oxygen concentrations and extracellular signals. Therefore, to effectively treat metastatic cancer, it is important to understand the strategies that cancer cells adopt during the metastatic process. Finally, we focus on studying the *"rewiring"* between healthy and tumor samples using BRCA as a model (Fig. 7). We used breast cancer since it is the only dataset that fits the criteria for a mutual

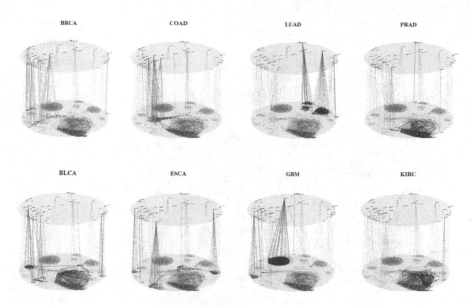

Fig. 6. Multilayer networks of IG-encoded proteins and their cancer-specific interactors. Protein-protein interactions were found for deregulated IGs in each cancer. The cancer-specific deregulated IGs that encode for proteins interact with exclusive groups of proteins in each cancer. Upper layer shows deregulated IGs across cancer, highlighting the cancer-specific upregulated (in red) and downregulated IGs (in blue). Links to the bottom layer connect IG-encoded proteins to proteins found in PPIs by STRING. Yellow links highlight interactions that can be affected by deregulation of IGs in a specific type of cancer. (Color figure online)

information analysis (at least 100 samples for each condition are required). A network of DE-IGs and their co-expressed genes was built and analyzed for each condition. Our results show that the tumor co-expression network is composed of a total of 62,462 interactions among 15,347 genes, while the healthy network has 13,037 genes with 45,941 interactions. There are only 9,615 interactions shared between the two network topologies. All the differences between these two networks are potential rewiring-caused co-expression interactions that may be part of the mechanism that BRCA cells follow to achieve the characteristics of their phenotype. For instance, as seen in Fig. 7, IGs like *RRS1, FNDC10, EPOP* and *NND* are highly affected by this *"rewiring"* behavior in cancer. According to differential expression and network analysis, we observe that histones play a key role in rewiring the co-expression mechanism from healthy to cancer tissue. Examples of this are *H2AW* and *H2BC8*, which are core histones that have only few healthy tissue-specific interactions that are almost lost during cancer development, which creates many new interactions with other genes.

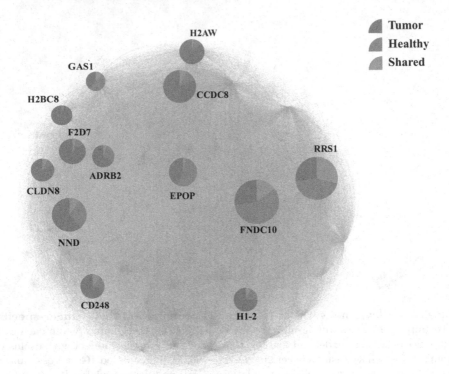

Fig. 7. Co-expression network of healthy and tumoral breast tissue. In this network, the co-expression of DE-IGs found in BRCA in healthy tissue (depicted by brown links) and breast tumor tissue (purple links) is shown. Shared co-expression relations are depicted with green links. Pie charts for IG nodes of a higher degree are shown. Each of them indicates the percentage of each type of interaction: in purple tumor-specific interaction, in brown healthy-specific interaction, in green conserved interaction. We can see the effect of rewiring by observing that a high percentage of interactions that are present in healthy tissue is not observed in tumor tissue. Moreover, many more new interactions emerge in the tumor tissue. (Color figure online)

3 Discussion

Here, we present a comprehensive analysis of differentially expressed Intronless genes (IGs) across eight cancers using a standardized bioinformatics pipeline. We uncover signature changes in gene expression patterns of IGs such as enhanced expression changes when compared to multi-exon genes, unique and shared transcriptional signatures among different cancer types, and association with specific protein-protein interactions that might have potential effects in post-transcriptional processes.

Our results show that IGs tend to have tissue-specific transcriptional profiles in normal tissues. In the case of cancers, the upregulated IGs were related to negative regulation of gene silencing and negative regulation of cell differentiation. Furthermore, we found a core set of chromatin-modifying genes such as *HASPIN* and Histone deacetylases (HDACs), located on chromosome 6, that were upregulated across all the analyzed cancers. We can observe in Fig. 4 that the expression of these genes in cancer differentiate from that of normal tissue in most of the cases, indicating deregulation in the diseased phenotype. On the other hand, the distinct expression among healthy tissues is evident, this may suggest that these histone-related genes present tissue-specific expression levels.

Although the pattern of DE-IGs was found to be not specifically shared across the different tumor types, highlighting their specificity modulating cancer signaling and proliferative pathways. For example, beta-catenin *Wnt* signaling was regulated in kidney and prostate tumors while GPCR signaling critical to inflammation [6,22,23] was regulated in gastrointestinal esophageal and colon adenocarcinomas, and *p53* pathways.

This biological behavior could be due to the tumor's cellular origin and heterogeneity [15], and tissue-specific expression [4,13]. Due to their remarkable heterogeneity, glioblastoma, prostate, and kidney tumors were found with a less representative shared behavior in the upregulation of IGs. Notably, glioma tumors are the malignant microenvironment where IGs showed the most characteristic gene expression regulation profile with 35 unique DE-IGs, more than 50% greater than other cancers. Notwithstanding, our results show that GBM represents the tumor where the IGs have more differential expression, specificity and concerted functional roles. This could be explained due to the "multiforme" nature of GBM that is driven by brain-specific gene expression as is further evidenced by least shared DE-IGs (79%) with other cancers [15].

In conclusion, IGs drive massive transcriptional rewiring, as observed in co-expression analysis, that may drive tumorigenesis and may represent novel therapeutic targets for gene therapy and in-silico drug design such as the popular GPCRs or emerging HDAC inhibitors [9]. Although we focus on 8 cancers based on their heterogeneity and aggressivity, we expect that using our analysis paradigm, IGs may also display characteristic expression profiles in other cancers.

4   Materials and Methods

4.1   Data Extraction and Curation for IG, and MEG Datasets

Data was extracted using Python scripts (https://github.com/GEmilioHO/intronless_genes), and the Application Programming Interface (API). Homo sapiens genome was assembled at a chromosome level and was accessed at the Ensembl REST API platform (http://rest.ensembl.org/ accessed using Python with the ensembl_rest package). The pipeline process was as follows: protein-coding genes with CDS identifiers for transcripts for all chromosomes were retrieved and classified into two datasets named "single-exon genes" (SEGs), and "multiple exon genes" (MEGs) depending on exon and transcript count. The SEG dataset was submitted to the Intron DB comparison (http://www.nextgenbioinformatics.org/IntronDB) to separate those with UTR introns and referred to as uiSEGs. The output of the pipeline was a third dataset containing only intronless genes (IGs). After data extraction, a manual curation step in the IG and uiSEG datasets was followed to discard incomplete annotated protein sequences and mitochondrial encoded proteins. The final IG dataset contained 940 protein-coding genes with only one exon and one transcript.

4.2   Gene Expression Profiles in Healthy Tissue Tissue

The 424 collected samples of IGs expression (see Sect. 4.4) were scattered in a plane (Fig. 2) by using t-SNE algorithm, which takes samples in a high dimensional space (defined as the expression level of the IGs) and returns a two-dimensional representation where two samples are close if their expression patterns are similar. This process was performed using the Scikit-learn python package [29].

4.3   Bipartite Network and Quantification of Shared and Unique DE-IGs

The bipartite network was constructed with upper nodes representing IGs and bottom nodes representing cancer types. We place links connecting each IG to the cancer types where such a gene is differentially expressed (DE). Nodes corresponding to genes in the bipartite network were sorted by degree aimed to visually identify intronless genes cancer-specific deregulated, and those whose expression is disrupted in most of the diseases.

Quantification of shared and cancer-specific deregulated IGs is computed by identifying the number of deregulated IGs in every pair of cancers, and dividing such quantity by the total of the disrupted genes in any of the compared diseases, this metric is known as the Jaccard similarity coefficient. The results are reported as a heatmap.

4.4   Data Source and Differential Expression Analysis Across Cancer

Currently, the (The Cancer Genome Atlas) TCGA possesses data for the study of 37 different tumor types. We selected 8 types of cancer primary malignant tumors and their respectives adjacent tissues for the present study. Gene expression data from patients for BRCA, BLCA, COAD, ESCA, GBM, KIRC, LUAD and PRAD cancers was downloaded from the NIH website (https://portal.gdc.cancer.gov/) using the TCGAbiolinks R package [7] with the following restriction criteria: samples types primary tumor and solid tissue normal (control); results of RNAseq experimental strategy; and workflow type HTSeq-counts format. Differential expression analysis was carried out using the TCGAutilis R package [25], indicating which of the obtained samples correspond to tumors and which to control, and establishing a filtering threshold of FDR = 0.05 and logFC = 1 (absolute value) to consider a gene significantly differentially expressed. For comparative analysis, normal healthy tissue data was obtained from the dataset gtex_v8 of the GTEx Portal in June 2020 by using the GTEx API. We found gene expression data for 424 IGs in 13 different tissues: bladder (21 samples), brain cerebellum (241), brain cortex (255), brain hippocampus (197), breast mammary tissue (459), colon sigmoid (373), colon transverse (406), esophagus gastroesophageal junction (375), esophagus mucosa (555), esophagus muscularis (515), kidney cortex (85), lung (578), and prostate (245).

4.5   Upregulation Significant Differences of IGs and MEGs Among Cancers

We obtained a dataset of all significantly upregulated IGs and MEGs for each type of cancer studied, in these datasets we compared IGs and MEGs gene expression employing Python scripts using a hypergeometric test. In order to find significant differences in upregulation between IGs and MEGs, the Wilcoxon test was carried out for each type of tumor. Additionally, to confirm that the datasets fulfilled all assumptions a Levene's test was performed for each individual cancer data set.

4.6   Functional Enrichment Analysis of Differentially Expressed IGs

The functional enrichment was conducted using the over-representation analysis of the functional assignment (ORA). Genes with differential expression up to one log2-fold change values were considered as up-regulated with a p- and q-value set at 0.05 and 0.10, respectively. First, the functional enrichment of the 338 differentially expressed human IG proteins (up-regulated and downregulated separately) was performed using all human IGs proteins as a background "universe" (selecting input as species: Homo sapiens, universe 940 human IGs). The comparative functional enrichment analyses were performed using Metascape (https://metascape.org/) for the biological process category, including KEGG and Reactome pathways. To delve insight into the role of specific IGs in the affected biological processes ORA was assessed to determine category barplots. The Circos software [20] was employed for data analysis and visualization.

4.7   DE-IGs PPI Network Construction and Protein Complex Identification

Network analysis and visualization were performed using python scripts (See repository https://gitlab.com/jarr.tecn/de-igs-cancers/-/tree/master) and the Gephi software. STRINGDB platform [32] was used to download physical and functional interactions data. The highest confidence scores (0.9) were filtered for this study, keeping the most probable interactions. Then, interactions were requested for the set of unique DE-IG of each cancer type, downloading relations between those genes, and also their interactors in the first shell (up to 50 interactors). All this data is assembled into a single network. Network metrics such as degree distribution, closeness, and betweenness centralities were computed using Python networkx library [14]. For protein complex detection, we assumed that highly clustered nodes in the PPI network correspond to protein complexes. Therefore, 27 protein complexes were predicted using the Louvain method to find communities implemented in Gephi.

4.8   BRCA Network Deconvolution

Co-expression networks were inferred using ARACNe-AP [21], a mutual information-based tool for gene regulatory network deconvolution. In this analysis, we used separated submatrices for healthy and tumor tissue and the list of every DE-IG in the BRCA dataset. For this study, a *p-value* of $1 \times 10\text{-}8$ was set up and 100 bootstraps were carried out. Then, they were consolidated into a single network, getting an inferred co-expression network for each condition.

Acknowledgments. We would like to thank Roddy Jorquera and Carolina González for fruitful discussions. Special thanks to Fernando Flores for helping with designing and constructing the co-expression network. We are also thankful to Carlos González for visualization and technical support.

This research was funded by Conacyt Ciencia Basica Project 254206. K.A.P (CVU:227919), J.A.R.R (CVU: 1147711), and O.Z.M (CVU: 1147042) received financial support from CONACyT. K.A.P is a current holder of a fellowship from the Fulbright Comexus García-Robles foundation.

Appendix

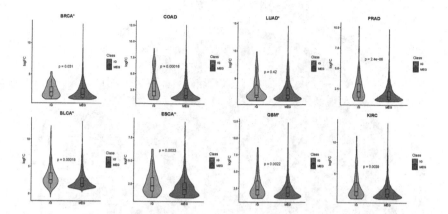

Fig. 8. Upregulated gene expression levels among cancers. A violin plot is shown for each cancer type, separated by gene type (IGs or MEGs); expression levels are shown on the y-axis using log2fold change values. The *p-value* for the Wilcoxon test is shown in order to indicate the statistical difference between both populations. Cancer types with asterisk represent those datasets that meet all the assumptions of the Wilcoxon test for this comparison. Levene's test was performed to measure shape similarity between IG and MEG groups. Only up-regulated genes log fold change equal or greater than 1 are selected.

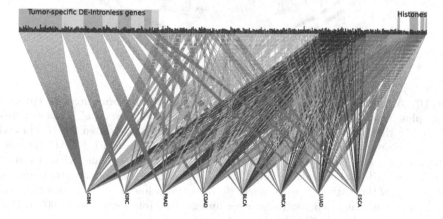

Fig. 9. IG-cancer type bipartite network. Intronless genes are shown in the upper group of nodes, while cancers are located in the bottom group. A link joining an IG and a cancer type indicates that the expression level of the IG is found disrupted in the linked cancer type. Upper nodes are color-sorted by their cancer specificity: genes deregulated in a single cancer are shown at the left in blue, followed by genes deregulated in two cancers, three cancers, and so on. At the right of the upper nodes, there is a single gene (*HASPIN*) whose expression is disrupted in all tissues. In general, all histones are deregulated in almost all cancers. (Color figure online)

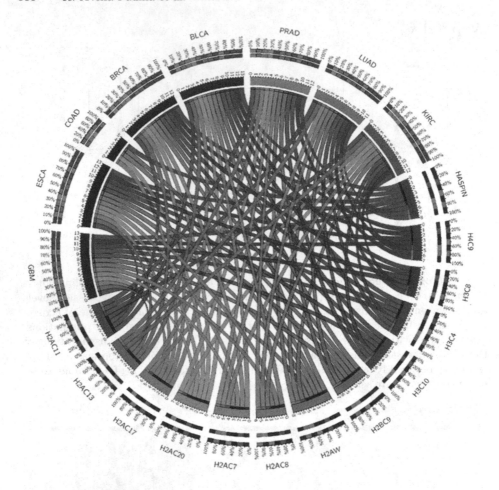

Fig. 10. A high conserved group of up-regulated histones across tumors.
Circa plot depicts the conservation of the deregulation of a group of histones across
cancers. In the upper hemisphere, the total number of histone-related DE-IGs in each
cancer is represented by an arc: GBM (yellow), ESCA (mustard), COAD (orange),
BRCA (dark orange), BLCA (red), PRAD (pink), LUAD (violet), and KIRC (purple).
Histone encoded IGs are depicted in the lower hemisphere. The inner arcs represent the
distribution of the histones in each cancer (0–13 genes), while the outside arcs represent
the tumors that have that particular gene upregulated (on a scale of 0–100%). Data
was obtained from the TCGA database. (Color figure online)

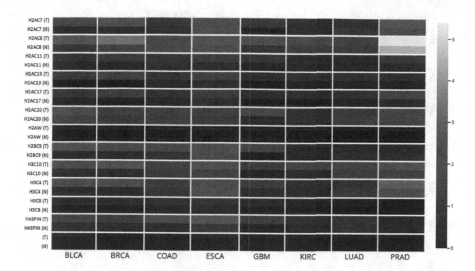

Fig. 11. Expression of histones among tumors and normal tissues. Heatmap depicting a comparison of the expression of the 13 histone-related genes identified with upregulated conserved expression among the studied cancers. The expression of each gene is shown for each type of tumor (T), as well as its expression in normal tissue from GTEx data (N). The color key from darker to light indicates low to high transcript per million (TPM) values.

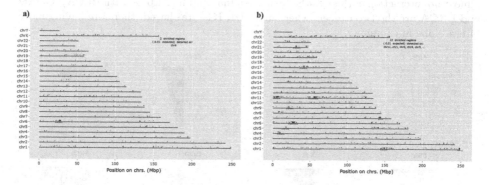

Fig. 12. Comparison of genome location of highly conserved DE-IGs. a) Genome location of IGs in the human genome. b) Genome location of highly conserved DE-IGs IGs encoded proteins are represented by red dots. The purple lines indicate regions where these genes are statistically enriched, compared to the density of genes in the background. The hypergeometric test is used to determine if the presence of the genes is significant. Essentially, the genes in each region define a gene set/pathway. The chromosomes may be only partly shown as the last gene's location to draw the line is used. Data was obtained from the ENSEMBL database.

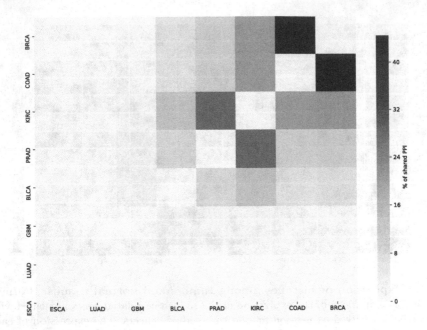

Fig. 13. Shared PPI heatmap. Color intensity indicates a higher percentage of shared PPI related to DE IGs of the compared diseases. LUAD does not share any DE IG interaction with other cancer types, and ESCA only one with BCLA. GMB shares almost no interaction with any other cancer (a maximum of 1.64% with BLCA). COAD and BRCA are the two cancers where more DE IGs are shared, therefore having more interactions in common.

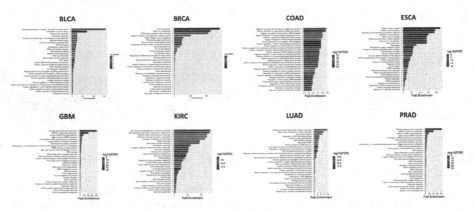

Fig. 14. PPI networks ontologies. FDR is calculated based on nominal *p-value* from the hypergeometric test. Fold Enrichment is defined as the percentage of genes in each cancer belonging to a pathway, divided by the corresponding percentage in the background. FDR indicats how likely the enrichment is by chance. In x-axis the Fold Enrichment indicates how drastically genes of a certain pathway are overrepresented. FDR values are depicted red (high) to blue (low). (Color figure online)

Table 1. Available expression data for each type of cancer

Cancer type	Primary solid tissue samples	Solid tissue normal samples
BLCA	414	19
BRCA	1102	113
COAD	478	41
ESCA	161	11
GBM	156	5
KIRC	538	72
LUAD	533	59
PRAD	498	52

Table 2. Gene Ontology (GO) enrichment of deregulated IGs for each cancer

Upregulated IGs									
GO	Description	Log10P							
		BLCA	BRCA	COAD	ESCA	GBM	KIRC	LUAD	PRAD
GO:0061077	Chaperone-mediated protein folding	0.00	0.00	0.00	−3.54	0.00	0.00	0.00	0.00
GO:0043523	Regulation of neuron apoptotic process	0.00	0.00	−3.41	0.00	0.00	0.00	0.00	0.00
GO:0000280	Nuclear division	0.00	0.00	0.00	0.00	0.00	0.00	−3.40	0.00
GO:0000904	Cell morphogenesis involved in differentiation	0.00	0.00	0.00	0.00	0.00	0.00	−3.68	0.00
GO:0030900	Forebrain development	0.00	0.00	0.00	0.00	0.00	0.00	−2.19	0.00
GO:0002009	Morphogenesis of an epithelium	0.00	0.00	0.00	0.00	0.00	−2.73	−3.17	0.00
R-HSA-420499	Class C/3 (Metabotropic glutamate/pheromone receptors)	0.00	0.00	0.00	0.00	0.00	−3.50	0.00	0.00
GO:0008285	Negative regulation of cell population proliferation	0.00	0.00	0.00	−2.68	−4.14	0.00	−2.46	0.00
GO:0007423	Sensory organ development	−2.60	0.00	0.00	0.00	−5.21	0.00	−2.38	0.00
hsa05022	Pathways of neurodegeneration - multiple diseases	0.00	0.00	0.00	0.00	−3.00	0.00	0.00	0.00
GO:0045669	Positive regulation of osteoblast differentiation	0.00	0.00	0.00	0.00	−3.79	0.00	0.00	0.00
GO:0098609	Cell-cell adhesion	0.00	0.00	0.00	0.00	−2.57	0.00	0.00	0.00
GO:0048732	Gland development	0.00	0.00	0.00	0.00	−4.33	0.00	0.00	0.00
GO:1903706	Regulation of hemopoiesis	0.00	−2.54	0.00	−2.08	−3.87	−2.13	0.00	0.00
GO:0045596	Negative regulation of cell differentiation	0.00	−2.02	−2.12	−2.33	−2.34	−2.23	0.00	0.00
GO:0016570	Histone modification	−2.34	−2.14	0.00	−2.37	−3.26	0.00	−2.19	−3.94
GO:0007517	Muscle organ development	−3.88	0.00	0.00	−2.74	−2.82	−2.20	−2.56	0.00
R-HSA-3214815	HDACs deacetylate histones	−10.77	−12.90	0.00	−9.84	−5.83	−9.04	−3.25	−15.15
R-HSA-3214858	RMTs methylate histone arginines	−5.62	−7.73	0.00	−7.63	−4.60	−8.42	−2.90	−8.98
GO:0006335	DNA replication-dependent chromatin assembly	−2.26	−4.69	0.00	−3.15	−2.65	−5.06	0.00	−4.32
Downregulated IGs									
GO	Description	Log10P							
		BLCA	BRCA	COAD	ESCA	GBM	KIRC	LUAD	PRAD
GO:0060341	Regulation of cellular localization	0.00	−3.03	−3.03	0.00	0.00	0.00	0.00	0.00
GO:0007423	Sensory organ development	0.00	−2.85	−2.85	0.00	0.00	0.00	0.00	0.00
GO:0042391	Regulation of membrane potential	0.00	−4.93	−4.93	0.00	0.00	0.00	0.00	0.00
WP236	Adipogenesis	0.00	−4.11	−4.11	0.00	0.00	0.00	0.00	0.00
WP4239	Epithelial to mesenchymal transition in colorectal cancer	0.00	−3.87	−3.87	0.00	0.00	−3.26	0.00	0.00
GO:0040007	Growth	0.00	−5.90	−5.90	0.00	0.00	−2.70	0.00	−3.01
GO:0030029	Actin filament-based process	−2.04	−2.08	−2.08	0.00	0.00	0.00	0.00	−4.23

(continued)

Table 2. (*continued*)

Upregulated IGs		Log10P							
GO	Description	BLCA	BRCA	COAD	ESCA	GBM	KIRC	LUAD	PRAD
GO:0010942	Positive regulation of cell death	−2.04	−2.08	−2.08	0.00	0.00	0.00	0.00	−3.01
GO:0090066	Regulation of anatomical structure size	−2.70	−8.04	−8.04	0.00	−3.49	0.00	−2.44	−2.57
R-HSA-373076	Class A/1 (Rhodopsin-like receptors)	0.00	−2.36	−2.36	0.00	−3.90	0.00	−4.62	−2.16
GO:0042752	Regulation of circadian rhythm	0.00	0.00	0.00	0.00	−2.77	0.00	0.00	0.00
GO:0009725	Response to hormone	0.00	0.00	0.00	0.00	−3.11	−2.70	0.00	0.00
GO:0043269	Regulation of ion transport	−2.70	0.00	0.00	0.00	−3.49	0.00	0.00	0.00
GO:0007193	Adenylate cyclase-inhibiting G protein-coupled receptor signaling pathway	−2.66	0.00	0.00	0.00	−3.01	0.00	−3.41	0.00
GO:1903530	Regulation of secretion by cell	−2.58	−2.62	−2.62	0.00	−3.30	0.00	0.00	0.00
GO:0008285	Negative regulation of cell population proliferation	0.00	0.00	0.00	0.00	0.00	−2.16	0.00	0.00
GO:2000027	Regulation of animal organ morphogenesis	0.00	0.00	0.00	0.00	0.00	−2.84	0.00	0.00
GO:0030162	Regulation of proteolysis	0.00	0.00	0.00	0.00	0.00	−3.98	0.00	0.00
GO:0009611	Response to wounding	0.00	0.00	0.00	0.00	0.00	0.00	0.00	nan
WP2877	Vitamin D receptor pathway	0.00	0.00	0.00	0.00	0.00	0.00	−3.04	0.00

Table 3. Cancer-specific deregulated genes

Cancer	Cancer-specific deregulated genes	Number of genes
BLCA	ZXDB, TRIL, ZXDA, FTHL17, OR52E8, LRRC10B, KRTAP2-3, ERVV-1, MAS1L, OXER1, HDGFL1	11
BRCA	CALML5, CCDC87, CAPZA3, GP5, ASCL4, MOCS3, PKDREJ, H4C12, MTRNR2L8, CCDC96, PIGM	11
COAD	CCDC85B, KRTAP3-3, PROB1, HNRNPA1L2, CCDC184, MAGEH1, GPR25, P2RY4, SPRR1A, KCNA10, DUSP21, KRTAP1-1, RTL8B	13
ESCA	LDHAL6B, IFNE, OR51B4, H2AC14, RNF225, CTAGE15	6
GBM	PPIAL4A, RAP2B, SOX3, ANKRD63, SPPL2C, ANKRD34C, ATOH7, CSNK1A1L, TTC30A, H3C14, OR14I1, IRS4, SF3B5, PPP1R2B, FFAR1, F8A3, KLLN, MTRNR2L10, PGAM4, KLHL34, USP27X, MTRNR2L6, SSTR4, EXOC8, H2AC19, ATP5MGL, GPR52, BLOC1S4, INSM2, C6orf226, TTC30B, CENPB, SFN, SRY, MC4R	35
KIRC	OR2A4, NPAP1, KRTAP5-8, TAS2R13, TSPYL6, PCDHB10, DDI1, CTAGE9, POU5F2, TAL2, ELOA2, TAS2R3, KLHDC7A, TAS2R30, RXFP4, KAAG1, MAGEF1, EBLN2	18
LUAD	GALNT4, FOXE1, THBD, TAF1L, PPP4R3C, MAP10	6
PRAD	PBOV1, KRTAP13-2, KBTBD13, RAB6D, KDM4E, H3C1, FRAT2, LENG9, CCDC54, CTAGE8, ATOH1, IER5, FOXB2, MARS2, SUMO4, H1-6	16

Table 4. Shared PPIs. The first column contains pairs of cancers, followed by the number of PPI related to unique DE IG of the compared diseases, finally, in the third column is shown the percentage of shared interactions. Note that two cancer may share more genes than other couples of genes but have a smaller percentage, this is due to a difference in the total number of interactions associated with the compared tumors, for example, PRAD and KIRK share 227 PPIs, representing a percentage of 28.64, while COAD and BRCA have less common interactions (145) but a bigger percentage: 44.61.

Cancers	Percentage of total interactions in compared cancers	Interactions shared
COAD, BRCA	44.62	145
KIRC, PRAD	28.65	277
KIRC, BRCA	21.32	132
KIRC, COAD	20.75	138
BLCA, KIRC	14.17	118
BRCA, PRAD	11.64	95
COAD, PRAD	10.94	95
BLCA, PRAD	9.35	95
BLCA, BRCA	7.48	43
BLCA, COAD	6.86	43
BLCA, GBM	1.65	14
PRAD, GBM	1.31	15
KIRC, GBM	0.91	9
BRCA, GBM	0.30	2
COAD, GBM	0.28	2
BLCA, ESCA	0.18	1

Table 5. Identified PPI complexes

Protein complex	Enriched functions	Cancers	FDR
CD40 signaling pathway	Regulation of protein monoubiquitination; Regulation of CD40 signaling pathway	BLCA	1.80E−02
Multi-multicellular organism process	Response to corticosterone;Multi-multicellular organism process; Myeloid leukocyte differentiation; Positive regulation of transcription by RNA polymerase II	BLCA	1.40E−05
Electron transport chain	Mitochondrial electron transport, ubiquinol to cytochrome c; Pons development; Pyramidal neuron development	BLCA	6.70E−03
Chromatid sister cohesion	Positive regulation of maintenance of sister chromatid cohesion; Positive regulation of maintenance of mitotic sister chromatid cohesion; Regulation of maintenance of sister chromatid cohesion	BLCA, PRAD	5.10E−05
Dendrite morphogenesis	Positive regulation of dendrite morphogenesis; Regulation of dendrite morphogenesis; Regulation of mitotic nuclear division	BRCA	5.70E−04
tRNA modifications	Protein urmylation; TRNA thio-modification; TRNA wobble position uridine thiolation	BRCA	1.50E−10
Protein localization to nucleolus	Ribosomal large subunit export from nucleus; Protein localization to nucleolus	BRCA	5.50E−03

(continued)

Table 5. (*continued*)

Protein complex	Enriched functions	Cancers	FDR
Wnt signaling pathways	Canonical Wnt signaling pathway; Wnt signaling pathway; Cell-cell signaling by wnt	KIRC, COAD, BRCA, BLCA, GBM	3.90E−19
Follicle-stimulating hormone signaling	Desensitization of G protein-coupled receptor signaling pathway by arrestin; Norepinephrine-epinephrine-mediated vasodilation involved in regulation of systemic arterial blood pressure; Follicle-stimulating hormone signaling pathway; Negative regulation of multicellular organism growth	BRCA	6.50E−05
GPI anchor	Preassembly of GPI anchor in ER membrane; GPI anchor biosynthetic process; GPI anchor metabolic process	BRCA	3.40E−08
Hormone-mediated apoptosis	Hormone-mediated apoptotic signaling pathway; Somatostatin receptor signaling pathway; Somatostatin signaling pathway; Positive regulation of T cell anergy	COAD	2.20E−02
Light stimulus	Wnt signaling pathway, calcium modulating pathway; Detection of light stimulus; Phototransduction	ESCA	1.10E−03
Vasodilatation	Glutamate catabolic process to 2-oxoglutarate; Response to transition metal nanoparticle; Negative regulation of collagen biosynthetic process; Regulation of angiotensin levels in blood; Angiotensin maturation	ESCA	2.10E−02
Protein folding	Regulation of cellular response to heat; Chaperone cofactor-dependent protein refolding; Inclusion body assembly; de novo protein folding	ESCA	4.10E−08
Heart morphogenesis	Coronary vein morphogenesis; Negative regulation of cell proliferation involved in heart valve morphogenesis; Mitral valve formation; Cardiac right atrium morphogenesis; Cell proliferation involved in heart valve morphogenesis	ESCA	9.90E−03
tRNA and NcRNA processing	TRNA 5 -leader removal; TRNA 5 -end processing; NcRNA 5 -end processing; Endonucleolytic cleavage involved in tRNA processing	ESCA	6.90E−03
Splicing	U2-type pre-spliceosome assembly; MRNA 3 -splice site recognition; Histone mRNA metabolic process; Spliceosomal snRNP assembly	GBM	9.40E−07
Habituation	Habituation; Nonassociative learning; Regulation of glycogen (starch) synthase activity	GBM	1.20E−02
Vesicle transport	Golgi to plasma membrane transport; Vesicle docking involved in exocytosis; Vesicle docking; Exocytic process	GBM	8.50E−07
Intraciliary transport	Intraciliary anterograde transport; Negative regulation of myotube differentiation; Intraciliary transport	GBM	1.90E−02
Regulation of feeding behavior	Positive regulation of feeding behavior; Regulation of feeding behavior; Positive regulation of behavior; Negative regulation of myotube differentiation	KIRC	5.00E−04
Transcription preinitiation complex assembly	RNA polymerase II preinitiation complex assembly; Transcription preinitiation complex assembly	BLCA, LUAD, GBM	2.90E−15
BMP signaling pathway	Pharyngeal system development; Endocardial cushion development; Positive regulation of bone mineralization; Negative regulation of BMP signaling pathway	LUAD	8.80E−12
Innate immune response	Innate immune response activating cell surface receptor signaling pathway; Innate immune response-activating signal transduction; Activation of innate immune response	LUAD	8.90E−13
Blood coagulation system	Positive regulation of blood coagulation; Positive regulation of hemostasis; Positive regulation of coagulation	LUAD	6.30E+00
Translation and postranslation processes	Translation; Peptide biosynthetic process; Amide biosynthetic process	PRAD	1.30E−03
Beta-catenin destruction complex	Negative regulation of type B pancreatic cell development; Superior temporal gyrus development; Beta-catenin destruction complex assembly; Regulation of type B pancreatic cell development; Beta-catenin destruction complex disassembly	PRAD	4.10E−02

References

1. Ademuwagun, I., et al.: Contributors. In: Forero, D.A., Patrinos, G.P. (eds.) Genome Plasticity in Health and Disease, pp. xi–xiv. Translational and Applied Genomics, Academic Press (2020). https://doi.org/10.1016/B978-0-12-817819-5.01002-0. https://www.sciencedirect.com/science/article/pii/B9780128178195010020

2. Amigo, J.D., et al.: The reprimo gene family: a novel gene lineage in gastric cancer with tumor suppressive properties. Int. J. Mol. Sci. **19**(7), 1862 (2018). https://doi.org/10.3390/ijms19071862. https://www.mdpi.com/1422-0067/19/7/1862

3. An, G., et al.: Cloning and characterization of UROC28, a novel gene overexpressed in prostate, breast, and bladder cancers1. Cancer Res. **60**(24), 7014–7020 (2000)

4. Aviña-Padilla, K., et al.: Evolutionary perspective and expression analysis of intronless genes highlight the conservation of their regulatory role. Front. Genet. **12**, 1101 (2021). https://doi.org/10.3389/fgene.2021.654256. https://www.frontiersin.org/article/10.3389/fgene.2021.654256

5. Bujko, M., Kober, P., Mikula, M., Ligaj, M., Ostrowski, J., Siedlecki, J.: Expression changes of cell-cell adhesion-related genes in colorectal tumors. Oncol. Lett. **9**, 2463–2470 (2015). https://doi.org/10.3892/ol.2015.3107. https://doi.org/10.3109/10428194.2010.514968. pMID 20919851

6. Chen, K., Bao, Z., hua Gong, W., Tang, P.C., Yoshimura, T., Wang, J.M.: Regulation of inflammation by members of the formyl-peptide receptor family. J. Autoimmun. **85**, 64–77 (2017). https://doi.org/10.1016/j.jaut.2017.06.012. https://pubmed.ncbi.nlm.nih.gov/28689639/

7. Colaprico, A., et al.: TCGAbiolinks: an R/Bioconductor package for integrative analysis of TCGA data. Nucleic Acids Res. **44**(8), e71–e71 (2015). https://doi.org/10.1093/nar/gkv1507

8. Croci, G.A., et al.: Reproducibility of histologic prognostic parameters for mantle cell lymphoma: cytology, Ki67, p53 and SOX11. Virchows Arch. **477**(2), 259–267 (2020). https://doi.org/10.1007/s00428-020-02750-7

9. Dai, E., Zhu, Z., Wahed, S., Qu, Z., Storkus, W., Guo, Z.S.: Epigenetic modulation of antitumor immunity for improved cancer immunotherapy. Mol. Cancer **20**, 141–152 (2021). https://doi.org/10.1186/s12943-021-01464-x. https://link.springer.com/article/10.1007/s10555-020-09944-0

10. Dai, J., Sultan, S., Taylor, S.S., Higgins, J.M.: The kinase haspin is required for mitotic histone H3 Thr 3 phosphorylation and normal metaphase chromosome alignment. Genes Dev. **19**, 472–488 (2005). https://doi.org/10.1101/gad.1267105

11. Denny, J.C., et al.: Variants near foxe1 are associated with hypothyroidism and other thyroid conditions: using electronic medical records for genome- and phenome-wide studies. Am. J. Hum. Genet. **89**(4), 529–542 (2011). https://doi.org/10.1016/j.ajhg.2011.09.008. https://www.sciencedirect.com/science/article/pii/S0002929711003983

12. Goh, K.I., Cusick, M.E., Valle, D., Childs, B., Vidal, M., Barabási, A.L.: The human disease network. Proc. Natl. Acad. Sci. **104**(21), 8685–8690 (2007). https://doi.org/10.1073/pnas.0701361104. https://www.pnas.org/doi/abs/10.1073/pnas.0701361104

13. Grzybowska, E.A.: Human intronless genes: functional groups, associated diseases, evolution, and mrna processing in absence of splicing. Biochem. Biophys. Res. Commun. **424**(1), 1–6 (2012). https://doi.org/10.1016/j.bbrc.2012.06.092. https://www.sciencedirect.com/science/article/pii/S0006291X12011874

14. Hagberg, A., Swart, P., Chult, D.: Exploring network structure, dynamics, and function using networkx, pp. 11–15 (2008)

15. Herrera-Oropeza, G.E., Angulo-Rojo, C., Gástelum-López, S.A., Varela-Echavarría, A., Hernández-Rosales, M., Aviña-Padilla, K.: Glioblastoma multiforme: a multi-omics analysis of driver genes and tumour heterogeneity. Interface Focus **11**(4), 20200072 (2021). https://doi.org/10.1098/rsfs.2020.0072. https://royalsocietypublishing.org/doi/abs/10.1098/rsfs.2020.0072

16. Huang, S.H., Lo, Y.S., Luo, Y.C., Tseng, Y.Y., Yang, J.M.: A homologous mapping method for three-dimensional reconstruction of protein networks reveals disease-associated mutations. BMC Syst. Biol. **12**(S2), 8685–8690 (2018)

17. Hung, M.S., et al.: Functional polymorphism of the ck2α intronless gene plays oncogenic roles in lung cancer. PLoS ONE **5**(7), 1–10 (2010). https://doi.org/10.1371/journal.pone.0011418

18. Inamdar, A.A., et al.: Mantle cell lymphoma in the era of precision medicine-diagnosis, biomarkers and therapeutic agents. Oncotarget **7**(30), 48692–48731 (2016). https://doi.org/10.18632/oncotarget.8961. https://www.oncotarget.com/article/8961/

19. Jao, T.M., et al.: PCDH10 exerts tumor-suppressor functions through modulation of EGFR/AKT axis in colorectal cancer. Cancer Lett. **499**, 290–300 (2021). https://doi.org/10.1016/j.canlet.2020.11.017. https://www.sciencedirect.com/science/article/pii/S0304383520306108

20. Krzywinski, M., et al.: Circos: an information aesthetic for comparative genomics. Genome Res. **19**, 1639–1645 (2009). https://doi.org/10.1101/gr.092759.109. https://genome.cshlp.org/content/19/9/1639

21. Lachmann, A., Giorgi, F.M., Lopez, G., Califano, A.: ARACNe-AP: gene network reverse engineering through adaptive partitioning inference of mutual information. Bioinformatics **32**(14), 2233–2235 (2016). https://doi.org/10.1093/bioinformatics/btw216

22. Le, Y., Murphy, P.M., Wang, J.M.: Formyl-peptide receptors revisited. Trends Immunol. **23**(11), 541–548 (2002). https://doi.org/10.1016/S1471-4906(02)02316-5. https://www.sciencedirect.com/science/article/pii/S1471490602023165

23. Liang, W., et al.: The contribution of chemoattractant GPCRs, formylpeptide receptors, to inflammation and cancer. Front. Endocrinol. **11**, 17 (2020). https://doi.org/10.3389/fendo.2020.00017. https://www.readcube.com/articles/10.3389/fendo.2020.00017

24. Liu, X.Y., et al.: Methylation of SOX1 and vim promoters in serum as potential biomarkers for hepatocellular carcinoma. Neoplasma **64**, 745–753 (2017). https://doi.org/10.4149/neo_2017_513

25. Marcel, R., Lucas, S., Sean, D., Levi, W.: TCGAutils: TCGA utility functions for data management (2021). https://bioconductor.org/packages/release/bioc/html/TCGAutils.html. R package version 1.14.4

26. Martin, A.M., Kulski, J.K., Witt, C., Pontarotti, P., Christiansen, F.T.: Leukocyte Ig-like receptor complex (LRC) in mice and men. Trends Immunol. **23**(2), 81–88 (2002). https://doi.org/10.1016/S1471-4906(01)02155-X. https://www.sciencedirect.com/science/article/pii/S147149060102155X

27. Ohki, R., et al.: Reprimo, a new candidate mediator of the P53-mediated cell cycle arrest at the G2 phase*. J. Biol. Chem. **275**(30), 22627–22630 (2000). https://doi.org/10.1074/jbc.C000235200. https://www.sciencedirect.com/science/article/pii/S0021925819661077

28. Pan, J., et al.: Lineage-specific epigenomic and genomic activation of oncogene HNF4A promotes gastrointestinal adenocarcinomas. Cancer Res. **80**(13), 2722–2736 (2020). https://doi.org/10.1158/0008-5472.CAN-20-0390

29. Pedregosa, F., et al.: Scikit-learn: machine learning in python. J. Mach. Learn. Res. **12**, 2825–2830 (2012). https://www.jmlr.org/papers/volume12/pedregosa11a/pedregosa11a.pdf

30. Sander, B., et al.: Mantle cell lymphoma—a spectrum from indolent to aggressive disease. Virchows Arch. **468**(3), 245–257 (2015). https://doi.org/10.1007/s00428-015-1840-6

31. Sander, B.: Mantle cell lymphoma: recent insights into pathogenesis, clinical variability, and new diagnostic markers. Semin. Diagn. Pathol. **28**(3), 245–255 (2011). https://doi.org/10.1053/j.semdp.2011.02.010. https://www.sciencedirect.com/science/article/pii/S0740257011000153. Seminars on Lymphomas, Part II

32. Szklarczyk, D., et al.: The STRING database in 2021: customizable protein-protein networks, and functional characterization of user-uploaded gene/measurement sets. Nucleic Acids Res. **49**(D1), D605–D612 (2020). https://doi.org/10.1093/nar/gkaa1074

33. Verhaegen, M.E., et al.: Merkel cell polyomavirus small T antigen initiates merkel cell carcinoma-like tumor development in mice. Cancer Res. **77**(12), 3151–3157 (2017). https://doi.org/10.1158/0008-5472.CAN-17-0035

34. Wang, F., et al.: Histone H3 Thr-3 phosphorylation by haspin positions aurora b at centromeres in mitosis. Science **330**(6001), 231–235 (2010). https://doi.org/10.1126/science.1189435. https://www.science.org/doi/abs/10.1126/science.1189435

35. Wang, Y.Y., et al.: Expression of SOX11 mRNA in mantle cell lymphoma and its clinical significance. Zhonghua xue ye xue za zhi = Zhonghua xueyexue zazhi **33**(7), 556–560 (2012). http://europepmc.org/abstract/MED/22967417

36. Xu, W., Li, J.Y.: SOX11 expression in mantle cell lymphoma. Leuk. Lymphoma **51**(11), 1962–1967 (2010). https://doi.org/10.3109/10428194.2010.514968. pMID 20919851

37. Yao, Z., et al.: The role of tumor suppressor gene SOX11 in prostate cancer. Tumour Biol. **8**(36), 6133–6138 (2015). https://doi.org/10.1007/s13277-015-3296-3. http://europepmc.org/abstract/MED/22967417

Author Index

Printed in the United States
by Baker & Taylor Publisher Services